The MIDI Manual

Third Edition

The MIDI Manual

A Practical Guide to MIDI in the Project Studio

Third Edition

David Miles Huber

Amsterdam Boston Heidelberg London New York Oxford Paris
San Diego San Francisco Singapore Sydney Tokyo

Acquisitions Editor: Catharine Steers
Publishing Services Manager: George Morrison
Project Manager: Paul Gottehrer
Assistant Editor: Terri Jadick
Marketing Manager: Christine Degon Veroulis
Cover Design: Eric DeCicco

Foca Press is an imprint of Elsevier
30 Corporate Drive, Suite 400, Burlington, MA 01803, USA
Linacre House, Jordan Hill, Oxford OX2 8DP, UK

Library of Congress Cataloging-in-Publication Data
Application submitted

British Library Cataloguing-in-Publishing Data
A catalogue record for this book is available from the British Library.

ISBN 13: 978-0-240-80798-0
ISBN 10: 0-240-80798-7

For information on all Focal Press publications
visit our website at www.books.elsevier.com

07 08 09 10 11 5 4 3 2 1

Printed in the United States America

Table of Contents

Foreword

The most amazing aspect of the book you're holding in your hands is that it talks about a technological subject that, while more than a decade and a half old, is still vital, growing, and relevant. Think back to other artifacts from the same era as MIDI'S early days: Is anyone still writing books about the Commodore 64, the joys of telecommunicating with 300-baud modems, getting the most out of a Mac Plus, or how to organize data on your 20-megabyte hard drive? Of course not. Technology has a certain ruthlessness; in the rush toward the new, the old is forgotten—cast away as an embarrassing reminder of what we once were.

But this hasn't happened with MIDI. Far from disappearing, MIDI has lived long and prospered. Once used by only a handful of computer-literate musicians, MIDI is now commonplace in computers around the world. It has become "the new sheet music," but for computer media rather than the printed page.

What accounts for MIDI's stubborn refusal not to yield to something new and different, and even evolve to newer heights? Part of the reason is that MIDI was born to good parents. Virtually every music-related hardware and software manufacturer in the early 1980s worked together to hammer out a standard: Americans, Japanese, and Europeans debated, refined, and tested until they had forged a truly worldwide phenomenon. They had the good sense to make MIDI inexpensive enough that there was no reason not to include it as part of musical hardware, which created the volume that made it worthwhile for other companies to develop MIDI-compatible widgets.

MIDI was also powerful and useful enough that it served a wide range of musical needs. But the story would have ended there if not for the MIDI Manufacturers Association, the "guardians" of the MIDI spec. Periodically, this organization drafts new proposals that take MIDI to another level. In the years since MIDI'S inception, it has become part of studio recording, post-production, video games, lighting control, Broadway shows, the Internet, and much more. All of these improvements have been done in an orderly, consensus-oriented fashion that has guaranteed the spec's universality and effectiveness.

Which brings us to this book. MIDI's evolution has meant that there has been a continuing need for books that explore this evolution. David Huber is an excellent guide to this world. He loves music, technology, and people, which puts him in a good position to communicate about technology to those who want to learn more about this "MIDI thing." Of course, *The MIDI Manual* includes info on MIDI basics, but more importantly, it discusses the many updates and enhancements that have occurred during the past 20+ years. While it's a fine place for beginners to get an overview, what's special about *The MIDI Manual* is that it's also "continuing education" for those who want to remain current about MIDI, and get a glimpse of where it's going to be heading in the future.

So sit back, relax, put on a background CD (which, odds are, had MIDI involved somewhere in its production), and find out more about the marvelous world of MIDI. Like a truly fine wine, it doesn't get older—it just gets better.

—Craig Anderton

1 What Is MIDI?

Simply stated, Musical Instrument Digital Interface (MIDI) is a digital communications language and compatible specification that allows multiple hardware and software electronic instruments, performance controllers, computers, and other related devices to communicate with each other over a connected network (Figure 1.1). MIDI is used to translate performance- or control-related events (such as playing a keyboard, selecting a patch number, varying a modulation wheel, triggering a staged visual effect, etc.) into equivalent digital messages and then transmit these messages to other MIDI devices where they can be used to control sound generators and other performance parameters. The beauty of MIDI is that its data can be recorded into a hardware device or software program (known as a sequencer), where it can be edited and transmitted to electronic instruments or other devices to create music or control any number of parameters.

In artistic terms, this digital language is an important medium that lets artists express themselves with a degree of flexibility and control that, before its inception, wasn't possible on an individual level. Through the transmission of this performance language, an electronic musician can create and develop a song or composition in a practical, flexible, affordable, and fun production environment.

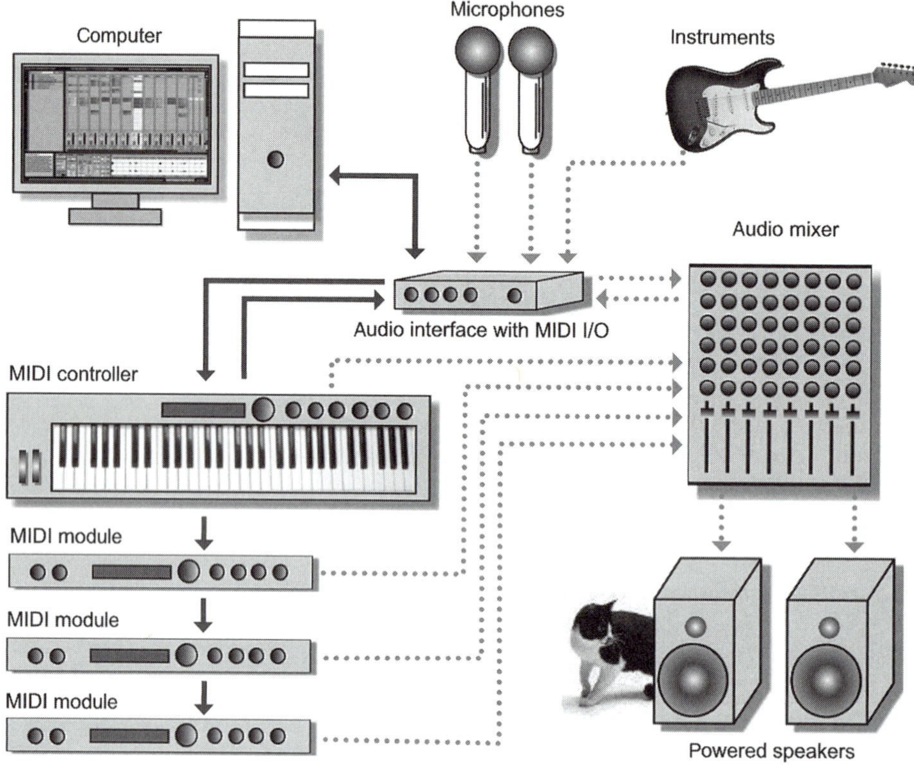

Figure 1.1. *Example of a typical MIDI system with the MIDI network connections.*

In addition to composing and performing a song, musicians can also act as techno-conductors, having complete control over a wide palette of sounds, their timbre (sound and tonal quality), overall blend (level, panning), and other real-time controls. MIDI can also be used to vary the performance and control parameters of electronic instruments, recording devices, control devices, and signal processors in the studio, on the road, or on the stage.

The term "interface" refers to the actual data communications link and software/hardware systems in a connected MIDI network. Through the use of MIDI, it is possible for all of the electronic instruments and devices within a network to be addressed through the transmission of real-time performance and control-related MIDI data messages throughout a system to multiple instruments and devices through a single data line (which can be chained from device to device). This is possible because a single data cable is capable of transmitting performance and control messages over 16 discrete channels. This simple fact allows electronic musicians to record, overdub, mix, and play back their performances in a working environment that loosely resembles the multitrack recording process. Once mastered, MIDI surpasses this analogy by allowing a composition to be edited, controlled, altered, and called up with complete automation and repeatability—providing production challenges and possibilities that are well beyond the capabilities of the traditional tape-based multitrack recording process.

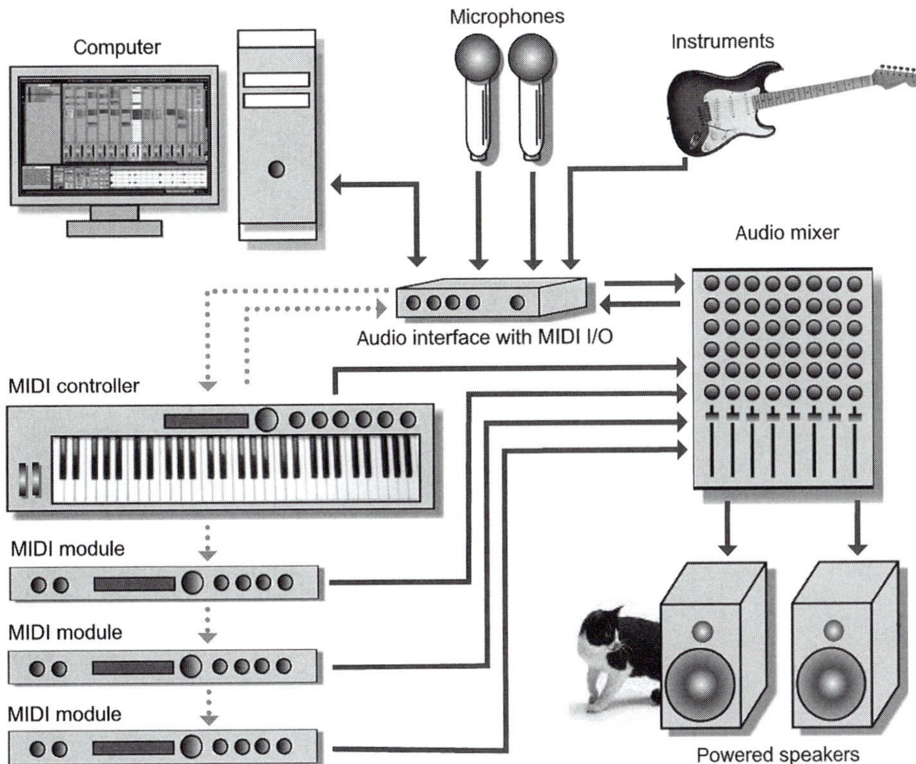

Figure 1.2. *Example of a typical MIDI system with the audio connections.*

What MIDI Isn't

For starters, let's dispel one of MIDI's greatest myths: *MIDI does not communicate audio nor can it create sounds!* It is a digital language that instructs a device or program to create, playback, or alter sounds. MIDI is a data protocol that communicates on/off triggering and a wide range of parameters to instruct an instrument or device to generate, reproduce, or control audio or production-related functions. Because of these differences, the MIDI data path and the audio routing paths are entirely separate from each another (Figure 1.2). Even if they digitally share the same transmission cable (such as through FireWire™ or USB), the actual data paths and formats are completely separate.

In short, MIDI communicates information that instructs an instrument to play or a device to carry out a function. It can be thought of as the dots on a player-piano roll—when we put the paper roll up to our ears, we hear nothing. When the cut-out dots pass over the sensors on a player piano, the instrument itself begins to make beautiful music. It's exactly the same with MIDI. A MIDI file or data stream is simply a set of instructions that pass down a wire in a serial fashion, but when an electronic instrument interprets the data we begin to hear sound.

A Brief History

In the early days of electronic music, keyboard synthesizers were commonly monophonic devices (capable of sounding only one note at a time) and often generated a thin sound quality. These limiting factors caused early manufacturers to look for ways to combine instruments together to create a thicker, richer sound texture. This was originally accomplished by establishing an instrument link that would allow a synthesizer (acting as a master controller) to directly control the performance parameters of one or more synthesizers (known as slave sound modules). As a result of these links, a basic control signal (known as control voltage, or CV) was developed (Figure 1.3).

This simple yet problematic system was based on the fact that when most early keyboards were played, they generated a DC voltage that could directly control another instrument's voltage-controlled oscillators (which affected the pitch of a sounding note) and voltage-controlled amplifiers (which affected the note's volume and on/off nature). Since many keyboards of the day generated a DC signal that ascended at a rate of 1 volt per octave (breaking each musical octave into 1/12-volt intervals), it was possible to use this standard control voltage as a master-reference signal for transmitting pitch information to other synths. In addition to a control voltage, this

Figure 1.3. *The late, great synth pioneer Bob Moog, who was outstanding in his field. (Photograph courtesy of Roger Luther; www.moogarchives.com.)*

standard required that a keyboard transmit a gate signal. This second signal was used to synchronize the beginning and duration times of each note. With the appearance of more advanced polyphonic synthesizers (which could generate more than one note at a time) and early digital devices, it was clear that this standard would no longer be the answer to system-wide control, and new standards began to appear on the scene (thereby creating the fun of having incompatible control standards). With the arrival of early drum machines and sequencing devices, standardization became even more of a dilemma.

Synchronization between these early devices was also problematic, as manufacturers would often standardize on different sync-pulse clock rates. Synchronizing incompatible systems could be extremely difficult, because they would lose sync over a very short period of time, rendering sync nearly impossible without additional sync-rate converters or other types of modifications. Because of this mess, Dave Smith and Chet Wood (then of Sequential Circuits, a now-defunct manufacturer of electronic instruments) began creating a digital electronic instrument protocol, which was named the Universal Synthesizer Interface (USI). As a result of this early protocol, equipment from different manufacturers could finally communicate directly (*e.g.*, a synth from one company finally worked with another company's sequencer). In the fall of 1981, USI was proposed to the Audio Engineering Society. During the following two years, a panel (which included representatives from the major electronic instrument manufacturers) modified this standard and adopted it under the name of Musical Instrument Digital Interface (MIDI Specification 1.0).

The strong acceptance of MIDI was largely due to the need for a standardized protocol and fast-paced advances in technology that allowed complex circuit chips and hardware designs to be manufactured cost effectively. It was also due, in part, to the introduction of Yamaha's popular DX-7 synthesizer in the winter of 1983, after which time keyboard sales began to grow at an astonishing rate.

With the adoption of this industry-wide standard, any device that incorporated MIDI into its design could transmit or respond to digital performance and control-related data conforming to the MIDI 1.0 specification. For the first time, any new device that conformed to the MIDI spec would integrate into an existing MIDI system and actually work … without any muss or fuss.

Over the course of time, new instruments came onto the market that offered improved sound and functional capabilities that led to the beginnings of software sound generators, samplers, and effects devices. With the eventual maturation of software instruments and systems that could emulate existing devices or create entirely an entirely new range of functions and sound, hardware controllers began to quickly spring onto the scene that made use of MIDI to communicate physical control movements into analogous moves in a program or plug-in software interface. In fact, this explosion of software emulation and control has breathed a new degree of life into the common, everyday use of MIDI.

Why Is MIDI?

Before MIDI, it was pretty much necessary to perform a musical passage in real time. Of course, there are a few exceptions to this statement. In earlier days, music could be created and re-created though the mechanical triggering of a musical device (music boxes and the aforementioned player piano come to mind). When tape-based recording came along in the middle part of the last century, it became possible to edit two or more problematic performances together into a single, good take. However, when it came to the encoding of a musical passage and then faithfully playing it back—while still being able to edit or alter the tempo, notes, and control variables of a performance—we were pretty much back in the horse-and-buggy days.

With the introduction of electronic music production and MIDI, a musical performance could be captured in the digital domain and then faithfully played back in a production-type environment that mimicked the traditional form and functions of multitrack recording. Basic tracks could be recorded one at a time, allowing a composition to be built up using various electronic instruments. But, here's the kicker: MIDI finally made it possible for a performance track to be edited, layered, altered, spindled, mutilated, and improved with relative ease and under completely automated computer control. If you played a bad note, fix it. If you want to change the key or tempo of a piece, change it. If you want to change the expressive volume of a phrase in a song, just do it! Even its sonic character (timbre) can be changed! These capabilities just hint at the power of MIDI!

This affordable potential for future expansion and increased control throughout an integrated production system has spawned the growth of an industry that's also very personal in nature. For the first time in music history, it is possible for an individual to cost-effectively realize a full-scale sound production on his or her own time. Because MIDI is a real-time performance medium, it is also possible to listen to and edit a production at every stage of its development, all within the comfort of one's own home or personal project studio.

I'd also like to address another issue that has sprung up around MIDI and electronic music production. With the introduction of drum machines, modern-day synths, samplers, and powerful hardware or software instruments, it is not only possible but also relatively easy to build up a composition using instrument voices that closely mimic virtually any instrument that can be imagined. In the early days, studio musicians spoke out against MIDI, saying that it would be the robot that would make them obsolete. Although there was a bit of truth to this, these same musicians are now using the power of MIDI to expand their own musical palate and create productions of their own. Today, MIDI is being used by many professional and nonprofessional musicians alike to perform an expanding range of production tasks, including music production, audio-for-video and film postproduction, and stage production. Such is progress.

MIDI in the Home

A vast number of electronic musical instruments, effects devices, computer systems, and other MIDI-related devices are currently available on the new and used electronic music market. This diversity lets us select the type of production system that best suits our own particular musical taste and production style. With the introduction of the large-scale integrated circuit chip (which allows complex circuitry to be quickly and easily mass produced), many of the devices that make up an electronic music system are affordable for almost every musician or composer, whether he or she is a working professional, aspiring artist, or beginning hobbyist (Figure 1.4).

One of the greatest benefits of a project or portable production system centers around the idea that an artist can select from a wide range of tools and toys to generate specific sounds—or to get

Figure 1.4. *Gettin' it all going in the bedroom studio. (Photograph courtesy of Yamaha Corporation of America; www.yamaha.com.)*

the particular sounds that he or she likes. This technology is often extremely powerful, as the components combine to create a vast palette of sounds and handle a wide range of task-specific functions. Such a system might include one or more keyboard synthesizers, synth modules, samplers, drum machines, a computer (with a digital audio workstation and sequencing package), effects devices, and audio mixing capabilities.

Systems like these are constantly being installed in the homes of working and aspiring musicians. Their size can range from taking up a corner of an artist's bedroom to being a larger system that's been installed in a dedicated project studio. All of these system types can be designed to handle a wide range of applications and have the important advantage of letting the artist produce his or her music in a comfortable environment whenever the creative mood hits. Such production luxuries, which would have literally cost an artist a fortune twenty years ago, are now within the reach of almost every working and aspiring musician.

MIDI on the Go

Of course, MIDI production systems can appear in any number of shapes and sizes and can be designed to match a wide range of production and budget needs. For example, a portable, all-in-one keyboard instrument (known as a MIDI workstation) often includes an integrated keyboard, polyphonic synthesizer, percussion sounds, built-in sequencer, and audio recording capabilities … all in a single hardware package. Laptops have hit the production scene big time, as they can combine software recording and production applications with portable keyboard

Figure 1.5. *Between takes. (Photograph courtesy of M-Audio, A Division of Avid Technology; www.m-audio.com.)*

controllers and audio interface devices to create a professional production system that lets us compose, produce, and mix in the studio or on the beach of a remote seaside island (Figure 1.5).

MIDI in the Studio

MIDI has also dramatically changed the sound, technology, and production habits of the recording studio (Figure 1.4). Before MIDI and the concept of the home project studio, the professional recording studio was one of the only production environments that allowed an artist or composer to combine instruments and sound textures into a final recorded product. Often, the process of recording a group in a live setting was (and still can be) an expensive and time-consuming process. This is due to the high cost of hiring session musicians and the high hourly rates that are charged for a professional studio—not to mention Murphy's studio law, which states that you'll always spend more time and money than you thought you ever could in an effort to capture that elusive "ideal performance."

Because of the digital audio workstation (DAW) and MIDI, much of the music production process can now be preplanned and rehearsed (or even totally produced and recorded) before you step into the studio (Figure 1.6). This out-and-out luxury has reduced the number of hours that are needed for laying down recorded tracks to a cost-effective minimum; for example, it is now commonplace for groups to record and produce entire albums in their own project studios. Once completed (or nearly completed), the group can either dump the tracks to tape or simply bring their entire set of MIDI and recorded audio tracks into the studio and lay the instrument tracks down to disc or tape. In a professional studio, the tracks can be sweetened into a polished state by adding vocals or other instruments. Finally, the tracks can then be professionally mixed down into a final product. In essence, through the use of careful planning and preproduction in the project studio, a project can be produced in a much more timely fashion (and hopefully on budget) than would otherwise be possible.

Figure 1.6. Steve Tushar, sound effects editor and musician, with his MC Media Application Controller running Nuendo. (Photograph courtesy of Andrew Wild, Euphonix, Inc.; www.euphonix.com.)

Figure 1.7. Euphonix System 5 console at Goldcrest Post in London. (Photograph by Patrick Denis, courtesy of Euphonix, Inc.; www.euphonix.com.)

MIDI in Audio-for-Visual and Film

Electronic music has long been an indispensable tool for the scoring and audio postproduction of television commercials, industrial videos, television shows, and full-feature motion picture sound tracks (Figure 1.7). For productions that are on a tight budget, entire scores are often written and produced in a project studio at a mere fraction of what it might cost to hire the musicians, a studio, and mixdown rooms. Even high-budget projects make extensive use of MIDI in the preproduction and production phases. Often, orchestral scores for such projects are composed, edited, and finessed as a MIDI version of the composer's score before the expensive orchestral tracks are finally recorded in the studio. Before MIDI, this simply wasn't possible. Once approved, the final MIDI score can be printed and distributed to the musicians before the session.

MIDI in Live Performance

Electronic music production and MIDI are also at home on the stage. Obviously, MIDI has played a crucial role in helping to bring live music to the masses. The ability to sequence rhythm and background parts in advance, chain them together into a single, controllable sequence (using a jukebox-type sequencing program), and then play them on stage has become an indispensable live-performance tool for many musicians. This technique is widely used by solo artists who have become one-man bands by singing and playing their guitar to a series of background sequences. Larger techno-groups commonly use extensive on-stage loop and power sequencing to drive instruments, lighting and visuals in ways that are staggeringly compelling.

Again, the power of MIDI lies in the fact that much of a performance can be composed and produced in advance of going on stage or on tour. With advances in digital audio, recorded sounds can be easily integrated into the performance. The integration of looping technology often allows for on-the-spot improvisation, adding a fresh and varied feel to the performance for those on stage and in the audience. In addition to communicating performance data, MIDI controllers can be used to vary a staggering range of control parameters in real time.

In addition to allowing for control over on-stage music performance, lighting, and preproduced sequencing, MIDI can play a strong role in the production and execution of on-stage lighting and special effects. Most modern-day lighting boards are equipped with a MIDI interface, allowing the lighting to be controlled from a linear sequencer over the course of a song or production—or, from a loop-based sequencer that allows for scene changes in a more interactive and on-the-spot manner (while still allowing scenes to be synchronized to the basic script, when needed).

MIDI and Visuals

Speaking of on-screen effects, the ability to offer control over a preprogrammed sequence or interactive loops has put MIDI directly into the driver's seat when it comes to on-stage visuals and video playback (Figures 1.8 and 1.9). Many music acts are beginning to integrate visuals into their band—so much so that VJs now stand alongside their bandmates on-stage, offering up compelling visuals that can be diced, sequenced, and scratched in forms that can instantly switch from being totally chaotic to being in perfect sync … and then back again.

MIDI and Multimedia

One of the "media" in multimedia is definitely MIDI. It often pops up in places that you might expect—and in others that might take you by surprise. With the advent of General MIDI (a standardized specification that makes it possible for any soundcard or GM-compatible device to play back a score using the originally intended sounds and program settings), it is possible (and common) for MIDI scores to be integrated into multimedia games, text documents, CD-ROMs, and even websites. Due to the fact that MIDI is simply a series of performance commands (unlike digital audio, which actually encodes the audio information), the media's data

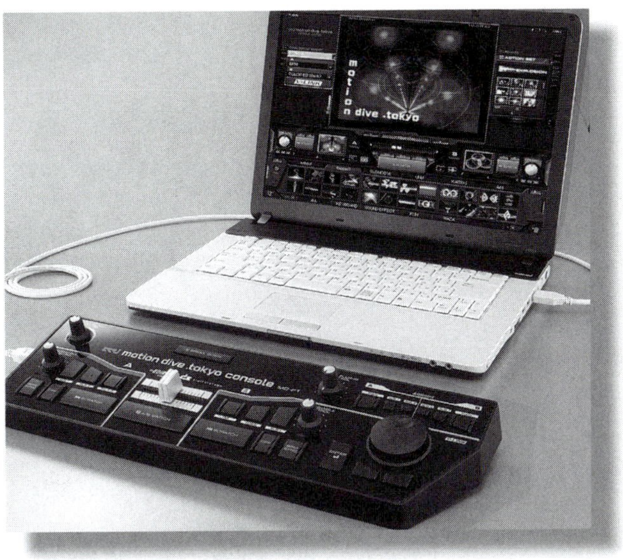

Figure 1.8. *The motion dive .tokyo performance package. (Photograph courtesy of Roland Corporation; www.rolandus.com/edirol.)*

Figure 1.9. *ArKaos VJ MIDI video authoring software. (Photograph courtesy of ArKaos S.A.; www.arkaos.net.)*

overhead requirements are extremely low. This means that almost no processing power is required to play MIDI, making it the ideal medium for playing real-time music scores while you are actively browsing text, graphics, or other media over the Internet. Truly, when it comes to weaving MIDI into the various media types, the sky (and your imagination) is the creative and technological limit.

MIDI on the Phone

With the integration of the General MIDI standard into various media devices, one of the fastest-growing MIDI applications, surprisingly, is probably comfortably resting in your pocket or purse

Figure 1.10. *One ringy-dingy … MIDI helps us to reach out and touch someone through ring tones.*

right now—the ring tone on your cell phone (Figure 1.10). The ability to use MIDI (and often digital soundfiles) to let you know who is calling has spawned an industry that allows your cell to be personalized and super fun. One of my favorite ring tone stories happened on Hollywood Boulevard in L.A. This tall, lanky man was sitting at a café when his cell phone started blaring out the "If I Only Had a Brain" sequence from *The Wizard of Oz.* It wouldn't have been nearly as funny if the guy didn't look a lot like the scarecrow character. Of course, everyone laughed.

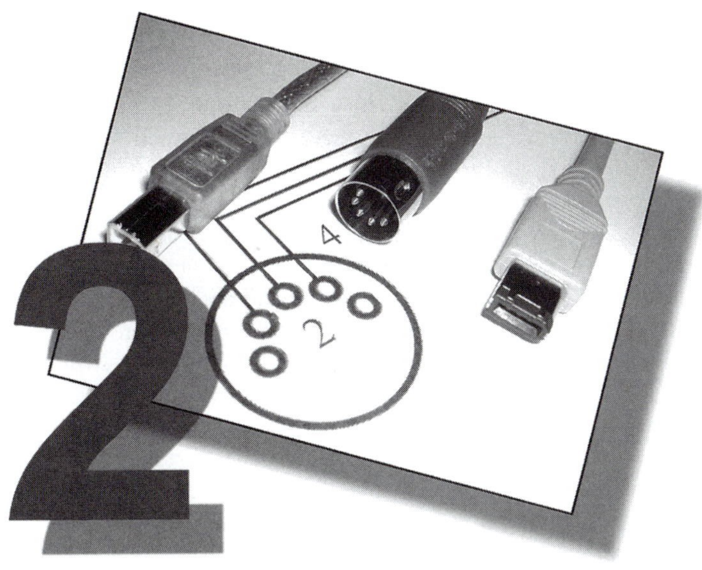

MIDI 1.0

The Musical Instrument Digital Interface is a digital communications protocol. That's to say, it is a standardized control language and hardware specification that makes it possible for electronic instruments, processors, controllers, and other device types to communicate performance and control-related data in real time.

Exploring the Spec

MIDI is a specified data format that must be strictly adhered to by those who design and manufacture MIDI-equipped instruments and devices. Because the format is standardized, you don't have to worry about whether the MIDI output of one device will be understood by the MIDI in port of a device that's made by another manufacturer. As long as the data ports say and/or communicate MIDI, you can be assured that the data (at least the basic performance functions) will be transmitted and understood by all devices within the connected system. In this way, the user need only consider the day-to-day dealings that go hand-in-hand with using electronic instruments, without having to be concerned with data compatibility between devices.

The Digital Word

One of the best ways to gain insight into how the MIDI specification works is to compare MIDI to a spoken language. As humans, we've adapted our communication skills to best suit our physical bodies. Ever since the first grunt, we've found that it's easiest for us to communicate with our vocal chords. And we've been doing it ever since. Over time, language developed by assigning a standardized meaning to a series of vocalized sounds (words). Eventually these words came to be grouped in such a way as to convey meanings that can be easily communicated ... and, finally, written. For example, in order to record the English language, a standard notation system was developed that assigned 26 symbols to specific sounds (letters of the alphabet) that, when grouped together, could communicate an equivalent spoken word (Figure 2.1). When these words are strung into complete sentences, a more complex form of communication is used that can convey information in a way that has a greater meaning (whether spoken or recorded onto paper or other medium). For example, the letters B, O, O, and K don't mean much when used individually; however, when grouped into a word, they refer to a physical media device that contains recorded data, which hopefully conveys messages that relate to a general theme. Changing a symbol in the word could change its meaning entirely; for example, changing the K to a T within the grouped word makes it refer to a tool that's better off worn on your feet than carried in a backpack. When placed into a sentence, grouped words can be used to convey complex messages with greater clarity (*e.g.*, Dude, is that a pirate book about pirate boots?).

Microprocessors and computers, on the other hand, are digitally based devices that obviously lack vocal chords and ears (although even that's changing). However, because they have the unique advantage of being able to process numbers at a very high rate, the obvious language of choice is the reception and transmission of digital data.

Unlike our base 10 system of counting, computers are limited to communicating with a binary system of 0's and 1's (or off and on). Like humans, computers group these binary digits (known as bits) into larger numeric "words" that represent and communicate specific information and instructions. Just as humans communicate using simple sentences, a computer can generate and respond to a series of related digital words that are understood by other digital systems or software programs (Figure 2.2).

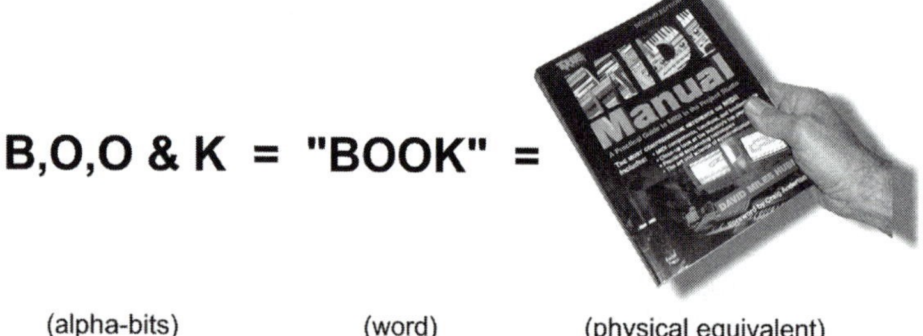

Figure 2.1. *Meaning is given to the alphabet letters B, O, O, and K when they're grouped into a word or placed into a sentence.*

(1001 0100) (0100 0001) (0101 1001)

Status Byte Data Byte #1 Data Byte #2

Figure 2.2. *Example of a digitally generated MIDI message.*

The MIDI Message

MIDI digitally communicates musical performance data between devices as a string of MIDI messages. These messages are traditionally transmitted through a standard MIDI line in a serial fashion at a speed of 31,250 bits/sec. Within a serial data transmission line, data travels in a single-file fashion through a single conductor cable (Figure 2.3a); a parallel data connection, on the other hand, is able to simultaneously transmit digital bits in a synchronous fashion over a number of wires (Figure 2.3b).

When using a standard MIDI cable, it's important to remember that data can only travel in one direction from a single source to a destination (Figure 2.4a). In order to make two-way communication possible, a second MIDI data line must be used to communicate data back to the device, either directly or thru the MIDI chain (Figure 2.4b).

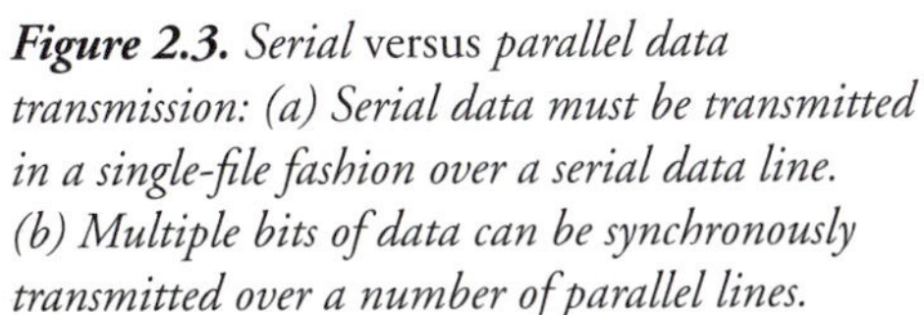

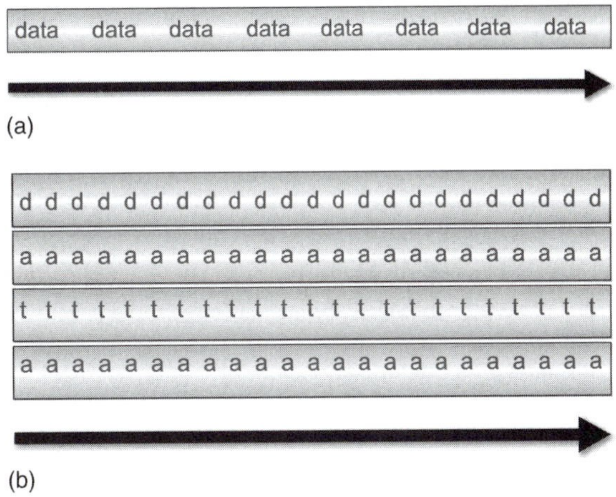

Figure 2.3. *Serial* versus *parallel data transmission: (a) Serial data must be transmitted in a single-file fashion over a serial data line. (b) Multiple bits of data can be synchronously transmitted over a number of parallel lines.*

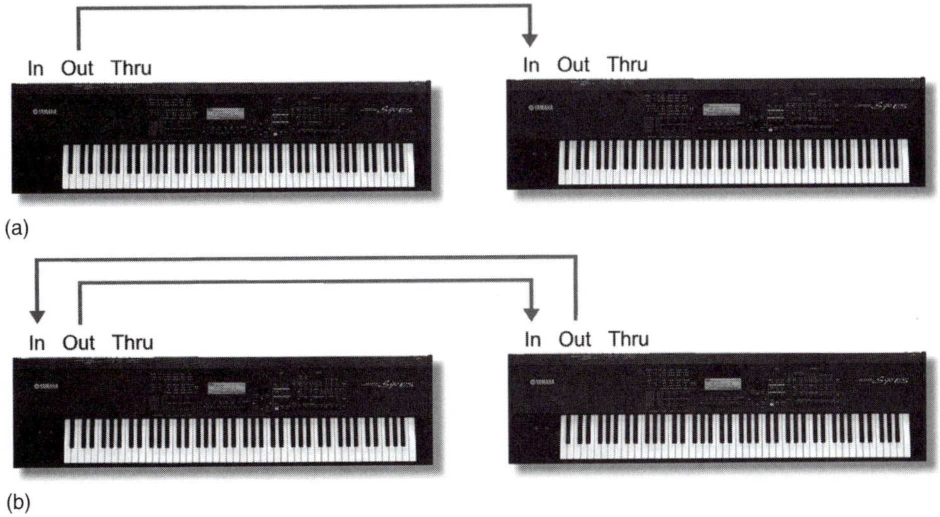

Figure 2.4. *MIDI data can only travel in one direction through a single MIDI cable: (a) data transmission from a single source to a destination; (b) two-way data communication using two cables.*

Figure 2.5. The most significant bit of a MIDI data byte is used to distinguish between a status byte (where MSB = 1) and a data byte (where MSB = 0).

MSB of a status byte is always "1"
(1SSS SSSS)
MSB of a data byte is always "0"
(0DDD DDDD)

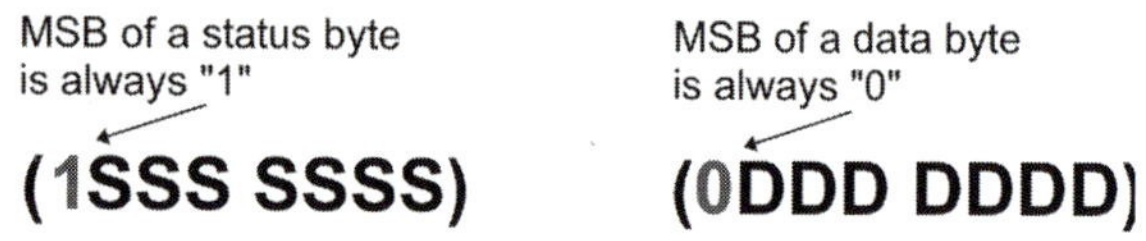

MIDI messages are made up of groups of 8-bit words (known as bytes), which are transmitted in a serial fashion to convey a series of instructions to one or all MIDI devices within a system.

Only two types of bytes are defined by the MIDI specification: the status byte and the data byte.

- A *status byte* is used to identify what type of MIDI function is to be performed by a device or program. It is also used to encode channel data (allowing the instruction to be received by a device that's set to respond to the selected channel).
- A *data byte* is used to associate a value to the event that's given by the accompanying status byte.

Although a byte is made up of eight bits, the most significant bit (MSB; the leftmost binary bit within a digital word) is used solely to identify the byte type. The MSB of a status byte is always 1, while the MSB of a data byte is always 0 (Figure 2.5). For example, a 3-byte MIDI Note-On message (which is used to signal the beginning of a MIDI note) in binary form might read as shown in Table 2.1. Thus, a 3-byte Note-On message of (10010100) (01000000) (01011001) will transmit instructions that would be read as: "Transmitting a Note-On message over MIDI channel #5, using keynote #64, with an attack velocity [volume level of a note] of 89."

Binary numbers (0's and 1's) are placed into groups (called binary words) that translate human concepts into an analogous language that's spoken by computers. For example, with regard to MIDI channels:

0000 = CH#1	0100 = CH#5	1000 = CH#9	1100 = CH#13
0001 = CH#2	0101 = CH#6	1001 = CH#10	1101 = CH#14
0010 = CH#3	0110 = CH#7	1010 = CH#11	1110 = CH#15
0011 = CH#4	0111 = CH#8	1011 = CH#12	1111 = CH#16

Table 2.1. Status and Data Byte Interpretation

1	*Status Byte*	*Data Byte 1*	*Data Byte 2*
Description	Status/channel #	Note #	Attack velocity
Binary data	(1001.0100)	(0100.0000)	(0101.1001)
Numeric value	(Note on/CH#5)	(64)	(89)

MIDI Channels

Just as a public speaker might single out and communicate a message to one individual in a crowd, MIDI messages can be directed to communicate information to a specific device or range of devices within a MIDI system. This is done by imbedding a channel-related nibble (4 bits) within the status/channel number byte (Figure 2.6). This makes it possible for performance or control information to be communicated to a specific device—or a sound generator within a device—that's assigned to a particular channel.

> Since this nibble is 4 bits wide, up to 16 discrete MIDI channels can be transmitted through a single MIDI cable or designated port.

Whenever a MIDI device, sound generator within a device, or program function is instructed to respond to a specific channel number, it will only respond to messages that are transmitted on that channel (*i.e.*, it ignores channel messages that are transmitted on any other channel). For example, let's assume that we're going to create a short song using a synthesizer that has a built-in sequencer (a device or program that's capable of recording, editing, and playing back MIDI data) and two other "synths" (Figure 2.7):

1. We could start off by recording a drum track into the master synth using channel 10 (many synths are preassigned to output drum/percussion sounds on this channel).
2. Once recorded, the sequence will transmit the notes and data over channel 10, allowing the synth's percussion section to played.

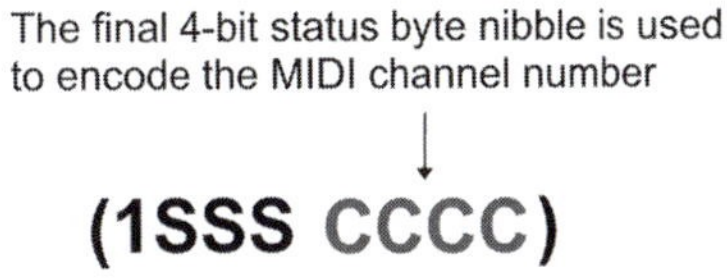

The final 4-bit status byte nibble is used to encode the MIDI channel number

↓

(1SSS CCCC)

Figure 2.6. *The least significant nibble (4 bits) of the status/channel number byte is used to encode channel number data, allowing up to 16 discrete MIDI channels to be transmitted through a single MIDI cable or designated port.*

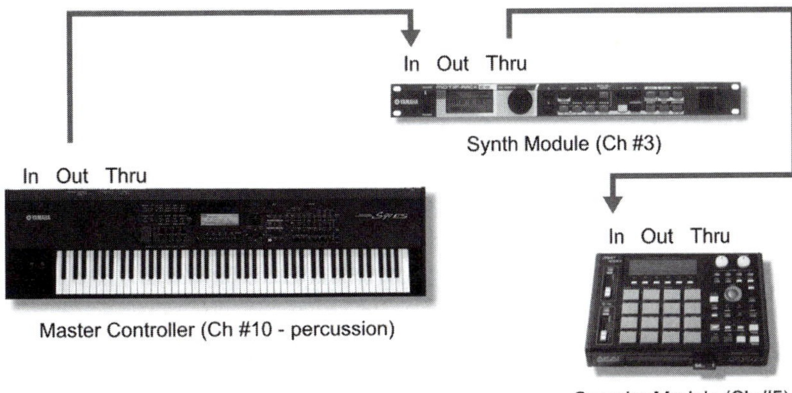

Figure 2.7. *MIDI setup showing a set of MIDI channel assignments.*

3. Next, we could set a synth module to channel 3 and instruct the master synth to transmit on the same channel (since the synth module is set to respond to data on channel 3, its generators will sound whenever the master keyboard is played). We can now begin recording a melody line into the sequencer's next track.
4. Playing back the sequence will then transmit data to both the master synth (percussion section) and the module (melody line) over their respective channels. At this point, our song is beginning to take shape.
5. Now we can set a sampler (or other instrument type) to respond to channel 5 and instruct the master synth to transmit on the same channel, allowing us to further embellish the song.
6. Now that the song's complete, the sequencer can play the musical parts to the synths on their respective MIDI channels—all in an environment that allows us to have complete control of volume, to edit, and to have a wide range of functions over each instrument. In short, we've created a true multichannel working environment.

It goes without saying that the above example is just but one of the infinite setup and channel possibilities that can be encountered in a production environment. It's often true, however, that even the most complex MIDI and production rooms will have a system … a basic channel and overall layout that makes the day-to-day operation of making music easier. This layout and the basic decisions that you might make in your own room are, of course, up to you. Streamlining a system to work both efficiently and easily will come with time, experience, and practice.

Auto Channelizing

Keeping track of the channels that are assigned to the various devices and tone generators can sometimes be a pain. As the previous paragraph suggests, it's only natural that you'll eventually come up with your own system for assigning MIDI channels to the various devices in your production studio. One of the ways to ease the pain of assigning channels throughout the studio is through the use of a channelizing feature that's built into most sequencers. In short, channelizing allows a sequencer to accept incoming data, regardless of its MIDI channel number, reassign the outgoing data to a selected track's channel, and then output it to the appropriate device. This can be best understood by trying it out for yourself:

Auto Channelizing

1. Plug a keyboard controller into a DAW or MIDI sequencer and make it the active MIDI input source.
2. Create a MIDI track and assign its channel and port numbers to one of your favorite devices (*e.g.*, transmitting on channel 1 to your favorite synth).
3. Create another MIDI track and assign its channel and port numbers to another of your favorite devices (*e.g.*, transmitting on channel 2 on another synth).

4. Create yet another MIDI track and assign its channel and port numbers to another of your favorite devices (*e.g.*, transmitting on channel 3 on yet another synth lead.
5. Next, place the first track into the Record Ready mode and play your keyboard. This should cause the MIDI data to route to channel 1, allowing the first device to begin playing.
6. Now, unselect the first track and place the second MIDI track into Record/Monitor Ready mode. This should cause the MIDI data to automatically route to channel 2, allowing the second device to begin playing.
7. Finally, place the third track into the Record Ready mode and play your keyboard. Does the output channel routing automatically change so the third device responds to the MIDI data by playing? If not, be patient. Read through the manual and check your cables. If at first you don't succeed ....

From this, you can see that the MIDI channel or port will automatically change to match that of the selected track—in effect, making MIDI channel assignment much easier and allowing you to stick to the task of making music.

MIDI Modes

Electronic instruments often vary in the number of sounds and notes that can be simultaneously produced by their internal sound generating circuitry. For example, certain instruments can only produce one note at a single time (known as a monophonic instrument), while others can generate 16, 32, and even 64 notes at once (these are known as polyphonic instruments). The latter type can easily play chords or more than one musical line on a single instrument at a time.

In addition, some instruments are only capable of producing a single generated sound patch (often referred to as a "voice") at any one time. Its generating circuitry could be polyphonic, allowing the player to lay down chords and bass or melody lines, but it can only produce these notes using a single, characteristic sound at any one time (*e.g.*, an electric piano, a synth bass, or a string patch). However, the vast majority of newer synths differ from this in that they're multi-timbral in nature, meaning that they can generate numerous sound patches at any one time (*e.g.*, an electric piano, a synth bass, and a string patch, as can be seen in Figure 2.8). That's to say that it is common to run across electronic instruments that can simultaneously generate a number of voices, each offering its own control over parameters (such as volume, panning, or modulation). Best of all, it is also common for different sounds to be assigned to their own MIDI channels, allowing multiple patches to be internally mixed within the device to a stereo output bus or independent outputs.

It should be noted that the word "patch" is a direct reference from earlier analog synthesizers, where patch chords were used to connect one sound generator block or processing function to another to create a generated sound.

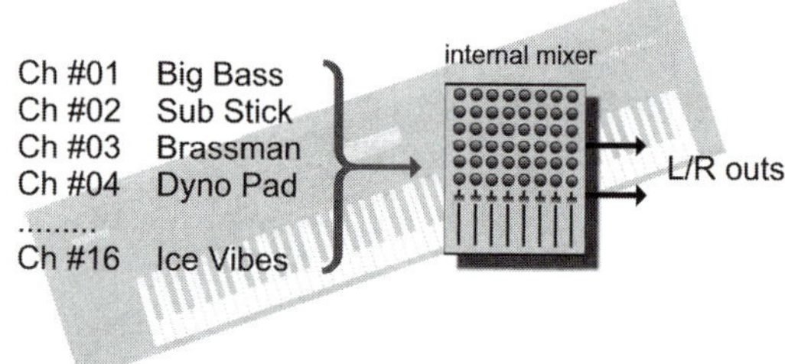

Figure 2.8. *Multi-timbral instruments are virtual bands-in-a-box that can simultaneously generate multiple patches, each of which can be assigned to its own MIDI channel.*

As a result of these differences between instruments and devices, a defined set of guidelines (known as MIDI reception modes) has been specified that allows a MIDI instrument to transmit or respond to MIDI channel messages in several ways. For example, one instrument might be programmed to respond to all 16 MIDI channels at one time, while another might be polyphonic in nature, with each voice being programmed to respond only to a single MIDI channel.

Poly/Mono

An instrument or device can be set to respond to MIDI data in either the poly mode or the mono mode. Stated simply, an instrument that's set to respond to MIDI data polyphonically will be able to play more than one note at a time. Conversely, an instrument that's set to respond to MIDI data monophonically will only be able to play a single note at any one time.

Omni On/Off

Omni on/off refers to how a MIDI instrument will respond to MIDI messages at its input. When Omni is turned on, the MIDI device will respond to all channel messages that are being received regardless of its MIDI channel assignment. When Omni is turned off, the device will respond only to a single MIDI channel or set of assigned channels (in the case of a multi-timbral instrument).

The following list and figures explain the four modes that are supported by the MIDI spec in more detail:

- *Mode 1-Omni On/Poly*—In this mode, an instrument will respond to data that's being received on any MIDI channel and then redirect this data to the instrument's base channel (Figure 2.9a). In essence, the device will play back everything that's presented at its input in a polyphonic fashion, regardless of the incoming channel designations. As you might guess, this mode is rarely used.
- *Mode 2-Omni On/Mono*—As in Mode 1, an instrument will respond to all data that's being received at its input, without regard to channel designations; however, this device will only be able to play one note at a time (Figure 2.9b). Mode 2 is used even more rarely than Mode 1, as the device can't discriminate channel designations and can only play one note at a time.

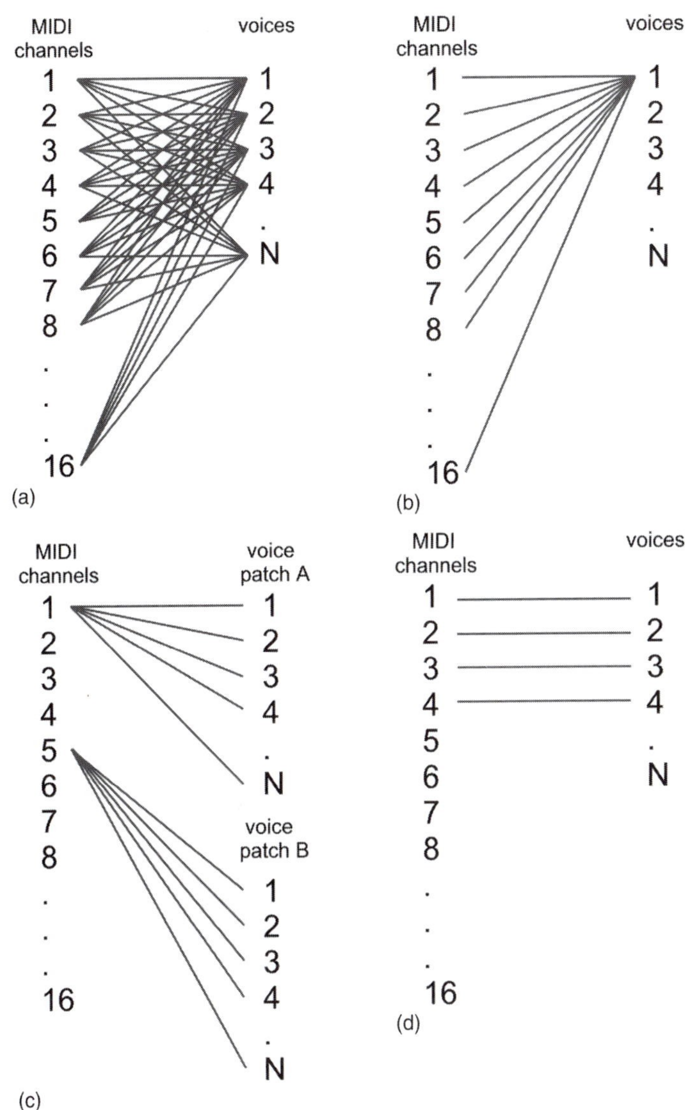

Figure 2.9. *Voice/channel assignments of the four modes that are supported by the MIDI spec: (a) omni on/poly; (b) omni on/mono; (c) omni off/poly; (d) omni off/mono.*

- *Mode 3-Omni Off/Poly*—In this mode, an instrument will only respond to data that matches its assigned base channel in a polyphonic fashion (Figure 2.9c). Data that's assigned to any other channel will be ignored. This mode is by far the most commonly used, as it allows the voices within a multi-timbral instrument to be individually controlled by messages that are being received on their assigned MIDI channels. For example, each of the 16 channels in a MIDI line could be used to independently play each of the parts in a 16-voice, multi-timbral synth.

- *Mode 4-Omni Off/Mono*—As with Mode 3, an instrument will be able to respond to performance data that's transmitted over a single, dedicated channel; however, each voice will only be able to generate one MIDI note at a time (Figure 2.9d). A practical example of this mode is often used in MIDI guitar systems, where MIDI data is monophonically transmitted over six consecutive channels (one channel/voice per string).

Base Channel

The assigned base channel determines which MIDI channel the instrument or device will respond to. For example, if a device's base channel is set to 1, it will respond to performance messages arriving on channel 1 while ignoring those arriving on channels 2 though 16.

Channel Voice Messages

Channel Voice messages are used to transmit real-time performance data throughout a connected MIDI system. They're generated whenever a MIDI instrument's controller is played, selected, or varied by the performer. Examples of such control changes could be the playing of a keyboard, pressing of program selection buttons, or movement of modulation or pitch wheels. Each Channel Voice message contains a MIDI channel number within its status byte, meaning that only devices that are assigned to the same channel number will respond to these commands. There are seven Channel Voice message types: Note-On, Note-Off, Polyphonic Key Pressure, Channel Pressure, Program Change, Pitch Bend Change, and Control Change.

Note-On Messages

A Note-On message is used to indicate the beginning of a MIDI note. It is generated each time a note is triggered on a keyboard, controller or other MIDI instrument (*i.e.*, by pressing a key, hitting a drum pad, or by playing a sequence). A Note-On message consists of 3 bytes of information (Figure 2.10):

- Note-on status/MIDI channel number
- MIDI pitch number
- Attack velocity value

The first byte in the message specifies a note-on event and a MIDI channel (1–16). The second byte is used to specify which of the possible 128 notes (numbered 0–127) will be sounded by an instrument. In general, MIDI note number 60 is assigned to the middle C key of an equally tempered keyboard, while notes 21 to 108 correspond to the 88 keys of an extended keyboard controller (Figure 2.11). The final byte is used to indicate the velocity or speed at which the key was pressed (over a value range that varies from 0 to 127). Velocity (Figure 2.12) is used to denote the loudness of a sounding note, which increases in volume with higher velocity values (although velocity can also be

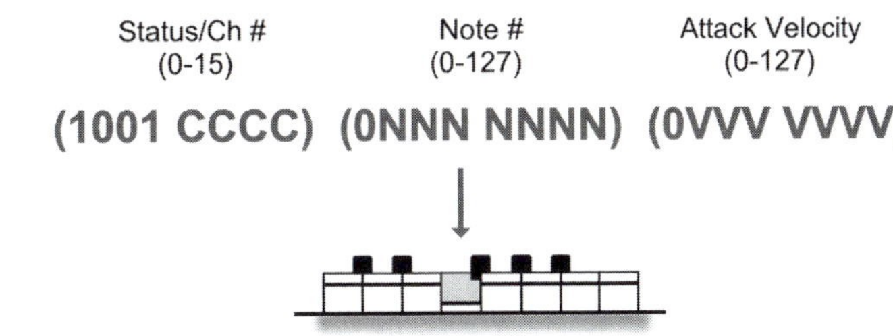

Figure 2.10. Byte structure of a MIDI Note-On message.

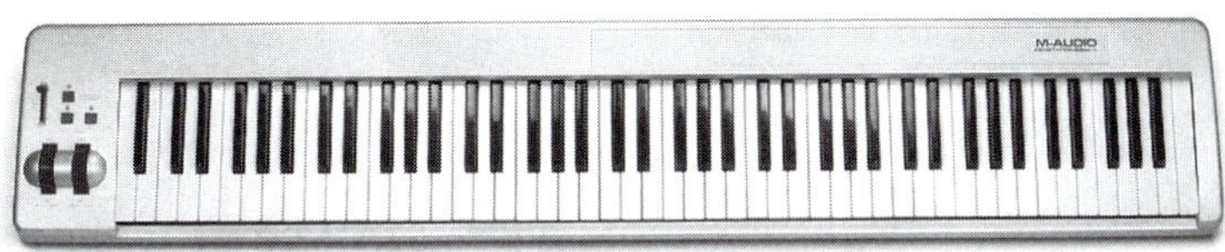

Figure 2.11. M-Audio Keystation 88es MIDI controller. (Courtesy of M-Audio, a division of Avid Technology, Inc.; www.m-audio.com.)

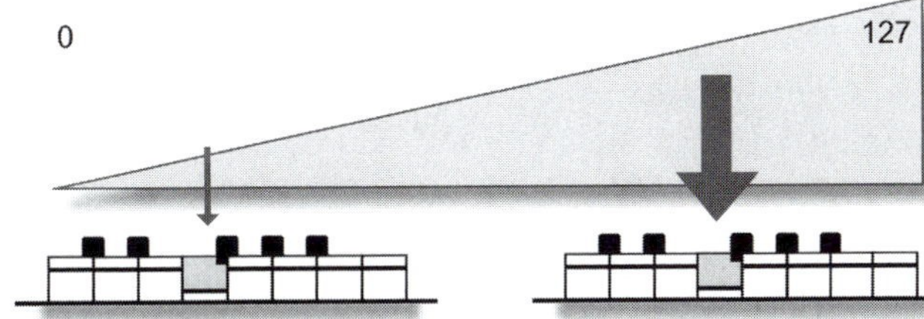

Figure 2.12. Velocity is used to communicate the volume or loudness of a note within a performance.

programmed to work in conjunction with other parameters, such as expression, control over timbre, or sample voice assignments).

Not all instruments are designed to interpret the entire range of velocity values (as with certain drum machines), and others don't respond dynamically at all. Instruments that don't support velocity information will generally transmit an attack velocity value of 64 for every note that's played, regardless of the how soft or hard the keys are actually being pressed. Similarly, instruments that don't respond to velocity messages will interpret all MIDI velocities as having a value of 64.

A Note-On message that contains an attack velocity of 0 (zero) is generally equivalent to the transmission of a Note-Off message. This common implementation tells the device to silence a currently sounding note by playing it with a velocity (volume) level of 0.

Note-Off Messages

A Note-Off message is used as a command to stop playing a specific MIDI note. Each Note-On message will continue to play until a corresponding Note-Off message for that note has been received.

In this way, the bare basics of a musical composition can be encoded as a series of MIDI note-on and note-off events. It should also be pointed out that a Note-Off message won't cut off a sound; it'll merely stop playing it. If the patch being played has a release (or final decay) slope, it will begin this stage upon receiving the message. As with the Note-On message, the note-off structure consists of 3 bytes of information (Figure 2.13):

- Note-off status/MIDI channel number
- MIDI note number
- Release velocity value

In contrast to the dynamics of attack velocity, the release velocity value (0–127) indicates the velocity or speed at which the key was released. A low value indicates that the key was released very slowly, whereas a high value shows that the key was released quickly. Although not all instruments generate or respond to MIDI's release velocity feature, instruments that are capable of responding to these values can be programmed to vary a note's speed of decay, often reducing the signal's decay time as the release velocity value is increased.

All Notes Off

On the odd occasion (often when you least expect it), a MIDI note can get stuck! This can happen when data drops out or a cable gets disconnected, creating a situation where a note receives a Note-On message but not a Note-Off message, resulting in a note that continues to plaaaaaaaaaaayyyyyyyyyy! Since you're often too annoyed or under pressure to take the time to track down which note is the offending sucka, it's generally far easier to transmit an All Notes Off message that silences everything on all channels and ports. If you get stuck in such a situation, pressing a "panic button" that might be built into your sequencer or hardware MIDI devices could save your ears and sanity.

Pressure (Aftertouch) Messages

Pressure messages (often referred to as "AfterTouch") occur after you've pressed a key and then decide to press down harder on it. For devices that can respond to (and therefore generally transmit)

Figure 2.13. Byte structure of a MIDI Note-Off message.

Status/Ch # (0-15)	Note # (0-127)	Release Velocity (0-127)
(1000 CCCC)	**(0NNN NNNN)**	**(0VVV VVVV)**

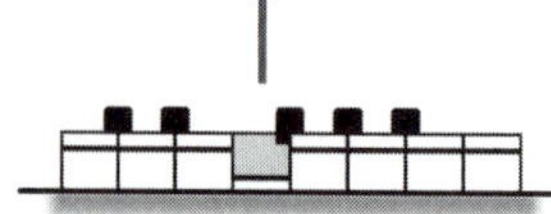

these messages, AfterTouch can often be assigned to such parameters as vibrato, loudness, filter cutoff, and pitch. Two types of Pressure messages are defined by the MIDI spec:

- Channel Pressure
- Polyphonic Key Pressure

Channel Pressure Messages

Channel Pressure messages are commonly transmitted by instruments that respond only to a single, overall pressure, regardless of the number of keys that are being played at any one time (Figure 2.14). For example, if six notes are played on a keyboard and additional aftertouch pressure is applied to just one key, the assigned parameter would be applied to all six notes. A Channel Pressure message consists of 3 bytes of information (Figure 2.15):

- Channel pressure status/MIDI channel number
- MIDI note number
- Pressure value

Polyphonic Key Pressure Messages

Polyphonic Key Pressure messages respond to pressure changes that are applied to the individual keys of a keyboard. That's to say that a suitably equipped instrument can transmit or respond to

single channel pressure message

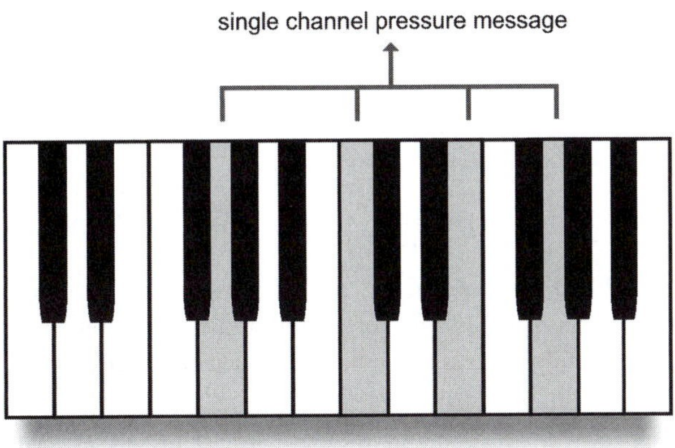

Figure 2.14. *Channel Pressure messages simultaneously affect all notes that are being transmitted over a MIDI channel.*

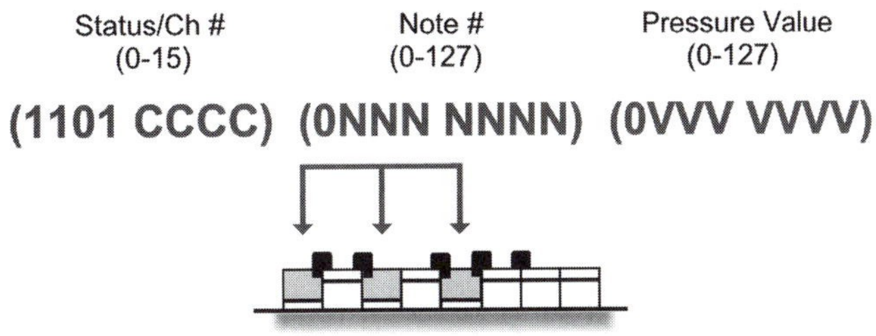

Status/Ch #
(0-15)
Note #
(0-127)
Pressure Value
(0-127)
(1101 CCCC) (0NNN NNNN) (0VVV VVVV)

Figure 2.15. *Byte structure of a MIDI Channel Pressure message.*

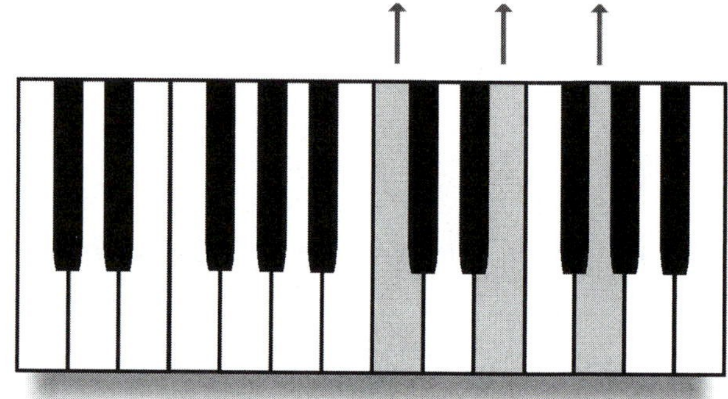

Figure 2.16. Individual Polyphonic Key Pressure messages are generated when additional pressure is applied to each key that's played.

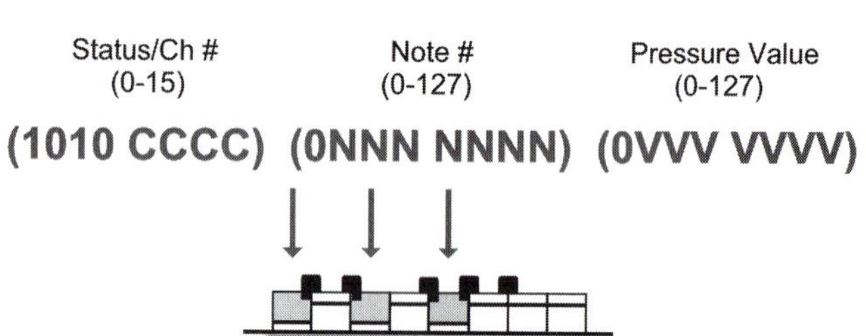

Figure 2.17. Byte structure of a MIDI Polyphonic Key Pressure message.

individual pressure messages for each key that's depressed (Figure 2.16). How a device responds to these messages will often vary from manufacturer to manufacturer (or can be assigned by the user); however, pressure values are commonly assigned to such performance parameters as vibrato, loudness, timbre, and pitch. Although controllers that are capable of producing polyphonic pressure are generally more expensive, it's not uncommon for an instrument to respond to these messages.

A Polyphonic Key Pressure message consists of 3 bytes of information (Figure 2.17):

- Polyphonic key pressure status/MIDI channel number
- MIDI note number
- Pressure value

Program Change Messages

Program Change messages are used to change the active program or preset number of a MIDI instrument or device. A preset is a user- or factory-defined number that actively selects a specific sound patch or system setup. Using this extremely handy message, up to 128 presets can be remotely selected from another device or controller; for example:

- A Program Change message can be transmitted from a remote keyboard or controller to an instrument, allowing sound patches to be remotely switched (Figures 2.18 and 2.19).

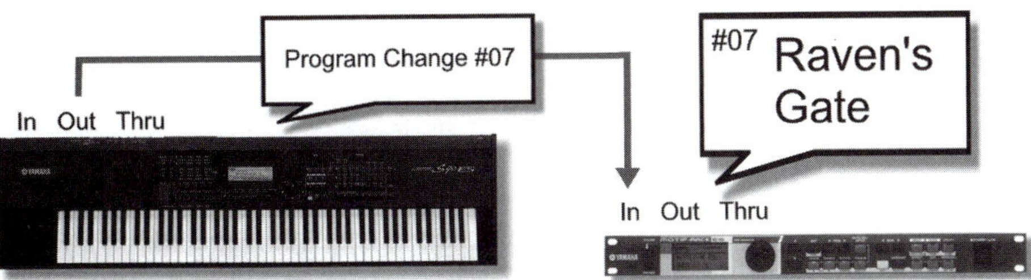

Figure 2.18. *Program Change messages can be used to change sound patches from a sequencer or from a remote controller.*

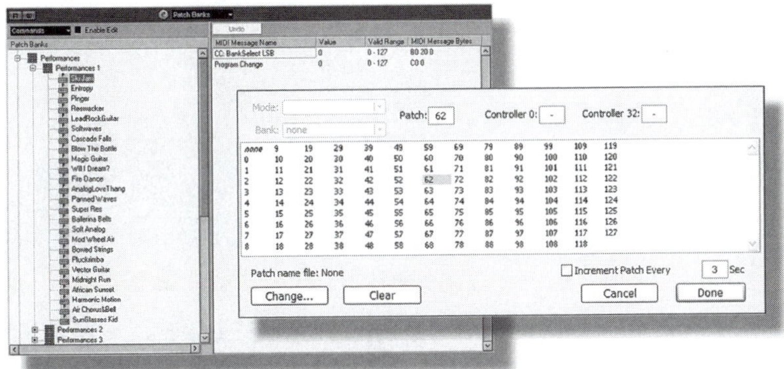

Figure 2.19. *Workstations and sequencer software systems will often allow patches to be recalled via Program Change messages. (Courtesy of Steinberg Media Technologies GmbH, A Division of Yamaha Corporation, www.steinberg.net, and Digidesign, A Division of Avid Technology, www.digidesign.com.)*

- Program Change messages could be programmed at the beginning of a sequence to instruct the various instruments or voice generators to set to the correct sound patch before playing.
- A Program Change message could be used to alter patches on an effects device, either in the studio or on stage.
- The list goes on …

A Program Change message (Figure 2.20) consists of 2 bytes of information:

- Program change status/MIDI channel number (1–16)
- Program ID number (0–127)

Pitch Bend Messages

Pitch bend sensitivity refers to the response sensitivity (in semitones) of a pitch-bend wheel or other pitch-bend controller (which, as you'd expect, is used to bend the pitch of a note upward

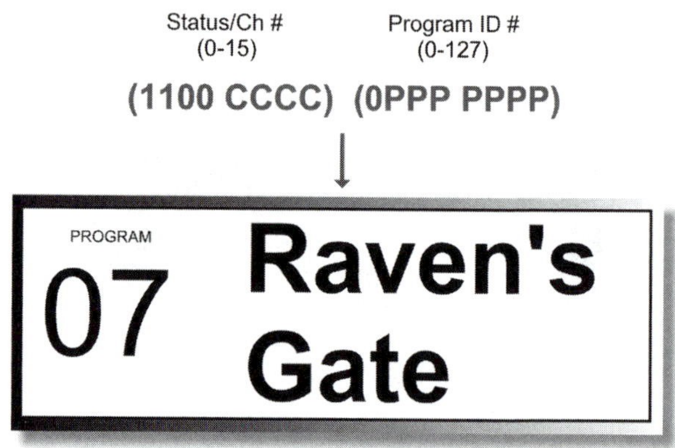

Figure 2.20. *Byte structure of a MIDI Program Change message.*

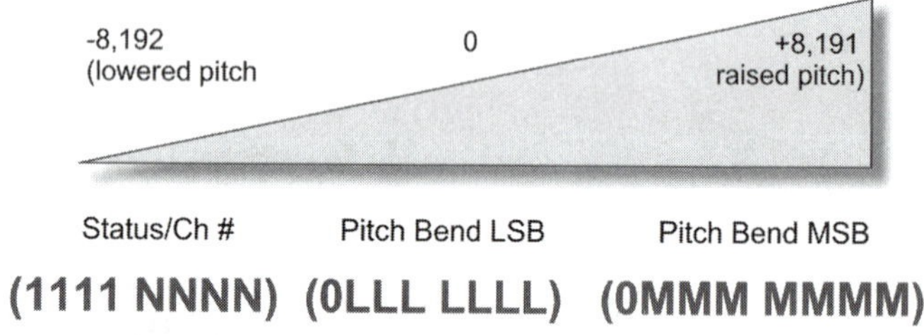

Figure 2.21. *Byte structure of a Pitch Bend message.*

Figure 2.22. *Pitch-bend wheel data value ranges.*

or downward). Since the ear can be extremely sensitive to changes in pitch, this control parameter is encoded using 2 data bytes, yielding a total of 16,384 steps. Since this parameter is most commonly affected by varying a pitch wheel (Figures 2.21 and 2.22), the control values range from −8192 to +8191, with 0 being the instrument's or part's unaltered pitch. Although the General MIDI spec recommends that the pitch-bend wheel have a range of ±2 half steps, most devices allows the pitch-bend range to be varied upwards to a full octave.

Control Change Messages

Control Change messages are used to transmit information to a device (either internally or through a MIDI line/network) that relates to real-time control over its performance parameters.

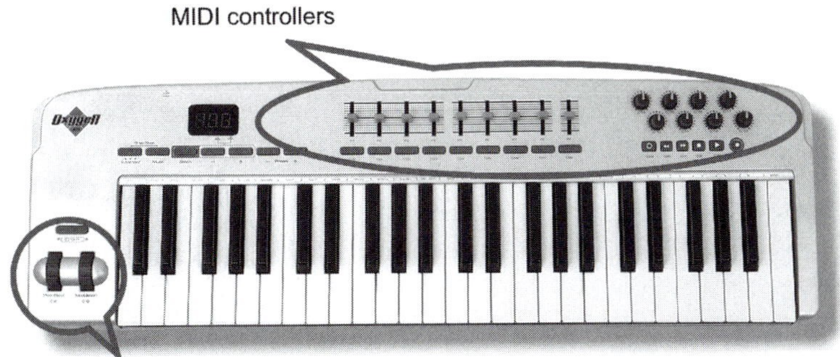

Figure 2.23. *M-audio controller. (Courtesy of M-Audio, A Division of Avid Technology, Inc.; www.m-audio.com.)*

Three types of Control Change messages can be transmitted via MIDI:

- *Continuous controllers*—Controllers that relay a full range of variable control settings (often ranging in value from 0 to 127, although, in certain cases, two controller messages can be combined in tandem to achieve a greater resolution)
- *Switch controllers*—Controllers that have either an "off" or an "on" state with no intermediate settings
- *Channel mode message controllers*—Controllers that range from controller numbers 120 through 127 and are used to set the note sounding status, instrument reset, local control on/off, all notes off, and MIDI mode status of a device or instrument

A single Control Change message or a stream of such messages is transmitted whenever controllers (such as foot switches, foot pedals, pitch-bend wheels, modulation wheels, or breath controllers) are varied in real time (Figure 2.23). Newer controllers and software editors often offer up a wide range of switched and variable controllers, allowing for extensive, user-programmable control over any number of device, voice, and mixing parameters in real time (Figure 2.24).

A Control Change message (Figure 2.25) consists of 3 bytes of information:

- Control change status/MIDI channel number (1–16)
- Controller ID number (0–127)
- Corresponding controller value (0–127)

Controller ID Numbers

As you can see, the second byte of the Control Change message is used to denote the controller ID number. This all-important value is used to specify which of the device's program or performance parameters are to be addressed. Before we move on to discuss the various control numbers

Figure 2.24. Alternate modulation/pitch and touch-pad controller devices. (Courtesy of Novation; www.novationmusic.com.)

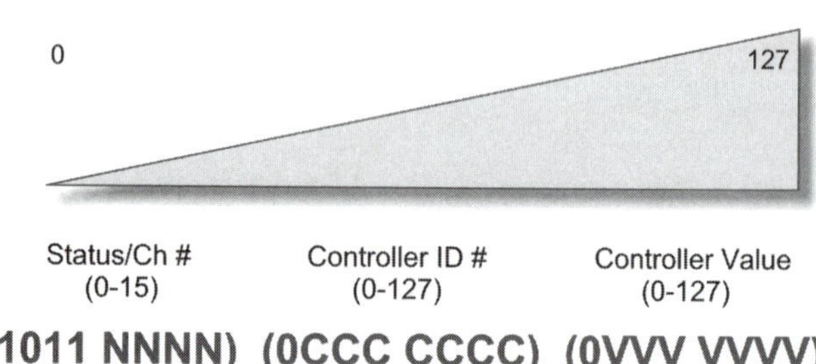

Figure 2.25. Byte structure of a Control Change message.

that are listed in the MIDI spec, it's best to take time out to discuss the values that are assigned to these parameters (*i.e.*, the third and subsequent bytes).

Controller Values

The third byte of the Control Change message is used to denote the controller's actual data value. This value is used to specify the position, depth, or level of a parameter. Here are a few examples as to how these values can be implemented to vary control and mix parameters:

- In the case of a variable control parameter that doesn't require that the settings be made in extremely fine increments, a 7-bit continuous controller allows for values over a 128-step range (with 0 being the minimum and 127 being the maximum value.
- The value range of the pan controller falls between 0 (hard left) and 127 (hard right), with a value of 64 representing a balanced center position (Figure 2.26).
- The value range of a switch controller is often 0 (off) and 127 (on) (Figure 2.27); however, it's not uncommon for a switching function to respond to continuous controller messages by recognizing the value range of 0 to 63 as being "off" and 64 to 127 as being "on."

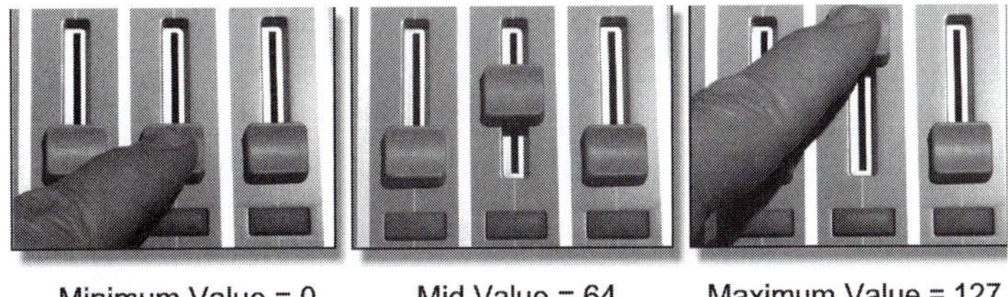

Figure 2.26. *Continuous controller data value ranges.*

Figure 2.27. *Switch controllers can be controlled by on/off buttons; however, switch functions also respond to a range of continuous controller values, such that 0 to 63 is seen as being "off" while 64 to 127 is seen as being "on."*

In certain cases, a single, 7-bit "course" message (128 steps) might not be enough to adequately manage a controller's resolution. For this reason, the MIDI spec allows the resolution of a number of parameters to be increased by adding an additional "fine" controller value message to the data stream, resulting in a resolution that yields an overall total of 16,384 discrete steps!

Explanation of Controller ID Parameters

The following sections detail the general categories and conventions for assigning controller numbers to an associated parameter, as specified by the 1995 update of the MMA (MIDI Manufacturers Association, www.midi.org). An overview of these controllers can be seen in Table 2.2. This is definitely an important chart to earmark, as these numbers will be an important guide to knowing or finding the right ID number that can help you on your path toward finding that perfect variable to make it sound right.

Bank Select

Numbers: 0 (coarse), 32 (fine)

Nowadays, it's common for an instrument or device to have more than 128 presets, patches, and general settings. Since MIDI Program Change messages can only switch between 128 settings, the Bank Select controller (sometimes called Bank Switch) can be used to switch between multiple groupings of 128 presets. As an example, a device that has 256 presets could be easily

Table 2.2. Controller ID Numbers, Outlining Both the Defined Format and Conventional Controller Assignments

	Control Number	*Parameter*
	14-Bit Controllers Coarse	*Most Significant Bit (MSB)*
0	Bank Select	0–127
1	Modulation Wheel or Lever	0–127
2	Breath Controller	0–127
3	Undefined	0–127
4	Foot Controller	0–127
5	Portamento Time	0–127
6	Data Entry MSB	0–127
7	Channel Volume (formerly Main Volume)	0–127
8	Balance	0–127
9	Undefined	0–127
10	Pan	0–127
11	Expression Controller	0–127
12	Effect Control 1	0–127
13	Effect Control 2	0–127
14	Undefined	0–127
15	Undefined	0–127
16–19	General Purpose Controllers 1–4	0–127
20–31	Undefined	0–127
	14-Bit Controllers Fine	*Least Significant Bit (LSB)*
32	LSB for Control 0 (Bank Select)	0–127
33	LSB for Control 1 (Modulation Wheel or Lever)	0–127
34	LSB for Control 2 (Breath Controller)	0–127
35	LSB for Control 3 (Undefined)	0–127
36	LSB for Control 4 (Foot Controller)	0–127
37	LSB for Control 5 (Portamento Time)	0–127
38	LSB for Control 6 (Data Entry)	0–127
39	LSB for Control 7 (Channel Volume, formerly Main Volume)	0–127
40	LSB for Control 8 (Balance)	0–127
41	LSB for Control 9 (Undefined)	0–127
42	LSB for Control 10 (Pan)	0–127
43	LSB for Control 11 (Expression Controller)	0–127
44	LSB for Control 12 (Effect Control 1)	0–127
45	LSB for Control 13 (Effect Control 2)	0–127
46–47	LSB for Control 14–15 (Undefined)	0–127
48–51	LSB for Control 16–19 (General Purpose Controllers 1–4)	0–127
52–63	LSB for Control 20–31 (Undefined)	0–127
	7-Bit Controllers	
64	Damper Pedal On/Off (Sustain)	<63 off, >64 on
65	Portamento On/Off	<63 off, >64 on

(Continued)

Table 2.2. (Continued)

	Control Number	*Parameter*
	7-Bit Controllers	*Most Significant Bit (MSB)*
66	Sustenuto On/Off	<63 off, >64 on
67	Soft Pedal On/Off	<63 off, >64 on
68	Legato Footswitch	<63 normal, >64 legato
69	Hold 2	<63 off, >64 on
70	Sound Controller 1 (default: Sound Variation)	0–127 LSB
71	Sound Controller 2 (default: Timbre/Harmonic Intens.)	0–127 LSB
72	Sound Controller 3 (default: Release Time)	0–127 LSB
73	Sound Controller 4 (default: Attack Time)	0–127 LSB
74	Sound Controller 5 (default: Brightness)	0–127 LSB
75	Sound Controller 6 (default: Decay Time; see MMA RP-021)	0–127 LSB
76	Sound Controller 7 (default: Vibrato Rate; see MMA RP-021)	0–127 LSB
77	Sound Controller 8 (default: Vibrato Depth; see MMA RP-021)	0–127 LSB
78	Sound Controller 9 (default: Vibrato Delay; see MMA RP-021)	0–127 LSB
79	Sound Controller 10 (default undefined; see MMA RP-021)	0–127 LSB
80–83	General Purpose Controller 5–8	0–127 LSB
84	Portamento Control	0–127 LSB
85–90	Undefined	—
91	Effects 1 Depth (default: Reverb Send Level)	0–127 LSB
92	Effects 2 Depth (default: Tremolo Level)	0–127 LSB
93	Effects 3 Depth (default: Chorus Send Level)	0–127 LSB
94	Effects 4 Depth (default: Celeste <Detune> Depth)	0–127 LSB
95	Effects 5 Depth (default: Phaser Depth)	0–127 LSB
	Controllers	
96	Data Increment (Data Entry +1)	—
97	Data Decrement (Data Entry –1)	—
98	Non-Registered Parameter Number (NRPN)	0–127 LSB
99	Non-Registered Parameter Number (NRPN)	0–127 MSB
100	Registered Parameter Number (RPN)	0–127 LSB
101	Registered Parameter Number (RPN)	0–127 MSB
102–119	Undefined	—
	Reserved for Channel Mode Messages	
120	All Sound Off 0	
121	Reset All Controllers	
122	Local Control On/Off (0 off, 127 on)	
123	All Notes Off	
124	Omni Mode Off	
125	Omni Mode On	
126	Poly Mode On/Off	
127	Poly Mode On	

divided into 2 banks of 128. Thus, if you wanted to call up preset 129, you would transmit a Bank Select message that would switch the system to its second bank and then follow this with a Program Change message that would select the first program in that bank. This parameter can be used on different channels to affect each part within a multi-timbral instrument.

Note: Bank Select can sometimes be used to switch between drum kits on multi-timbral devices that offer up percussion.

MOD Wheel

Numbers: 1 (coarse), 33 (fine)

This message is used to set the modulation (or MOD) wheel value. When modulation is used, a vibrato effect is introduced, often by introducing an low-frequency oscillator (LFO) effect. Although a fine modulation message can be used to give additional control, many units will often ignore the presence of this message. This parameter can be used on different channels to affect each part within a multi-timbral instrument.

Breath Control

Numbers: 2 (coarse), 34 (fine)

Breath Control is often used by wind players to add expression to a performance. As such, this message can be user assigned to control a wide range of parameters (such as aftertouch, modulation, or expression). This parameter can be used on different channels to affect each part within a multi-timbral instrument.

Foot Pedal

Numbers: 4 (coarse), 36 (fine)

This controller often makes use of a foot pedal, which uses a potentiometer to continuously control such user-assignable parameters as aftertouch, modulation, and expression. Although a foot controller can vary in value between 0 and 127, if it's being used as a foot switch, values ranging from 0 to 63 will be interpreted as being "off," while those ranging from 64 to 127 will be interpreted as being "on." This parameter can be used on different channels to affect each part within a multi-timbral instrument.

Portamento Time

Numbers: 5 (coarse), 37 (fine)

Portamento can be used to slide one note into another, rather than having the second note immediately follow the first. This value is used to determine the rate of slide (in time) between the two notes. This parameter can be used on different channels to affect each part within a multi-timbral instrument.

Data Entry

Numbers: 6 (coarse), 38 (fine)

This message value can be used to vary a Registered or Non-Registered Parameter. Registered Parameter Numbers (RPNs) and Non-Registered Parameter Numbers (NPRNs) are used to effect parameters that are specific to a particular MIDI device or aren't controllable by a defined

Control Change message. For example, Pitch Bend Sensitivity and Fine Tuning parameters are RPN parameters. Varying these values can be used to remotely alter these settings.

Note: Certain devices aren't able to control RPN or NRPN messages using this parameter. In such a case, it's often possible to directly assign a defined parameter to this message type.

Channel Volume (formerly known as Main Volume)

Numbers: 7 (coarse), 39 (fine)

This parameter affects the volume level of a device or individual musical parts (in the case of a multi-timbral instrument). This parameter is independent of individual note velocities and allows for the overall part or device volume that's transmitted over a channel to be altered.

Note: Fine resolution messages (controller 39) are rarely supported by most instruments and mixing devices. If a nonsupporting device receives these messages, they will simply be ignored.

Stereo Balance

Numbers: 8 (coarse), 40 (fine)

A Stereo Balance controller is used to vary the relative levels between two independent sound sources (Figure 2.28). As with the balance control on a stereo preamplifier, this controller is used to set the relative left/right balance of a stereo signal (unlike pan, which is used to place a mono signal within a stereo field). The value range of this controller falls between 0 (full left sound source) and 127 (full right sound source), with a value of 64 representing an equally balanced stereo field.

Pan

Numbers: 10 (coarse), 42 (fine)

A Pan controller is used to position the relative balance of a single sound source between the left and right channels of a stereo sound field (Figure 2.29). The value range of this controller falls between 0 (hard left) and 127 (hard right), with a value of 64 representing a balanced center position. Pan is a common parameter, in that it can be used to internally mix all of the parts of a multi-timbral or sound mixing device to a single, bused output.

Expression

Numbers: 11 (coarse), 43 (fine)

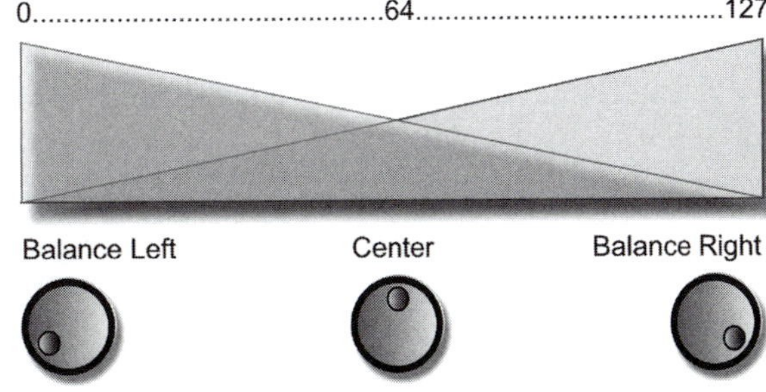

Figure 2.28. *Stereo Balance controller data value ranges and corresponding settings.*

Figure 2.29. Pan controller data value ranges.

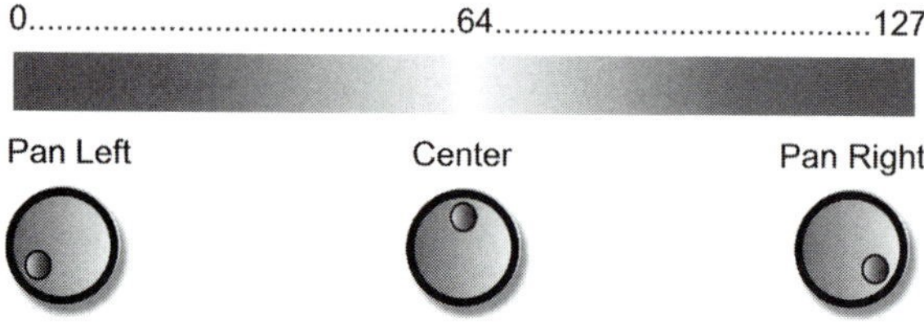

An Expression controller is used to accent the existing level settings of a MIDI instrument or device (*e.g.*, when a crescendo or decrescendo is needed). This control works by dividing a part's overall volume into 128 steps (or 16,384 if fine resolution is used), thereby making it possible to alter level over the course of a passage, without having to change the velocity levels of each note within that passage. Whenever the expression control is set to its maximum value (*i.e.*, 127 in coarse resolution), the output level will be set to the part's actual main volume setting. As the values decrease, the volume settings will proportionately decrease; that is, a value of 64 will reduce the signal to 50% of its full level, and 0 will reduce the output to 0% of its channel volume level (effectively turning the gain all the way down). It should be noted that the fine controller byte (43) is often ignored on most receiving systems.

Effect Control 1 and 2

Numbers: 12 and 13 (coarse), 44 and 45 (fine)

These controllers can be used to vary any parameter relating to an effects device (such as the reverb decay time for a reverb unit that's built into a GM sound module). This parameter can be used on different channels to affect each part within a multi-timbral instrument.

Note: Controllers 91 through 95 can be used to vary overall effect level, chorus, and phaser.

General Purpose Controllers

Numbers: 16 through 19 (coarse), 48 and 51 (fine)

These controllers are open to use for varying any parameter that can be assigned to them.

Note: Fine-resolution General Purpose messages are rarely supported by most instruments and mixing devices.

Hold Pedal

Number: 64

This control is used to sustain (hold) notes that are currently being played. When on, the notes will continue to play, even though the keys have been released, and Note Off messages will be

> The remaining controller parameters are capable only of being encoded using a single, 7-bit "coarse" message that will yield a total of 128 continuous steps or can be translated in value ranges that represent a switched on/off state.

transmitted only after the hold pedal has been switched off. This parameter can be used on different channels to affect each part within a multi-timbral instrument.

Portamento On/Off

Number: 65

This control determines whether the portamento effect is on or off. This parameter can be used on different channels to affect each part within a multi-timbral instrument.

Sustenuto Pedal

Number: 66

This control works much like a Hold Pedal message, with the exception that it will only sustain notes that are currently being played (*i.e.*, Note On messages have been transmitted, but their respective Note Off messages haven't). This has the effect of holding keys that were initiated when the pedal was pressed while not holding keys that are subsequently played during the sustain time. For example, this pedal could be used to hold an initial chord while not holding a melody that's being played around it. This parameter can be used on different channels to affect each part within a multi-timbral instrument.

Soft Pedal

Number: 67

When on, the volume of any played note is lowered. This parameter can be used on different channels to affect each part within a multi-timbral instrument.

Legato Pedal

Number: 68

This control is used to create a legato (or smooth transition) effect between notes. This is often achieved by not sounding the attack portion of the part's voltage-controlled amplifier (VCA) envelope. This effect can be used to simulate the phrasing of wind and brass players, as well as guitar pull-offs and hammer-ons (where secondary notes are not picked). This parameter can be used on different channels to affect each part within a multi-timbral instrument.

Hold 2 Pedal

Number: 69

Unlike the other Hold Pedal controller, this control doesn't permanently sustain a sounded note until the musician releases the pedal. It is used to lengthen the release time of a played note, effectively increasing the note's fade-out time by lengthening the VCA's release time.

Sound Variation (Sound Controller 1)

Number: 70

This control affects any parameter that's associated with the reproduction of sound. Sound Controller 1 can be used to alter or tune a soundfile during playback by varying its sample rate. This parameter can be used on different channels to affect each part within a multi-timbral instrument.

Sound Timbre (Sound Controller 2)

Number: 71

This control affects any parameter that's associated with the reproduction of sound. Sound Controller 2 can be used as a brightness control by varying the envelope of a voltage-controlled filter (VCF).

Note: VCA and VCF parameters can also be varied through the use of other controllers.

Sound Release Time (Sound Controller 3)

Number: 72

This controls the length of time that it takes a sound to fade out by altering the VCA's release time envelope. This parameter can be used on different channels to affect each part within a multi-timbral instrument.

Note: VCA and VCF parameters can also be varied through the use of other controllers.

Sound Attack Time (Sound Controller 4)

Number: 73

This controls the length of time that it takes a sound to fade in, by altering the VCA's attack time. This parameter can be used on different channels to affect each part within a multi-timbral instrument.

Note: VCA and VCF parameters can also be varied through the use of other controllers.

Sound Brightness (Sound Controller 5)

Number: 74

This parameter acts as a timbral brightness control by varying the VCF's cutoff filter frequency. This parameter can be used on different channels to affect each part within a multi-timbral instrument.

Note: VCA and VCF parameters can also be varied through the use of other controllers.

Sound Controller 6, 7, 8, 9, and 10

Numbers: 75, 76, 77, 78, and 79

These five additional controllers can be freely used to affect any assignable parameter that's associated with the audio production circuitry of a sound module. This parameter can be used on different channels to affect each part within a multi-timbral instrument.

General Purpose Buttons

Numbers: 80, 81, 82, and 83

These four General Purpose buttons can be freely used to affect any assignable on/off parameter within a device. This parameter can be used on different channels to affect each part within a multi-timbral instrument.

Portamento Control

Number: 84

A Portamento Control message (in conjunction with a MIDI note number) is used to indicate the starting note of a portamento (whereby one note slides into the next rather than having the second note immediately follow the first). The slide rate will be set by the Portamento Time (Controller 5), and the current status of the Portamento On/Off (Controller 65) is ignored.

Effects Level

Number: 91

This control is used to control the effect level or wet/dry balance for an instrument or part (often referring to its reverb or delay level). For devices with built-in effects, this parameter can be used on different channels to affect each part within a multi-timbral instrument.

Tremolo Level

Number: 92

This control is used to control the tremolo level within an instrument or part. This parameter can be used on different channels to affect each part within a multi-timbral instrument.

Chorus Level

Number: 93

By combining two identical (and often slightly delayed) signals that are slightly detuned in pitch from one another, another effect known as chorusing can be created. Chorusing is an effects tool that's often used by guitarists, vocalists, and other musicians to add depth, richness, and harmonic structure to their sound. This control is used to control the chorus level or wet/dry balance for an instrument or part. This parameter can be used on different channels to affect each part within a multi-timbral instrument.

Celeste Level

Number: 94

This control is used to control the celeste (detune) level for an instrument or part. This parameter can be used on different channels to affect each part within a multi-timbral instrument.

Phaser Level (Set it on stun, Scotty!)

Number: 95

This control is used to control a phasing effect level or wet/dry balance for an instrument or part. This parameter can be used on different channels to affect each part within a multi-timbral instrument.

Data Button Increment

Number: 96

This control causes a parameter (most often a Registered or Non-Registered Parameter) to increase its current value by 1.

Data Button Decrement

Number: 97

This control causes a parameter (most often a Registered or Non-Registered Parameter) to decrease its current value by one.

Non-Registered Parameter Number (NRPN)

Numbers: 99 (coarse), 98 (fine)

This control determines which Data Button or Data Entry controller will be incremented or decremented. The assignment of this parameter is entirely up to the manufacturer and doesn't have to be registered with the International MIDI Association (IMA).

Registered Parameter Number (RPN)

Numbers: 101 (coarse), 100 (fine)

This control determines which Data Button or Data Entry controller will be incremented or decremented. The functional parameter assignments and their designations are determined by the IMA.

All Sound Off

Number: 120

This control is used to turn off all sounding notes that were turned on by received Note On messages but which haven't yet been turned off by respective Note Off messages.

All Controllers Off

Number: 121

This control is used to reinitialize all of the controllers (continuous, switch, and incremental) within one or more receiving MIDI instruments or devices to their standard, power-up default state. Upon receipt of this message, all switched parameters will be turned off and continuous controllers will be set to their minimum values. This parameter can be used on different channels to affect each part within a multi-timbral instrument.

Local Keyboard On/Off

Number: 122

The Local Keyboard (Local Control) On/Off message is used to disconnect the controller of a MIDI instrument from its own internal voice generators (Figure 2.30). This feature is useful for turning a keyboard instrument into a master controller by disconnecting its keyboard from its sound generating circuitry. For example, when the instrument's local control is switched off, a synth can be used to output MIDI to a sequencer or other devices without having to listen to its own internal sounds.

With the Local turned off, you could easily route MIDI through your computer/sequencer and back into the device's own internal voice generators. In short, the local control feature splits your instrument into two parts: a master controller for playing other instruments in the system and a performance instrument that can be viewed as any other instrument in the MIDI setup. It's a setup that effectively makes the system more flexible while reducing setup and performance conflicts.

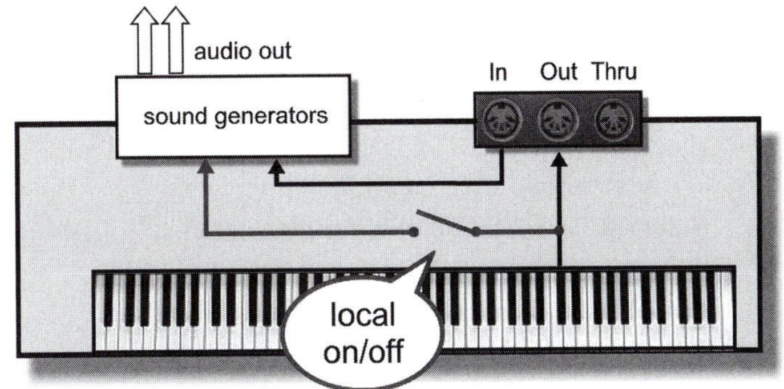

Figure 2.30. *The Local Keyboard On/Off function is used to disconnect a device's performance controller from its internal sound generators.*

It should be noted that turning the Local switch to "off" is important within a sequencer/DAW production setup, as this allows the respective devices, instruments, and parts within a setup to be heard without listening to sounds that would otherwise be playing from the master synth. Of course, if you're using a MIDI controller this setting doesn't apply, as it has no internal sound generators. A Local Control message consists of 2 bytes of information: a MIDI channel number and a Local Control on/off status byte.

All Notes Off

Number: 123

Occasionally, a MIDI instrument will receive a Note-On message, and (by some technical glitch) the following Note-Off message is somehow ignored or not received. This unfortunate event often results in a stuck note that continues to sound until a Note-Off message is received for that note. As an alternative to frantically searching for the right note-off key on the right MIDI channel, an All Notes Off "panic message" can be transmitted that effectively turns off all 128 notes on all channels and ports. Often, a MIDI interface or sequencer will include a button that can globally transmit this message throughout the connected system.

Omni Mode Off

Number: 124

Omni Mode Off refers to how an instrument will respond to MIDI messages at its input. Upon the reception of an Omni Mode Off message, a MIDI instrument or device will only respond to a single MIDI channel or set of assigned channels (in the case of a multi-timbral instrument).

Omni Mode On

Number: 125

Omni Mode On refers to how an instrument will respond to MIDI messages at its input. Upon the reception of an Omni Mode On message, a MIDI instrument or device will respond to all channel messages that are being received, regardless of its MIDI channel assignment.

Monophonic Operation
Number: 126

Upon receiving a Mono Mode On message, a MIDI instrument will assign individual voices to the consecutive MIDI channels, starting from the lowest currently assigned or base channel. The instrument will then be limited to playing only one note at a time on each MIDI channel, even if it's capable of playing multiple notes at any one time.

Polyphonic Operation
Number: 127

Upon receiving a Poly Mode On message, a MIDI instrument will assign voices to the consecutive MIDI channels, starting from the lowest currently assigned or base channel. The instrument will then respond to MIDI channels polyphonically, allowing the device to play more than one note at a time over a given channel or number of channels.

System Messages

As the name implies, System messages are globally transmitted to every MIDI device in the MIDI chain. This is accomplished because MIDI channel numbers aren't addressed within the byte structure of a System message. Thus, any device will respond to these messages, regardless of its MIDI channel assignment. The three System message types are System-Common messages, System Real-Time messages, and System-Exclusive messages.

System-Common Messages

System-Common messages are used to transmit MIDI time code, song position pointer, song select, tune request, and end-of-exclusive data messages throughout the MIDI system or 16 channels of a specified MIDI port.

MTC Quarter-Frame Messages
MIDI time code (MTC) provides a cost effective and easily implemented way to translate SMPTE (a standardized synchronization time code) into an equivalent code that conforms to the MIDI 1.0 spec. It allows time-based codes and commands to be distributed throughout the MIDI chain in a cheap, stable, and easy-to-implement way. MTC Quarter-Frame messages are transmitted and recognized by MIDI devices that can understand and execute MTC commands. A grouping of eight quarter frames is used to denote a complete time-code address (in hours, minutes, seconds, and frames), allowing the SMPTE address to be updated every two frames. Each Quarter-Frame message contains 2 bytes. The first is a quarter-frame common header, while the second byte contains a 4-bit nibble that represents the message number (0–7). A final nibble is used to encode the time field (in hours, minutes, seconds, or frames). More in-depth coverage of MIDI time code can be found in Chapter 11.

Song Position Pointer Messages

As with MIDI time code, the song position pointer (SPP) lets you synchronize a sequencer, tape recorder, or drum machine to an external source from any measure position within a song. The SPP message is used to reference a location point in a MIDI sequence (in measures) to a matching location within an external device. This message provides a timing reference that increments once for every six MIDI clock messages (with respect to the beginning of a composition). Unlike MTC (which provides the system with a universal address location point), SPP's timing reference can change with tempo variations, often requiring that a special tempo map be calculated in order to maintain synchronization. Because of this fact, SPP is used far less often than MIDI time code. SPP messages are generally transmitted while the MIDI sequence is stopped, allowing MIDI devices equipped with SPP to chase (in a fast-forward motion) through the song and lock to the external source once relative sync is achieved. More in-depth coverage of the SPP can be found in Chapter 11.

Song Select Messages

Song Select messages are used to request a specific song from a drum machine or sequencer (as identified by its song ID number). Once selected, the song will thereafter respond to MIDI Start, Stop, and Continue messages.

Tune Request Messages

The Tune Request message is used to request that a MIDI instrument initiate its internal tuning routine (if so equipped).

End of Exclusive Messages

The transmission of an End of Exclusive (EOX) message is used to indicate the end of a System-Exclusive message. In-depth coverage of System-Exclusive messages will be discussed later in this chapter and in Chapter 4.

System Real-Time Messages

Single-byte System Real-Time messages provide the precise timing element required to synchronize all of the MIDI devices in a connected system. To avoid timing delays, the MIDI specification allows System Real-Time messages to be inserted at any point in the data stream, even between other MIDI messages (Figure 2.31).

Timing Clock Messages

The MIDI Timing Clock message is transmitted within the MIDI data stream at various resolution rates. It is used to synchronize the internal timing clocks of each MIDI device within the system and is transmitted in both the start and stop modes at the currently defined tempo rate. In the early days of MIDI, these rates (which are measured in pulses per quarter note [ppq]) ranged from 24 to 128 ppq; however, continued advances in technology have brought these rates up to 240, 480, or even 960 ppq.

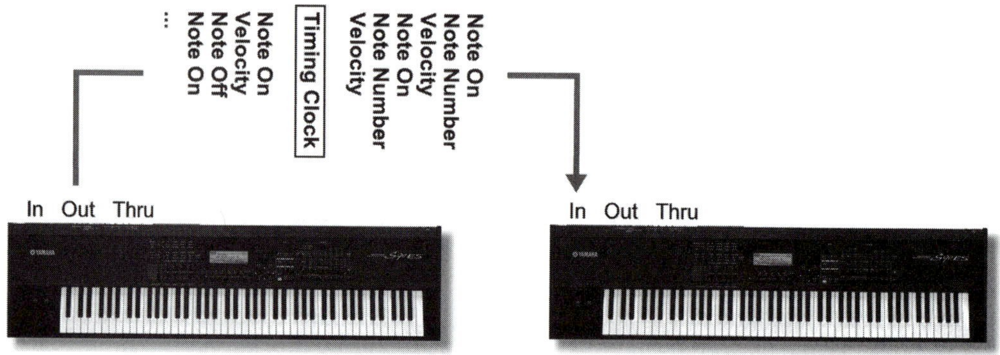

Figure 2.31. *System Real-Time messages can be inserted within the byte stream of other MIDI messages.*

Start Messages

Upon receipt of a timing clock message, the MIDI Start command instructs all connected MIDI devices to begin playing from their internal sequences initial start point. Should a program be in midsequence, the start command will reposition the sequence back to its beginning, at which point it will begin to play.

Stop Messages

Upon receipt of a MIDI Stop command, all devices within the system will stop playing at their current position point.

Continue Messages

After receiving a MIDI Stop command, a MIDI Continue message will instruct all connected devices to resume playing their internal sequences from the precise point at which it was stopped.

Active Sensing Messages

When in the Stop mode, an optional Active Sensing message can be transmitted throughout the MIDI data stream every 300 milliseconds. This instructs devices that can recognize this message that they're still connected to an active MIDI data stream.

System Reset Messages

A System Reset message is manually transmitted in order to reset a MIDI device or instrument back to its initial power-up default settings (commonly mode 1, local control on, and all notes off).

System-Exclusive Messages

The System-Exclusive (SysEx) message lets MIDI manufacturers, programmers, and designers communicate customized MIDI messages between MIDI devices. These messages give

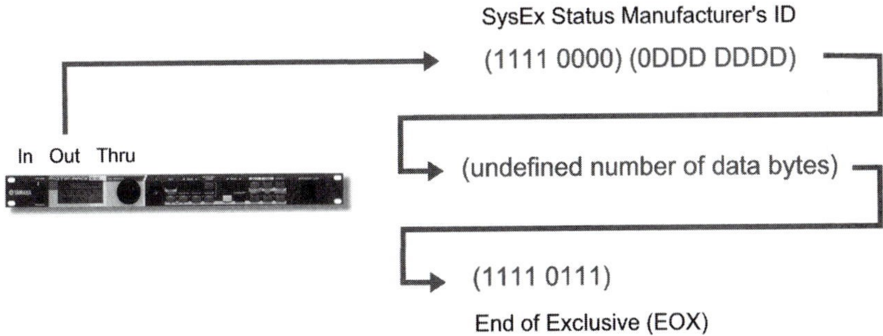

Figure 2.32. System-exclusive data (1-ID byte format).

manufacturers, programmers, and designers the freedom to communicate any device-specific data of an unrestricted length as they see fit. SysEx data is commonly used for the bulk transmission and reception of program/patch data, sample data, and real-time control over a device's parameters. The transmission format of a SysEx message (Figure 2.32) as defined by the MIDI standard includes a SysEx status header, manufacturer's ID number, any number of SysEx data bytes, and an EOX byte. Upon receiving a SysEx message, the identification number is read by a MIDI device to determine whether or not the following messages are relevant. This is easily accomplished, because a unique 1- or 3-byte ID number is assigned to each registered MIDI manufacturer. If this number doesn't match the receiving MIDI device, the ensuing data bytes will be ignored. Once a valid stream of SysEx data is transmitted, a final EOX message is sent, after which the device will again begin responding to incoming MIDI performance messages. A detailed practical explanation of the many uses (and wonders) of SysEx can be found in the synthesizer section of Chapter 4, as well as in the patch editor section of Chapter 6. I definitely recommend that you check these out, because SysEx is one of the most cost-effective and powerful tools that an electronic musician can have. It's definitely well worth the reading!

Universal Nonreal-Time System Exclusive

Universal Nonreal-Time SysEx data is a protocol that's used to communicate control and nonreal-time performance data. It's currently used to intelligently communicate a data-handshaking protocol that informs a device that a specific event is about to occur or that specific data is about to be requested. It is also used to transmit and receive universal sample-dump data or to transmit MIDI time-code cueing messages. A universal Nonreal-Time SysEx message consists of 4 or 5 bytes that includes two sub-ID data bytes that identify which nonreal-time parameter is to be addressed. It is then followed by a stream of pertinent SysEx data.

Universal Real-Time System Exclusive

Currently, two Universal Real-Time SysEx messages are defined. Both of them relate to the MTC synchronization code (which is discussed in detail in Chapter 10). These include full message data (relating to a SMPTE address) and user-bit data.

Running Status

Within the MIDI 1.0 specification, special provisions have been made to reduce the need for conveying redundant MIDI data. This mode, known as Running Status, allows a series of consecutive MIDI messages that have the same status byte type to be communicated without repeating the same status byte each time a MIDI message is sent. For example, we know that a standard MIDI message is made up of both a status byte and one or more data bytes. When using running status, a series of Pitch Bend messages that have been generated by a controller would transmit an initial status and data byte message, followed only by a series of related data (pitch-bend level) bytes, without the need for including redundant status bytes. The same could be said for Note-On, Note-Off, or any other status message type. Although the transmission of Running Status messages is optional, all MIDI devices must be able to identify and respond to this data transmission mode.

The Hardware

In addition to the huge number of electronic MIDI Instruments that are currently on the market, a vast array of supporting MIDI hardware systems also exists for the purpose of connecting, interfacing, distributing, processing, and diagnosing MIDI data. These systems are used to integrate all of the individual tools and toys into a working environment that will hopefully be designed to be powerful, cost effective and easy to use.

System Interconnection

As a data transmission medium, MIDI is relatively unique in the world of sound production in that it's able to pack 16 discrete channels of performance, controller, and timing information and then transmit it in one direction, using data densities that are economically small and easily to manage. In this way, it's possible for MIDI messages to be communicated from a specific source (such as a keyboard or MIDI sequencer) to any number of devices within a connected network over a single MIDI data chain. In addition, MIDI is flexible enough that multiple MIDI data lines can be used to interconnect devices in a wide range of possible system configurations; for example, multiple MIDI lines can be used to transmit data to instruments and devices over 32, 48, 128, or more discrete MIDI channels!

The MIDI Cable

A MIDI cable (Figure 3.1) consists of a shielded, twisted pair of conductor wires that has a male 5-pin DIN plug located at each of its ends. The MIDI specification currently uses only three of the five pins, with pins 4 and 5 being used as conductors for MIDI data and pin 2 to connect the cable's shield to equipment ground. Pins 1 and 3 are currently not in use, although the next section describes an ingenious system for powering devices through these pins that's known as MIDI phantom power. The cables use twisted cable and metal shield groundings to reduce outside interference, such as radiofrequency interference (RFI) or electrostatic interference, both of which can serve to distort or disrupt the transmission of MIDI messages.

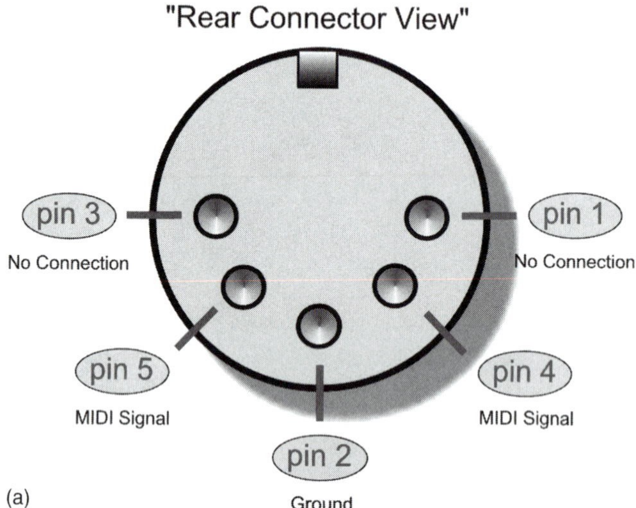

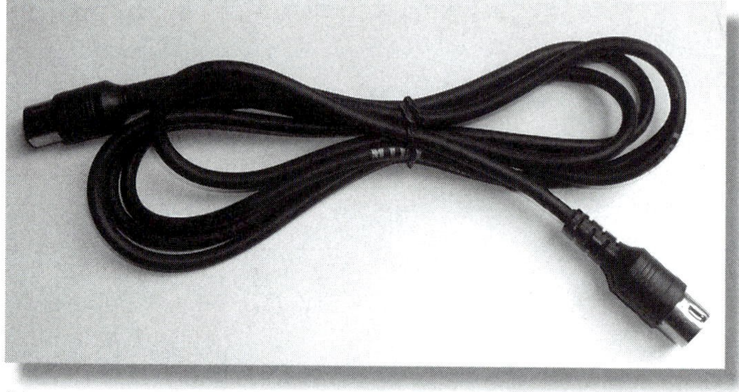

Figure 3.1. *The MIDI cable: (a) wiring diagram; (b) standard-length MIDI cable.*

MIDI Pin Description

- Pin 1 is not used in most cases; however, it can be used to provide the V− (ground return) of a MIDI phantom power supply.
- Pin 2 is connected to the shield or ground cable, which protects the signal from radio and electromagnetic interference.
- Pin 3 is not used in most cases; however, it can be used to provide the V+ (+9 to +15 V) of a MIDI phantom power supply.
- Pin 4 is a MIDI data line.
- Pin 5 is a MIDI data line.

MIDI cables come prefabricated in lengths of 2, 6, 10, 20, and 50 feet and can commonly be obtained from music stores that specialize in MIDI equipment. To reduce signal degradations and external interference that tends to occur over extended cable runs, 50 feet is the maximum length specified by the MIDI spec. (As an insider tip, I found that RadioShack® is a great source for picking up shorter MIDI cables at a fraction of what you'd sometimes spend at a music store.)

MIDI Phantom Power

In December of 1989, Craig Anderton (who wrote the forward to this book) wrote an article in *Electronic Musician* proposing an idea for allowing a source device to provide a standardized 12-V DC power supply to instruments and MIDI devices directly through pins 1 and 3 of a basic MIDI cable. Although pins 1 and 3 are technically reserved for possible changes in future MIDI applications (which never really came about), over the years several forward-thinking manufactures (and project enthusiasts) have begun to implement MIDI phantom power (Figure 3.2) directly into their studio and on-stage systems.

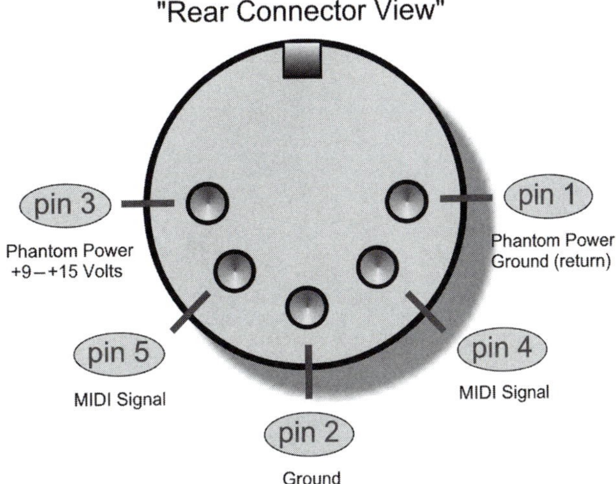

Figure 3.2. *MIDI phantom power wiring diagram.*

For the more adventurous types who aren't afraid to get their soldering irons hot, I've included a basic schematic for getting rid of your wall wart and powering a system through the MIDI cable itself. It should be noted that not all MIDI cables connect all five pins. Some manufacturers use only three wires to save on manufacturing costs, thus it's best to make sure what you've got before you delve into your own personal project. Of course, neither the publisher nor I extend any warranty to anyone regarding damage that might occur as a result of this mod. Think it out carefully, have a steady hand, and don't blow anything up!

On a personal note, I've retrofitted my own MIDI guitar rig, so the wall wart's far off-stage, eliminating the need for a nearby power plug (which I'd probably kick into the next state in the middle of a gig).

Wireless MIDI

Several companies have begun to manufacture wireless MIDI transmitters (Figures 3.3 and 3.4) that allow a battery-operated MIDI guitar, wind controller, etc., to be footloose and fancy free on-stage and in the studio. Working at distances of up to 500 feet, these battery-powered transmitter/receiver systems introduce very low delay latencies and can be switched over a number of radiochannel frequencies.

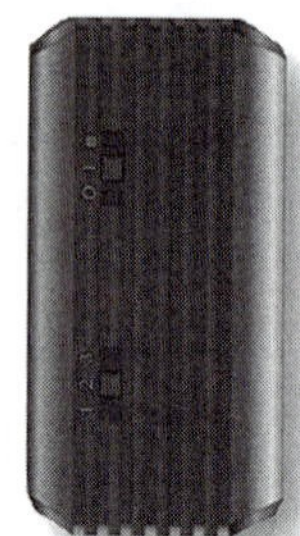

Figure 3.3. *CME's WIDI-X8 "C" wireless MIDI interface allows up to 84 wireless interface pairs to be running at the same time at a distance of up to 80 M. (Courtesy of CME, www.cme-pro.com.)*

Figure 3.4. *MidAir 25 25-key wireless USB MIDI controller and receiver. (Courtesy of M-Audio, A Division of Avid Technology, Inc.; www.m-audio.com.)*

MIDI Jacks

MIDI is distributed from device to device using three types of MIDI jacks: MIDI In, MIDI Out, and MIDI Thru (Figure 3.5). These three connectors use 5-pin DIN jacks as a way to connect MIDI Instruments, devices, and computers into a music or production network system. As a side note, it's nice to know that these ports[NCS1] (as strictly defined by the MIDI 1.0 spec.) are optically isolated to eliminate possible ground loops that might occur when connecting numerous devices together.

MIDI In Jack

The MIDI In jack receives messages from an external source and communicates this performance, control, and timing data to the device's internal microprocessor, allowing an instrument to be played or a device to be controlled. More than one MIDI In jack can be designed into a system to provide for MIDI merging functions or for devices that can support more than 16 channels (such as a MIDI interface). Other devices (such as a controller) might not have a MIDI In jack at all.

MIDI Out Jack

The MIDI Out jack is used to transmit MIDI performance and control messages, or SysEx from one device to another MIDI Instrument or device. More than one MIDI Out jack can be designed into a system, giving it the advantage of controlling and distributing data over multiple MIDI paths using more than just 16 channels (*i.e.*, 16 channels × *N* MIDI port paths).

MIDI Thru Jack

The MIDI Thru jack retransmits an exact copy of the data that's being received at the MIDI In jack. This process is important, because it allows data to pass directly through an instrument or device to the next device in the MIDI chain. Keep in mind that this jack is used to relay an exact copy of the MIDI In data stream, which isn't merged with data being transmitted from the MIDI Out jack.

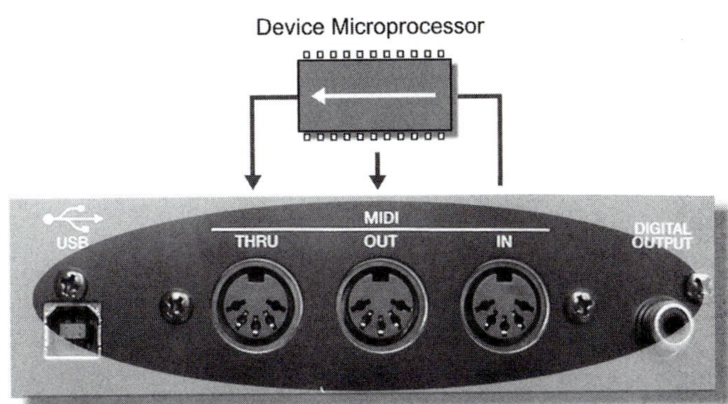

Figure 3.5. *MIDI In, Out and Thru ports, showing the device's signal path routing.*

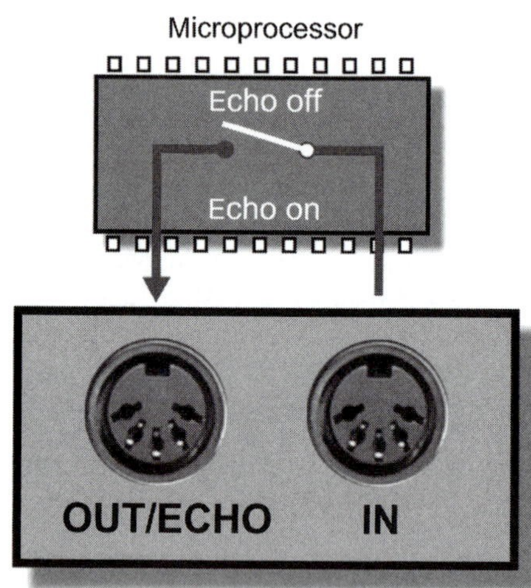

Figure 3.6. *MIDI echo configuration.*

MIDI Echo

Certain MIDI devices may not include a MIDI Thru jack at all. Some of these devices, however, may give the option of switching the MIDI Out between being an actual MIDI Out jack and a MIDI Echo jack (Figure 3.6). As with the MIDI Thru jack, a MIDI echo option can be used to retransmit an exact copy of any information that's received at the MIDI In port and route this data to the MIDI Out/Echo jack. Unlike a dedicated MIDI Out jack, the MIDI Echo function can often be selected to merge incoming data with performance data that's being generated by the device itself. In this way, more than one controller can be placed in a MIDI system at one time. It should be noted that, although performance and timing data can be echoed to a MIDI Out/Echo jack, not all devices are capable of echoing SysEx data.

Typical Configurations

Although electronic studio production equipment and setups are rarely alike (or even similar), there are a number of general rules that make it easy for MIDI devices to be connected into a functional network. These common configurations allow MIDI data to be distributed in the most efficient and understandable manner possible.

As a primary rule, there are only two valid ways to connect one MIDI device to another within a MIDI chain (Figure 3.7):

- Connecting the MIDI Out jack of a source device (controller or sequencer/computer) to the MIDI In of a second device in the chain
- Connecting the MIDI Thru jack of the second device to the MIDI In jack of the third device in the chain, and following this same Thru-to-In convention until the end of the chain is reached

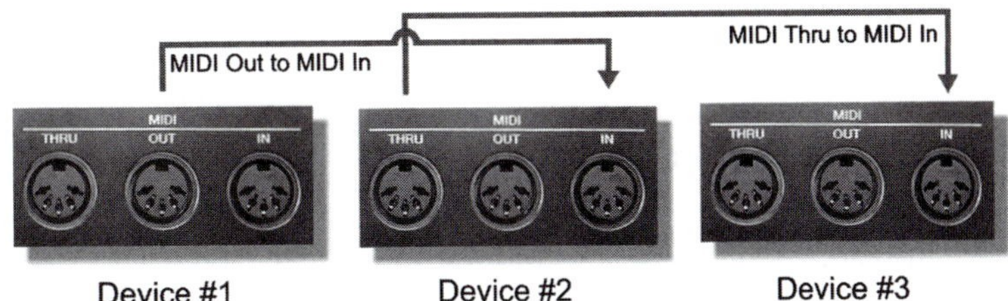

Figure 3.7. *The two valid means of connecting one MIDI device to another.*

The Daisy Chain

One of the simplest and most common ways to distribute data throughout a MIDI system is the daisy chain. This method relays MIDI data from a source device (controller or sequencer/computer) to the MIDI In jack of the next device in the chain (which receives and acts upon this data). This next device then relays an exact copy of the incoming data out to its MIDI Thru jack, which is then relayed to the next device in the chain, and so on through the successive devices. In this way, up to 16 channels of MIDI data can be chained from one device to the next within a connected data network—and it's precisely this concept of transmitting multiple channels through a single MIDI line that makes the whole concept work! Let's try to understand this system better by looking at a few examples.

Figure 3.8a shows a simple (and common) example of a MIDI daisy chain whereby data flows from a controller (MIDI Out jack of the source device) to a synth module (MIDI In jack of the second device in the chain). An exact copy of the data that flows into the second device is then relayed to its MIDI Thru jack to another synth (MIDI In jack of the third device in the chain). From the section on MIDI channels in Chapter 2, it shouldn't be hard to understand that if our controller is set to transmit on MIDI channel 3 the second synth in the chain (which is set to channel 2) will ignore the messages and not play, while the third synth (which is set to channel 3) will be playing its heart out. The moral of this story is that, although there's only one connected data line, a wide range of instruments and channel voices can be played in a surprisingly large number of combinations—all by using individual channel assignments along a daisy chain.

Another example (Figure 3.8b) shows how a computer can easily be designated as the master source within a daisy chain so a sequencing program can be used to control the entire playback and channel routing functions of a daisy-chained system. In this situation, the MIDI data flows from a master controller/synth to the MIDI In jack of a computer's MIDI interface (where the data can be played into, processed, and rechannelized through a MIDI sequencer). The MIDI Out of the interface is then routed back to the MIDI In jack of the master controller/synth (which receives and acts upon this data). The controller then relays an exact copy of this incoming data out to its MIDI Thru jack (which is then relayed to the next device in the chain) and so on, till the end of the chain is reached. When we stop and think about it, we can see that the controller is essentially used as a "performance tool" for entering data into the MIDI sequencer, which is then used to communicate this data out to the various instruments throughout the connected MIDI chain.

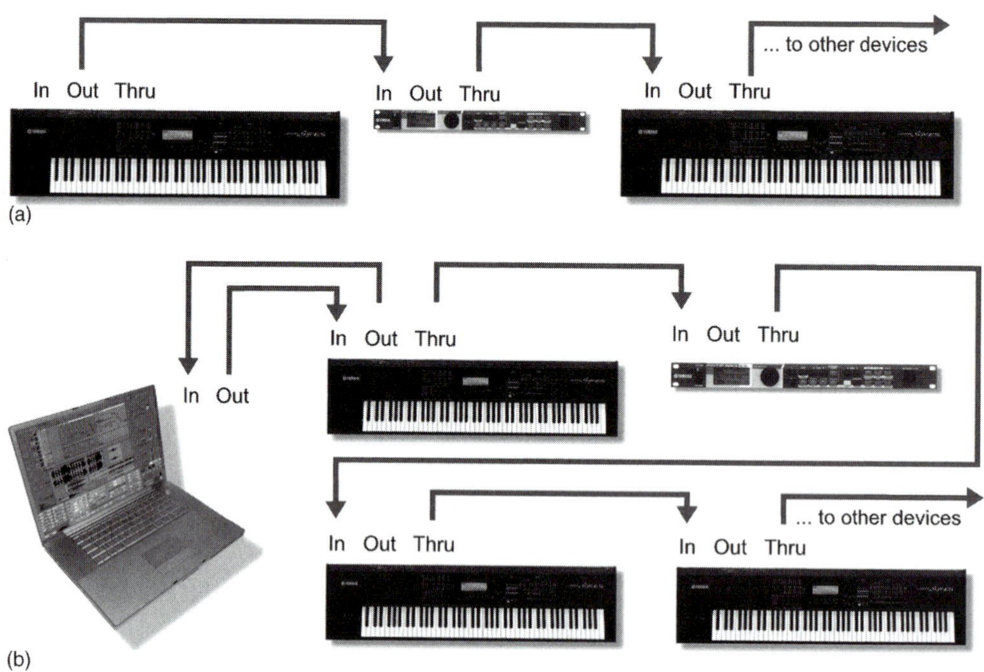

Figure 3.8. *Example of a connected MIDI system using a daisy chain: (a) typical daisy chain hookup; (b) example of how a computer can be connected into a daisy chain.*

The Multiport Network

Another common approach to routing MIDI throughout a production system involves distributing MIDI data through the multiple 2-, 4-, and 8-In/Out ports that are available on the newer multiport MIDI interfaces or through the use of multiple MIDI interfaces (typically these are USB devices).

> *Note:* Although the distinction isn't overly important, you might want to keep in mind that a MIDI "port" is a virtual data path that's processed through a computer, whereas a MIDI "jack" is the physical connection to the device itself.

In larger, more complex MIDI systems, a multiport MIDI network (Figure 3.9) offers several advantages over a single daisy chain path. One of the most important is its ability to address devices within a complex setup that requires more than 16 MIDI channels. For example, a 2 × 2 MIDI interface that has two-independent In/Out paths is capable of simultaneously addressing up to 32 channels (*i.e.*, port A 1–16 and port B 1–16), whereas an 8 × 8 port is capable of addressing up to 128 individual MIDI channels.

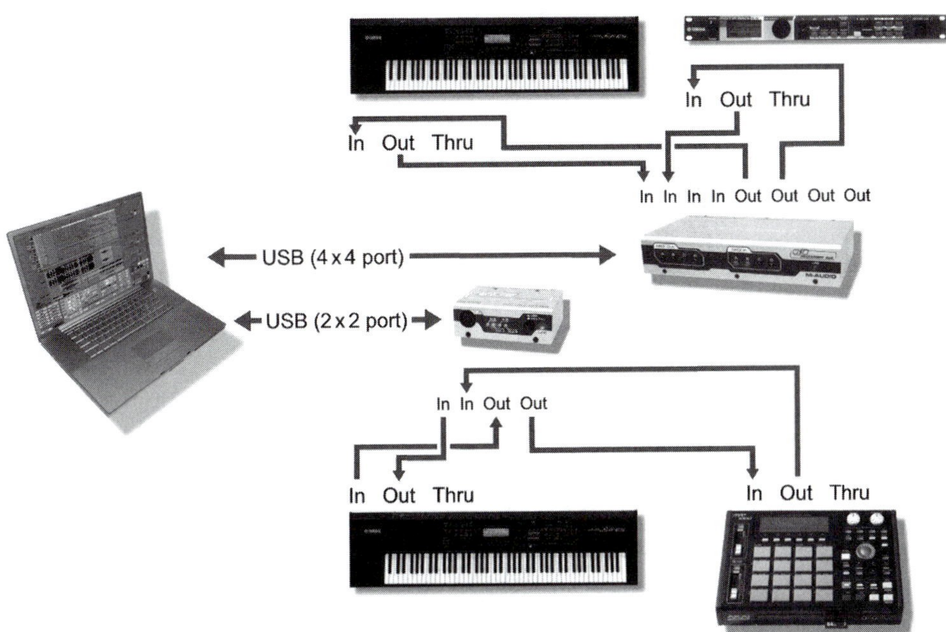

Figure 3.9. *Example of a multiport network using two MIDI interfaces.*

This type of network of independent MIDI chains has a number of advantages. As an example, port A might be dedicated to three instruments that are set to respond to MIDI channels 1 to 6, channel 7, and finally channel 11, whereas port B might be transmitting data to two instruments that are responding to channels 1–4 and 5–10 and port C might be communicating SysEx MIDI data to and from a MIDI remote controller for a digital audio workstation (DAW).

In this modern age of audio interfaces, multiport MIDI interfaces, and controller devices that are each fitted with MIDI ports, it's a simple matter for a computer to route and synchronously communicate MIDI data throughout the studio in any number of ingenious and cost-effective ways.

MIDI and the Personal Computer

Besides the coveted place of honor in which most electronic musician's hold their instruments, the most important device in a MIDI system is undoubtedly the personal computer. Through the use of software programs and peripheral hardware, the computer is often used to control, process, and distribute information relating to music performance and production from a centralized, integrated control position.

Essentially, the computer is a high-speed digital processing engine that can perform a wide range of work-, production-, or fun-related tasks. It's generally assembled from such building block components as a power supply, motherboard/central processing unit (CPU), random access memory (RAM), hard disks, and CD/DVD optical drives.

Most computers are designed with several hardware slots that let you add expansion cards to the system so as to perform specific tasks that can't be handled by the computer's own hardware. Peripherals are also the name of the game in the 21st century. They use such communications protocols as Universal Serial Bus (USB) and FireWire™ standards to integrate your computer to the world of digital audio, MIDI, video, the Internet, scanning, networking, and tons of amazing things that haven't been invented yet.

Software, that all-important binary computer-to-human interface, gives us countless options for performing all sorts of tasks such as music sequencing, data routing, digital audio production, word processing, or music printing. Thanks to the countless software and shareware options that are available for handling and processing most types of media data, it's nice (but sometimes frustrating) to have the luxury of being able to choose the company, look, feel, and programming style that best suits our personal working style and habits. When you get right down to it, the fact that a computer can be individually configured to best suit the individual's own needs has been one of the driving forces behind the digital age. It's turned the computer into a powerful "digital chameleon" that can change its form and function to fit the task at hand.

For the most part, two computer types are used in modern-day music production: the PC and the Mac. In truth, each brings its own particular set of advantages and disadvantages to personal computing, although their differences have greatly dwindled over the years. My personal take on the matter (a subject that's not even important enough to debate), is that it's a dual-platform world. The choice is yours and yours alone to make. Having said this, here are a few guidelines that can help you to choose which platform will best suit your needs:

- Which platform are you currently most familiar and comfortable with?
- Which platform and software is most used (and possibly demanded) by your clients?
- Which platform is most used by your friends or associates?
- What software and hardware do you already have and are familiar with?
- Which one do you like or feel best suits your needs?

Bottom line, do your research well and then realize that the choice is up to you.

Like I said, it's a dual-platform world. Many professional software and hardware systems can work on either platform. As I write this, my music collaborator is fully Mac and I'm fully PC, and it doesn't affect our production styles at all (Figure 3.10). Coexistence isn't much of a problem, either. Living a dual-platform existence can give you the edge of being familiar with both systems, which can be downright handy in a sticky production pinch.

The Mac

One computer type that's widely accepted by music professionals is the Macintosh® family of computers from Apple® Computer, Inc. The Mac® offers up a graphical user interface (GUI) that uses graphic icons and mouse-related commands in a friendly environment that lets you move, expand, tile, or stack windowed applications on the system's monitor. Due to the

Figure 3.10. *PC and Mac performance toys. (Courtesy of David Miles Huber, www.davidmileshuber.com, and Marcell Marias, www.marcellmarias.com.)*

rigid hardware design constraints that are required by the Mac's operating system (OS), a tight integration between the system's hardware and software exists that give many an "it just works" feeling.

The PC

Due to its cost effectiveness, the Windows® OS, large amount of available software, and its sheer numbers in both the home and business community, the Microsoft® Windows®-based personal computer (PC) clearly dominates the marketplace (musical or otherwise). Unlike the Mac®, which is made by one manufacturer, the PC's general specifications were licensed out to the industry at large. Because of this, countless manufactures make compatible systems that can be factory or user assembled and upgraded using standard, off-the-shelf hardware components. Like the Mac® OS, Windows® is a sophisticated, graphic-based, multitasking environment that can run multiple task-based applications at a time. With the introduction of Microsoft's (XP)erience and their 64-bit Vista™ OS, the differences between computer types with regard to hardware, software, networking, and peripheral integration have blurred to the point where we can truly proclaim that it really is a dual-platform world.

Portability

In this digital age, it goes without saying that both the Mac and PC are available in portable laptop configurations that can be easily taken on the road (Figures 3.11). With their increased power, portability, and inherent cool factor, these small, lightweight powerhouses often rival their desktop counterparts in CPU, graphics capabilities, hard disk space, and CD-/DVD-ROM drive capabilities. With the extensive development of USB and FireWire™ production peripherals, the laptop has been turned into an on-the-road, battery-operated computer that can dish out serious music production power with the big boys.

Figure 3.11. *Not just for on-the-go … a laptop can also act as an important second computer for e-mail and additional production tasks.*

Super-Portability

iPods, pocket PCs, and players—oh, my! For some, the concept of laptop portability is just too cumbersome for their on-the-go world. Such measures call for extreme portability, often in the form of an iPod® or other MP3 player to keep us in touch with the music, blogs, and podcasts of today. With the introduction of the personal digital assistant (PDA), it's become a simple matter to:

- Listen to music.
- Record a music ditty that's in your head.
- Make a cell phone call.
- View a document.
- Make an appointment.
- Surf the web.
- Take a picture.

all from one hand-held device (Figure 3.12). Heck, you can even download software that'll let you do basic MIDI sequencing in the palm of your hand on the summit of Mount Cucamonga.

Connecting to the Peripheral World

An important event in the evolution of personal computing has been the maturation of hardware and processing peripherals. With the development of the USB (www.usb.org) and FireWire™

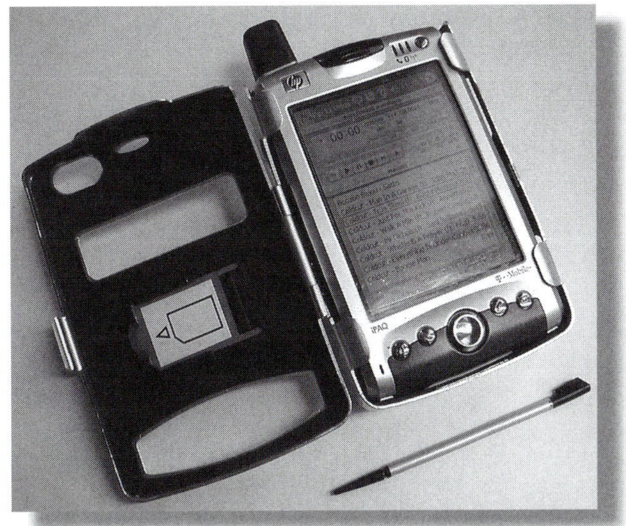

Figure 3.12. *Look, it's phone, it's a player … it's everything you need to stay connected!*

(www.1394ta.org) protocols, hardware devices such as mice, keyboards, cameras, audio interfaces, MIDI interfaces, CD and hard drives, MP3 players, and even portable fans can be plugged into an available port without any need to change frustrating hardware settings or open up the box. External peripherals are generally hardware devices that are designed to do a specific task or range of production tasks. For example, an audio interface is capable of translating analog audio (and often MIDI, control, and other media) into digital data that can be understood by the computer. Other peripheral devices can perform such useful functions as printing, media interfacing (video and MIDI), scanning, memory card interfacing, portable hard disk storage … the list could literally fill pages.

Peripheral Tips

- When receiving a new or used peripheral device, it's always best to check the web for the latest drivers. Even when a CD is included in the box, it could easily be out of date.
- Most peripheral drivers should be loaded *before* the hardware device has been plugged in. As always, it's best to consult the manual for installation details.

USB

In recent computer history, few protocols for interconnecting devices to a host computer have affected our lives like the Universal Serial Bus (USB). In short, USB is an open specification for connecting external hardware devices to the personal computer, as well as a special set of protocols for automatically recognizing and configuring them. The first of the following two speeds are supported by USB 1.0, but all three are supported by USB 2.0:

- USB 1.0 (1.5 Mbits/second)—A low speed for the attachment of low-bandwidth peripherals (such as a joystick or mouse)
- USB 1.0 (12 Mbits/second)—For the attachment of devices that require a higher throughput (such as data transfer, soundcards, digitally compressed video cameras, and scanners)
- USB 2.0 (480 Mbits/second)—For high-throughput and fast transfer of the above applications

The basic characteristics of USB include the following:

- Up to 127 external devices can be added to a system without having to open up the computer. As a result, the industry is moving toward a "sealed-case" or "locked-box" approach to computer hardware design.
- Newer operating systems will often automatically recognize and configure a basic USB device that's shipped with the latest device drivers.
- Devices are "hot pluggable," meaning they can be added (or removed) while the computer is on and running.
- The assignment of system resources and bus bandwidth is transparent to the installer and end user.
- USB connections allow data to flow bidirectionally between the computer and the peripheral.
- USB cables can be up to 5 meters in length (up to 3 meters for low-speed devices) and include two twisted pairs of wires, one for carrying signal data and the other pair for carrying a DC voltage to a bus-powered device. Those that use less than 500 milliamps (1/2 amp) can get their power directly from the USB cable's 5-V DC supply, while those having higher current demands will have to be externally powered.
- Standard USB cables have two types of connectors at each end; for example, a cable between the PC and a device would have an "A" plug at the PC (root) connection and a "B" plug for the device's receptacle.
- Cable distribution and daisy-chaining are done via a USB data hub (Figure 3.13). These devices act as a traffic cop in that they cycle through the various USB inputs in a sequential fashion, routing the data into a single data output line.

Note: It should be kept in mind that not all USB hubs are the same. Basic chipset and design differences can cause problems when passing data to and from a MIDI or audio interface. Since a hub can act as a bottleneck that reduces communication between such high-speed devices, it's often best to connect USB MIDI and audio interfaces directly to a USB port on the computer.

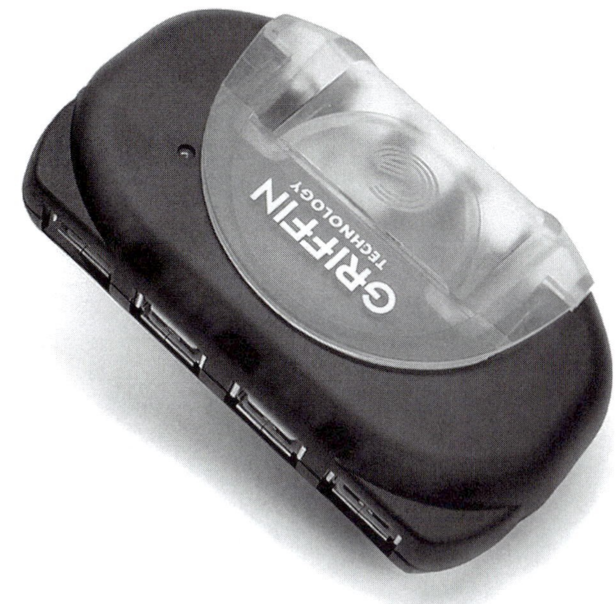

Figure 3.13. *Example of a four-port USB hub. (Courtesy of Griffin Technology, www.griffintechnology.com.)*

FireWire™

Originally created in the mid-1990s by Apple (and later standardized as IEEE-1394), the FireWire™ protocol is similar to the USB standard in that it uses a twisted-pair wiring to communicate bidirectional, serial data within a hot-swappable, connected chain. Unlike USB (which can handle up to 127 devices per bus), up to 63 devices can be connected within a connected FireWire™ chain. FireWire™ supports two speed modes:

- FireWire™ 400 or IEEE-1394a (400 Mbits/second)—Capable of delivering data over cables up to 4.5 meters in length, FireWire™ 400 is ideally for communicating large amounts of date to such devices as hard drives, video camcorders, and audio interface devices.
- FireWire™ 800 or IEEE-1394b (800 Mbits/second)—FireWire™ 800 is capable of communicating large amounts of data over cables up to 100 meters in length. When using fiberoptic cables, lengths in excess of 90 meters can be achieved in situations that require long-haul cabling (such as sound stages and studios).

Unlike USB, compatibility between the two modes is mildly problematic, as FireWire™ 800 ports are configured differently from their earlier predecessor and therefore require adapter cables to ensure compatibility.

Networking

Beyond the concept of connecting external devices to a single computer, another concept hits at the heart of the connectivity age—networking. The ability to set up and make use of a local area network (LAN) can be extremely useful in the home, studio, or office in that it can be used to link multiple computers with various data, platforms, and OS types. In short, a network can be set up

Figure 3.14. Local area network (LAN) connections. (a) Data may be shared between independent computers in a home or workplace LAN environment. (b) Computer terminals may be connected to a centralized server, allowing data to be stored, shared, and distributed from a central location.

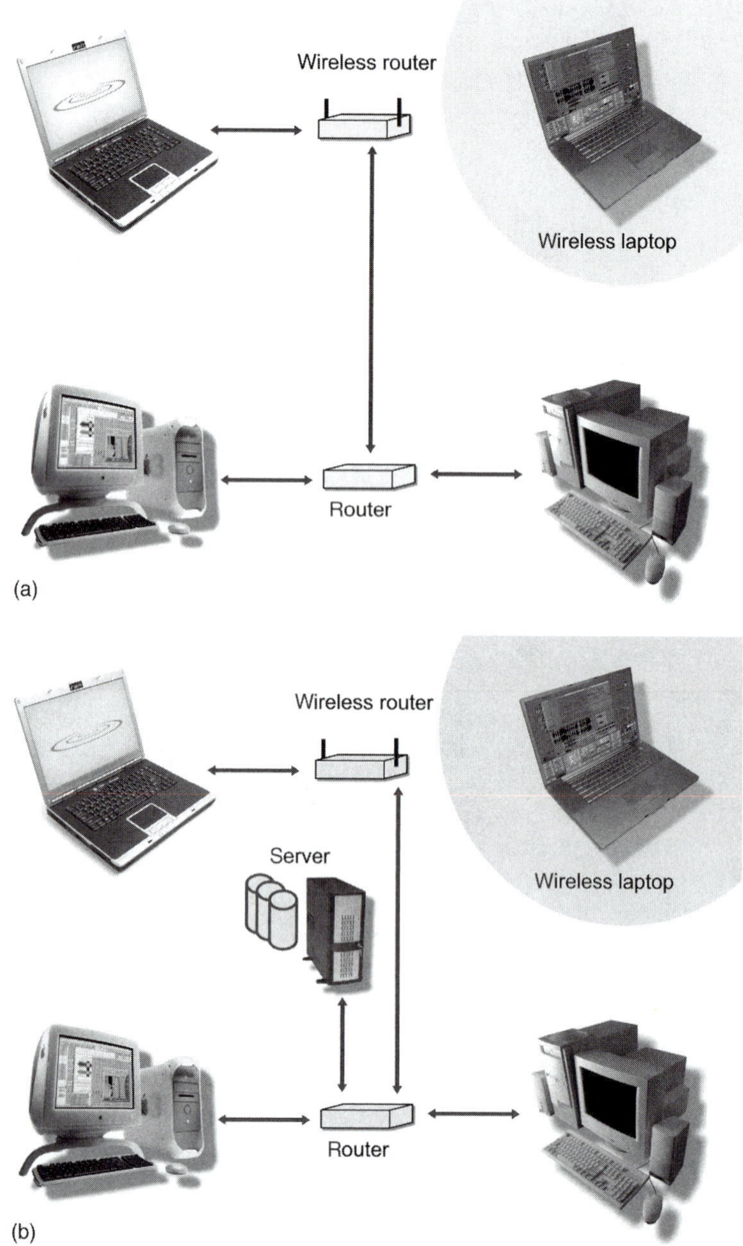

in a number of different ways, with varying degrees of complexity and administrative levels; however, there are basically two common ways that data can be handled over a LAN (Figure 3.14):

- The first is a system whereby the data that's shared between linked computers resides on the respective computers and is communicated back and forth in a decentralized manner.

- The second makes use of a centralized computer (called a server) that's basically an array of high-capacity hard drives that is used to store all of the data that relates to the everyday production aspects of a facility. Often, such a system will have a redundant set of drives that actually clone the entire system on a moment-to-moment basis, as a safety backup procedure. Alternatively (and in some cases, in addition to), a set of backup tapes may be made on a daily basic for extra insurance and archival purposes.

No matter what level of complexity is involved, some of the more common uses for a network connection include:

- *Sharing files*—Within a connected household, studio, or business, a LAN can be used to share files, soundfiles, video images—virtually anything—throughout the connected facility. This means that various production rooms, studios, and offices can simultaneously share and swap data or mediafiles in a way that's often transparent to the users.
- *Shared web connection*—One of the cooler aspects of using a LAN is the ability to share an Internet connection over the network from a single computer or server. The ability to connect from any computer with ease is just another reason why you should strongly consider wiring your studio or house with LAN connections.
- *Archiving and backup*—In addition to the benefits of archiving and backing up data with a server system, even the simplest LAN can be a true lifesaver. For example, let's say that we need to make a backup DVD of a session but don't have the time to tie up our production DAW. In this situation, we could simply burn the disc on another computer that's connected to the network and continue working away, without interruption.
- *Accessing soundfiles and sample libraries*—It goes without saying that sound and sample files can be easily accessed from any connected computer. Hey! If you're wireless (or have a long enough cable), go out to the pool and soak up the sun while getting your work done!

On a final note, those who are unfamiliar with networking are urged to learn about this powerful and easy-to-use data distribution and backup system. For a minimal investment in cables, hubs, and educational reading, you might be surprised at the time- and trouble-saving benefits that will be almost instantly be realized.

The MIDI Interface

Although computers and electronic instruments both communicate using the digital language of 1's and 0's, computers simply can't understand the language of MIDI without the use of a device that translates the serial messages into a data structure that computers can comprehend. Such a device is known as the MIDI interface. A wide range of MIDI interfaces currently exists that can be used with most computer systems and OS platforms. For the casual and professional musician, interfacing MIDI into a production system can be done in a number of ways. Probably the most common way to access MIDI In, Out, and Thru jacks is on modern-day USB or

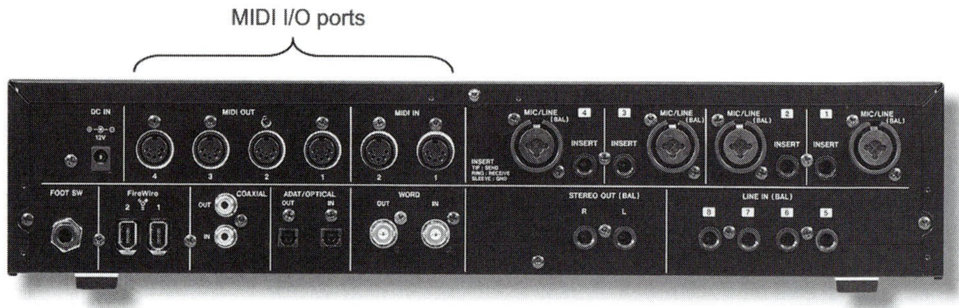

Figure 3.15. *Most audio interface devices include MIDI I/O ports. (Courtesy of TASCAM, www.tascam.com.)*

Figure 3.16. *M-Audio MIDISPORT 1 × 1 MIDI interface. (Courtesy of M-Audio, A Division of Avid Technology, Inc.; www.m-audio.com.)*

FireWire™ audio interface or controller surface (Figures 3.15). It's become a common matter for portable devices to offer 16 channels of I/O (on one port), while multichannel interfaces often include multiple MIDI I/O ports that can give you access to 32 or more channels.

The next option is to choose a USB MIDI interface that can range from simpler devices that include a single port (16 I/O channels, as shown in Figure 3.16) to a multiport system that can easily handle up to 128 channels over eight I/O ports. The multiport MIDI interface (Figure 3.17) is often the device of choice for most professional electronic musicians that require added routing and synchronization capabilities. These rack-mountable USB devices can be used to provide eight independent MIDI Ins and Outs to easily distribute MIDI data through separate lines over a connected network.

In addition to distributing MIDI data, these systems often includes driver software that can route and process MIDI data throughout the MIDI network. For example, a multiport interface could be used to merge together several MIDI Ins (or Outs) into a single data stream, filter out specific MIDI message types (used to block out unwanted commands that might adversely change an instrument's sound or performance), or rechannel data being transmitted on one

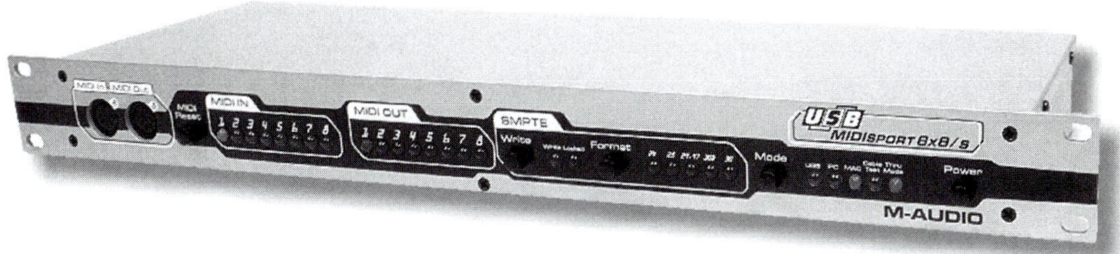

Figure 3.17. *M-Audio MIDISPORT 8 × 8/s MIDI interface. (Courtesy of M-Audio, A Division of Avid Technology, Inc.; www.m-audio.com.)*

MIDI channel or port to another channel or port (thereby allowing the data to be recognized by an instrument or device).

Another important function that can be handled by most multiport interfaces is synchronization. Synchronization (sync, for short) allows other, external devices (such as DAWs, video decks, and other media systems) to be simultaneously played back using the same timing reference. Interfaces that includes sync features will often read and write SMPTE time code, convert SMPTE to MIDI time code (MTC), and allow recorded time-code signals to be cleaned up when copying code from one analog device to another (jam sync). Further reading on synchronization can be found in Chapter 11.

In addition to the above interface types, a number of MIDI keyboard controllers and synth instruments have been designed with MIDI ports and jacks built right into them. For those getting started, this useful and cost-saving feature makes it easy to integrate your existing instruments into your DAW and sequencing environment.

The Keyboard Controller

The MIDI keyboard controller (Figures 3.18 through 3.20) is a keyboard device that's expressly designed to control hard-/software synths, samplers, modules, and other devices within a connected MIDI production system. These systems range from being USB-powered devices that are so portable that you could easily work on your latest project while passing time in an airport or in flight (you know, one of those great trips when you can spread out) to being full-sized keyboard controllers that have a wide array of variable controllers.

A keyboard controller will generally have no internal tone generators or sound-producing elements. Instead, its design has been optimized for control over performance and software control parameters by offering such features as a performance keyboard (that can range from a 2-octave, 25-key to a full 88-key surface), pitch and modulation control, any number of parameter knobs and controller surfaces (for real-time control over software and device parameters), drum/sample trigger pads, and even mixing and transport capabilities (which can be used to directly control a DAW program).

Figure 3.18. Novation ReMOTE 25SL MIDI controller/keyboard. (Courtesy of Novation Digital Music Systems, Ltd.; www.novationmusic.com.)

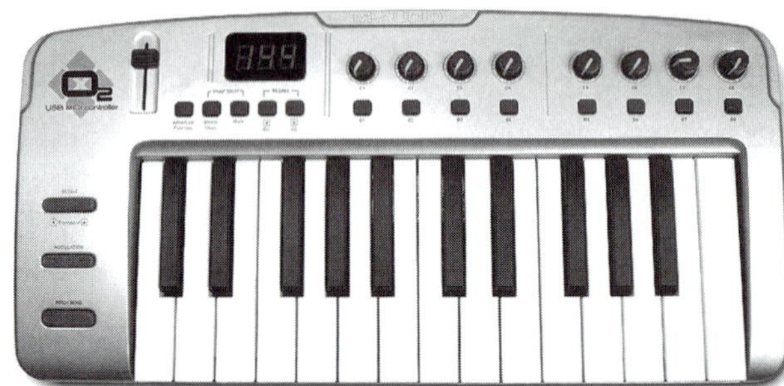

Figure 3.19. M-Audio 02 MIDI controller. (Courtesy of M-Audio, A Division of Avid Technology, Inc.; www.m-audio.com.)

Figure 3.20. CME UF5 controller. (Courtesy of Central Music Company, www.cme-pro.com; distributed by Yamaha Corporation of America, www.yamaha.com.)

Foot Controllers

Foot controllers (Figure 3.21) are sturdy little floor switch devices that are meant to be stomped on to provide any number of performance- and control-related functions, such as sustain, general device control (as defined by the user, software, or hardware) and record punch-in/-out (for use with DAWs and many types of recording systems). These devices are generally comprised of a simple non-latching momentary switch mechanism, cord, and quarter-inch plug. It should be noted that some devices require that the switch be closed (make an electrical connection) when

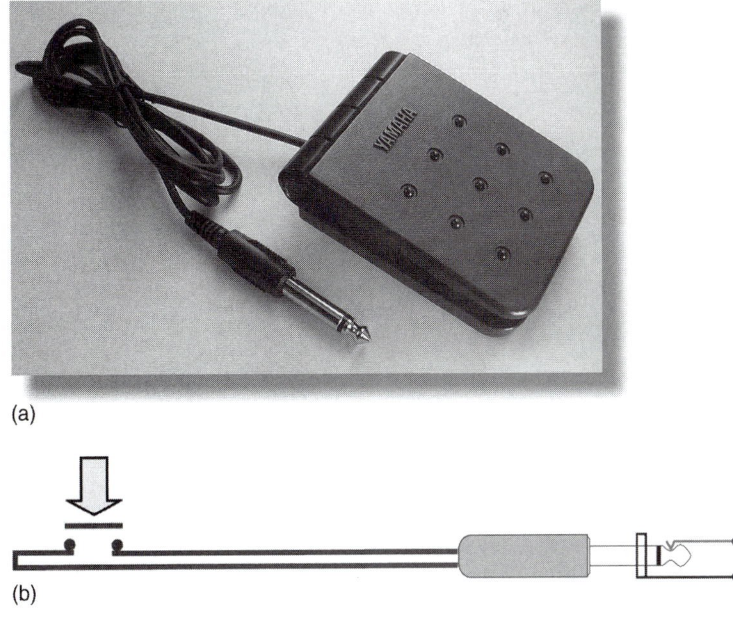

Figure 3.21. *Foot controller: (a) sustain pedal; (b) electrical diagram of a foot controller where the switch is normally open.*

Figure 3.22. *EX-P Universal Expression Controller Pedal. (Courtesy of M-Audio, A Division of Avid Technology, Inc.; www.m-audio.com.)*

stomped on, while others must be open (broken electrical connection) in order to operate. Other, smarter systems are actually able to sense which type of switch it is and adjust themselves accordingly. As always, check your manual before heading out to the local music store to buy one. Expression controllers (Figure 3.22), on the other hand, use a variable resistor to provide variable control over such MIDI controller parameters as volume, modulation, panning, filter sweep, etc. As with the foot controller, device polarity could be an issue. Again, check your manual before heading out to your local music store.

Figure 3.23. *Kid Beyond on stage with his voice, computer, and stomp box. (Courtesy of Kid Beyond, www.kidbeyond.com.)*

MIDI foot controller/stomp boxes offer up a range of functions that can include several MIDI assignable switches and variable expression controls that are sturdy enough to be stomped on and otherwise kicked around. A number of these devices include on-board effects or even an integrated audio interface. Most, however, can be used to control any number of MIDI parameters and devices, such as MIDI effects, music scene changes, stage light setups, effects loops, punch-in/-outs—you name it! Kid Beyond (Figure 3.23) uses this type of on-stage stomp box to control his amazing vocal beat effects via Ableton's Live loop/effects performance software (see Chapter 7). If you get a chance, don't pass up one of his performances!

The Audio Interface

An important peripheral that deserves careful consideration when putting together a computer-based production system is the digital audio interface. These devices might have a single, dedicated purpose, or they might be multifunctional in nature (being able to handle audio, MIDI, and control-related functions) In either case, their main purpose is to act as a bridge between the outside world of analog audio/media production and the computer's inner digital world (Figures 3.24 and 3.25).

A digital audio interface might be designed directly into the computer's circuitry, plug into the system via a USB or FireWire™ connection, or be designed as a card that plugs into a hardware slot. It could be a 2-in × 2-out (2 × 2) device that's limited in its sample-rate and bit-depth option. Or, it might be multichannel (8 × 8 or higher) and capable of handling rates up to 96 kHz/24 bits or higher.

Unless you buy a software package that has been designed to operate with a specific piece of hardware (as is the case with certain Digidesign® software), a vast number of audio interface types and options are available. When building a production system, it's important to weigh their differences and capabilities with patience and care. Taking the time to research the purchase of an audio interface and match it to your needs is definitely time well spent—the system and the dollars you'll save could be your own!

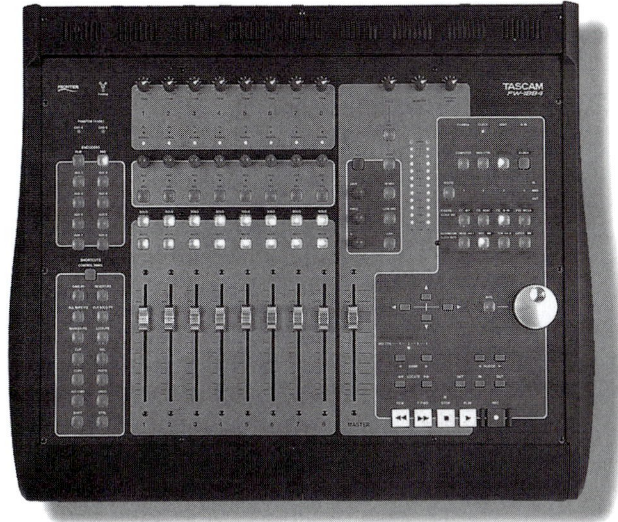

Figure 3.24. *TASCAM FW-1804 FireWire™ audio/MIDI interface. (Courtesy of TASCAM, www.tascam.com.)*

Figure 3.25. *M-Audio ProjectMix I/O 18 × 14 audio interface and control surface. (Courtesy of M-Audio, A Division of Avid Technology, Inc.; www.m-audio.com.)*

DAW Hardware Controllers

Often, one of the more common complaints that some people have against the digital audio workstation environment (particularly when relating to the use of on-screen mixers) is the lack of a hardware controller that gives the user access to hands-on controls. In recent years, this has been addressed by major manufacturers and third-party companies in the form of a hardware DAW controller interface (Figures 3.26 and 3.27).

These controllers generally mimic the design of an audio mixer in that they offer slide or rotary gain faders, pan pots, solo/mute, and channel select buttons, with the added bonus of a full transport remote. A channel select button is used to actively assign a specific channel to a section that contains a series of pots and switches that relate to equalization, effects, and dynamic functions (often being grouped over eight channel strips at a time). By switching between the banks (*e.g.*, 1–8, 9–16, 17–24), any number of the grouped inputs can be accessed by the virtual mixer. These devices also often include functional "soft keys" that can be programmed to give quick and easy access to the software's more commonly used program keys. Controller commands are most commonly transmitted between the controller and audio editor via

Figure 3.26. *Mackie Control Universal. (Courtesy of Loud Technologies, Inc.; www.mackie.com.)*

Figure 3.27. *Digidesign® ICON integrated console. (Courtesy of Digidesign, A Division of Avid Technology, Inc.; www.digidesign.com.)*

device-specific MIDI SysEx messages, which are communicated using existing USB or FireWire™ protocols.

Many of these systems are offered as an all-in-one, multifunctional device that often includes a multichannel audio interface, MIDI interface ports, monitor capabilities, and full controller and mix functions (including motorized long-throw faders). Again, due to the wide array of options and general functionality, it's wise to take time to research your personal and current equipment needs so as to make a choice that fits your studio, personal style, and budget.

MIDI Diagnostic Tools

Although software-based tools exist for analyzing MIDI data from a computer, a number of hardware tools can also be used to detect and diagnose messages as they travel over a single MIDI line or throughout a MIDI production studio or live stage setting. Such tools can be used for detecting the presence of MIDI data or for troubleshooting specific MIDI message types.

Probably the simplest and most practical MIDI diagnostic tool has been designed into almost every electronic instrument. This exists in the form of a MIDI In activity light-emitting diode (LED) that indicates the presence of incoming or outgoing MIDI data. These handy lights have fortunately been designed into the front panel of most MIDI instrument and interfaces, making it easy to see MIDI activity on all In and Out ports.

In certain situations, you might not be able to visually check for the presence of MIDI. This occurs if: (1) the instrument doesn't have a data indicator, or (2) activity indicators on an instrument won't light up if it's receiving data on a MIDI channel to which it isn't assigned. In either case, you could look at the LEDs on your multiport interface or you could buy or build a simple MIDI activity light that will let you know instantly if data activity is happening on that MIDI line. Do-it-yourself enthusiasts can tackle this option by going down to RadioShack®; picking up a single male 5-pin DIN plug, an 8.2-kOhm 1/4-watt resistor, and an LED of your favorite color choice; and following the simple diagram shown in Figure 3.28. On a fun note from the past, Ensoniq Corp. used to sell the Ensoniq Spider. This fun, plastic spider had two LED eyes that lit up whenever MIDI data was present … I'm sorry that I couldn't find a picture.

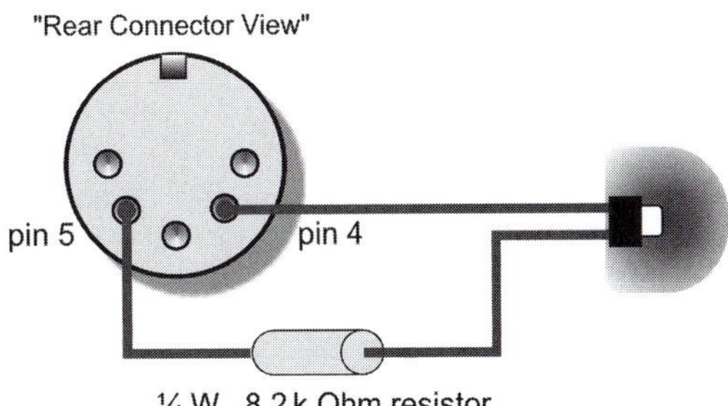

Figure 3.28. *Diagram detailing how to make your own MIDI activity light.*

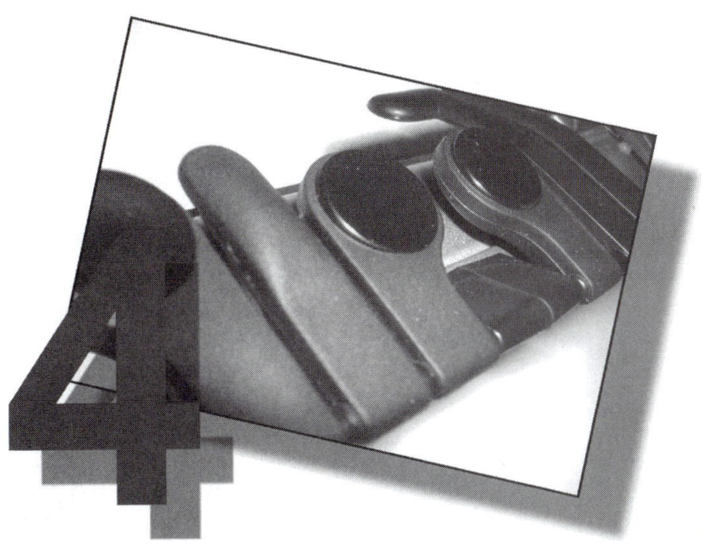

Electronic Instruments

Since their inception in the early 1980s, MIDI-based electronic instruments have played a central and important role in the development of music technology and production. These devices (which fall into almost every instrument category), along with the advent of cost-effective analog and digital audio recording systems, have probably been the most important technological advances to shape the industry into what it is today. In fact, the combination of these technologies has turned the personal project studio into one of the most important driving forces behind modern-day music production.

Inside the Toys

Although electronic instruments often differ from one another in looks, form, and function, they almost always share a common set of basic building block components (Figure 4.1), including the following:

- *Central processing units (CPU)*—CPUs are one or more dedicated computing devices (often in the form of a specially manufactured microprocessor chip) that contain all of

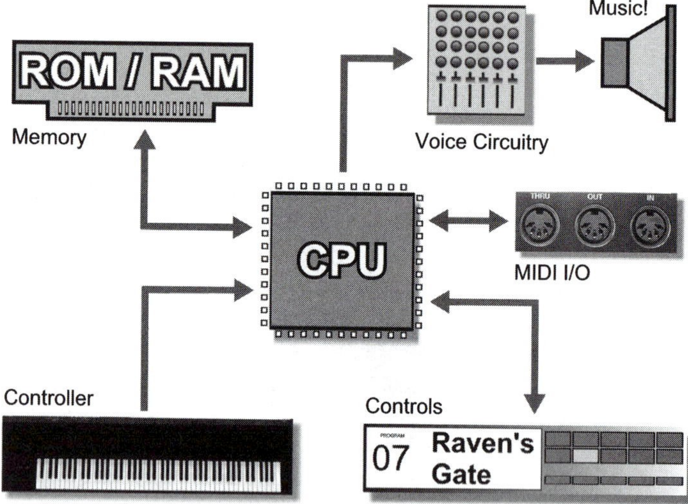

Figure 4.1. *The basic building blocks of an electronic musical instrument.*

the necessary instructional brains to control the hardware, voice data, and sound-generating capabilities of the entire instrument or device.

- *Performance controllers*—These include such interface devices as music keyboards, knobs, buttons, drum pads, and/or wind controllers for inputting performance data directly into the electronic instrument in real time or for transforming a performance into MIDI messages. Not all instruments have a built-in controller. These devices (commonly known as modules) contain all the necessary processing and sound-generating circuitry; however, the idea is to save space in a cramped studio by eliminating redundant keyboards or other controller surfaces.
- *Control panel*—The control panel is the all-important human interface of data entry controls and display panels that let you select and edit sounds and route and mix output signals, as well as control the instrument's basic operating functions.
- *Memory*—Digital memory is used for storing important internal data (such as patch information, setup configurations, and/or digital waveform data). This digital data can be encoded in the form of either read-only memory (ROM; data that can only be retrieved from a factory-encoded chip, cartridge, or CD/DVD-ROM) or random access memory (RAM; memory that can be read from and stored to a device's resident memory, cartridge, hard disk, or recordable media).
- *Voice circuitry*—Depending on the device type, this section can chain together digital processing "blocks" to either generate sounds (voices) or process and reproduce digital samples that are recorded into memory for playback according to a specific set of parameters. In short, it's used to generate or reproduce a sound patch, which can then be processed, amplified, and heard via speakers or headphones.
- *Auxiliary controllers*—These are external controlling devices that can be used in conjunction with an instrument or controller. Examples of these include foot pedals

(providing continuous-controller data), breath controllers, and pitch-bend or modulation wheels. Some of these controllers are continuous in nature, while others exist as a switching function that can be turned on and off. Examples of the latter include sustain pedals and vibrato switches.

- *MIDI communications ports*—These data ports and physical jacks are used to transmit and/or receive MIDI data.

Generally, no direct link is made between each of these functional blocks; the data from each of these components is routed and processed through the instrument's CPU. For example, should you wish to select a certain sound patch from the instrument's control panel, the control panel could be used to instruct the CPU to recall all of the waveform and sound-patch parameters from memory that are associated with the particular sound. These instructional parameters would then be used to modify the internal voice circuitry, so that when a key on the keyboard is pressed, the sound generators will output the desired patch's note and level values.

Of course, most if not all of an instrument's functional components can be used to access or communicate performance, patch, and system setup information via MIDI. A few of the many examples include:

- Transmission of performance and control-related data between devices, throughout a connected network.
- Transmission of real-time control parameters via MIDI and/or system exclusive (SysEx) messages.
- Bulk transmission of device patch and system parameters via system exclusive (SysEx) messages.

The remainder of this chapter will focus on the various types of MIDI instruments and controller devices that are currently available on the market. These devices can be grouped into such categories as keyboards, percussion, MIDI guitars and strings, woodwind instruments, and controlling devices.

It almost goes without saying that, over the years, model numbers, manufacturers, communications protocols, system applications, and production styles will come and go. The job of any self-respecting music professional and techno-geek will be to keep up with the latest developments as technology progresses, changes, and goes in and out of style. Keeping your mind open to new ways of working—through reading, surfing the Web, and getting your hands dirty experimentally—is a sure-fire way to stay fresh and youthful in the passionate pursuit of your career or hobby.

Keyboards

By far, the most common instruments that you'll encounter in almost any MIDI production facility will probably belong to the keyboard family (Figure 4.2). This is due, in part, to the fact that keyboards were the first electronic music devices to gain wide acceptance; also, MIDI was initially

Figure 4.2. *The OpenSynth NeKo™ (Gen2) production station makes use of the latest in computer technology. (Courtesy of Open Labs, Inc.; www.openlabs.com.)*

developed to record and control many of their performance and control parameters. The two basic keyboard-based instruments are the synthesizer and the digital sampler.

The Synth

A *synthesizer* (or *synth*) is an electronic instrument that uses multiple sound generators, filters, and oscillator blocks to create complex waveforms that can be combined into countless sonic variations. These synthesized sounds have become a basic staple of modern music and range from those that sound "cheesy" to ones that realistically mimic traditional instruments ... and all the way to those that generate other-worldly, ethereal sounds that literally defy classification.

Synthesizers generate sounds using a number of different technologies or program algorithms. The earliest synthesizers were analog in nature and generated sounds using a technology known as *frequency modulation* (FM) *synthesis*. Today, however, modern FM synthesis is usually implemented entirely in the digital domain.

FM synthesis techniques generally make use of at least two signal generators (commonly referred to as "operators") to create and modify a voice. Often, this is carried out by generating a signal that modulates or changes the tonal and amplitude characteristics of a base carrier signal. More sophisticated FM synths use up to four or six operators per voice, and these generators will also often use filters and variable amplifier types to alter a signal's characteristics into a sonic voice that can be complex and totally unique in nature.

Another basic technique that's used to create sounds is *wavetable synthesis*. This technique works by storing small segments of digitally sampled sound into a memory media. Various sample-based and synthesis techniques make use of looping, mathematical interpolation, pitch shifting, and digital filtering to create extended and richly textured sounds that use a surprisingly small amount of sample memory, allowing hundreds if not thousands of samples and sound variations to be stored in a single device or program.

These sample-based systems are often called *wavetable synthesizers* because the large number of pre-recorded samples that are encoded within the instrument's memory can be thought of as a "table" of sound waveforms that can be looked up and used when needed. Once selected, a vast range of modification parameters (such as sample mixing, envelope, pitch, volume, pan, and modulation) can be called up from the device's patch memory to control a sample's overall sound character.

Additive synthesis makes use of combined waveforms that are generated, mixed, and varied in level over time to create new timbres that are composed of multiple and complex harmonics that,

Figure 4.3. A6 Andromeda analog-modeled digital synthesizer. (Courtesy of Alesis, www.alesis.com.)

Figure 4.4. Bass Station analogue bass synth. (Courtesy of Novation Digital Music Systems, Ltd.; www.novationmusic.com.)

like the waveforms, vary over time. Subtractive synthesis makes extensive use of filtering to alter and subtract overtones from a generated waveform (or series of waveforms). For example, such a device could start with a square or sawtooth waveform that, with the use of filters, could be altered to approximate an acoustic instrument. These generated sounds can also be filtered and changed in level over time to more closely approximate a desired sound.

Of course, synths come in all shapes and sizes and use a wide range of patented synthesis techniques for generating and shaping complex waveforms, in a polyphonic fashion using 16, 32, or even 64 simultaneous voices (Figures 4.3 and 4.4). In addition, many synths often include a percussion section that can play a full range of drum and "perc" sounds, in a number of styles. Reverb and other basic effects are also commonly built into the architecture of these devices, reducing the need for using extensive outboard effects when being played on-stage or out of the box. Speaking of "out of the box," a number of synth systems are referred to as being "workstations." Such beasties are designed (at least in theory) to handle many of your basic production needs (including basic sound generation, MIDI sequencing, effects, etc.)—all in one neat little package.

Synth Modules

Synthesizers are also commonly designed into 19-inch rack-mountable or table-top systems (Figure 4.5). These devices, which are known as *synth modules,* often contain all of the features of

Figure 4.5. *Yamaha MOTIF-RACK ES synth. (Courtesy of Yamaha Corporation of America, www.yamaha.com.)*

Figure 4.6. *Generic soundcard for the PC.*

a standard synthesizer, except they don't have a keyboard controller. This space-saving feature means that more synths can be placed into your system rack and can be controlled from a master keyboard controller or sequencer without cluttering up your system with redundant keyboards.

Soundcard Synths

Although it would be easy to overlook these small hard- and software wonders, by far the greatest number of installed synthesizers have been designed into the computer's own operating system or onto a generic PC soundcard (Figure 4.6). These devices, which can be found in almost every home, generate sounds using a simple form of digitally controlled FM synthesis. Both the soft- and hardware synth systems almost always conform to the General MIDI specification, which has universally defined the overall patch and drum-sound structure so that a MIDI file will be uniformly played by all such synths with the correct instrument voicing and levels. Further information about General MIDI can be found in Chapter 10.

Software Synthesis and Sample Resynthesis

Since wavetable synths derive their sounds from prerecorded samples that are stored in a digital memory media, it logically follows that these sounds can also be easily stored onto hard disk (or any other media), where they can then be loaded into the RAM memory of a personal computer.

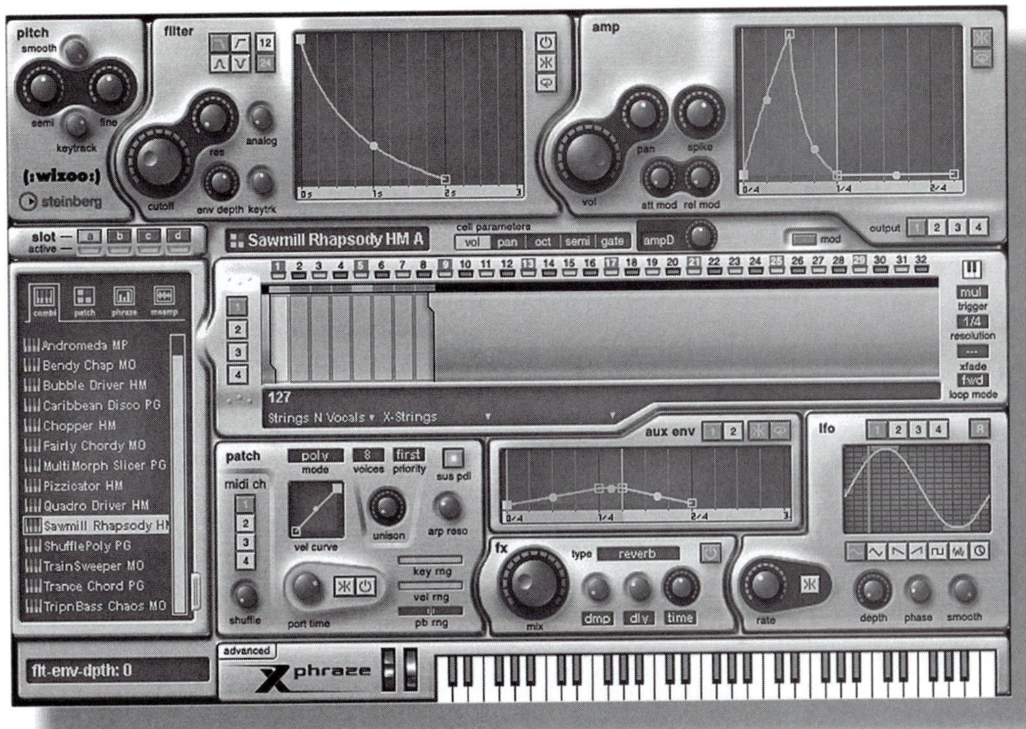

Figure 4.7. *Steinberg Xphraze synth plug-in. (Courtesy of Steinberg Media Technologies GmbH, A Division of Yamaha Corporation; www.steinberg.net.)*

This process of downloading wavetable samples into a computer and then manipulating these samples is used to create what is known as a *virtual synth* (or software synthesizer) (Figures 4.7 and 4.8).

Another virtual synth type lets you create a sound-generating or wavetable synth using on-screen, digital building blocks. In short, *modular software synthesis* works by linking various signal processing modules in a chain or parallel fashion to generate or modify a sound. These modules consist of such traditional synthesis building blocks as oscillators, voltage-controlled amplifiers, and voltage-controlled filters. These can then be mixed and processed to alter the signal's overall content and harmonic structure into almost any texture or synthesized sound that could possibly be imagined. Because the system exists in software, a newly created sound patch can be saved to disk for later recall.

As you might expect, the depth and capabilities of a software synth depend on the quality of the program and its generation techniques, wavetable signal quality, sample rate, and overall processing techniques. These can range from a simple General MIDI software synthesizer to professional-level software synthesizers that let you import, edit, and combine wavetable sounds with an amazing degree of ease and control.

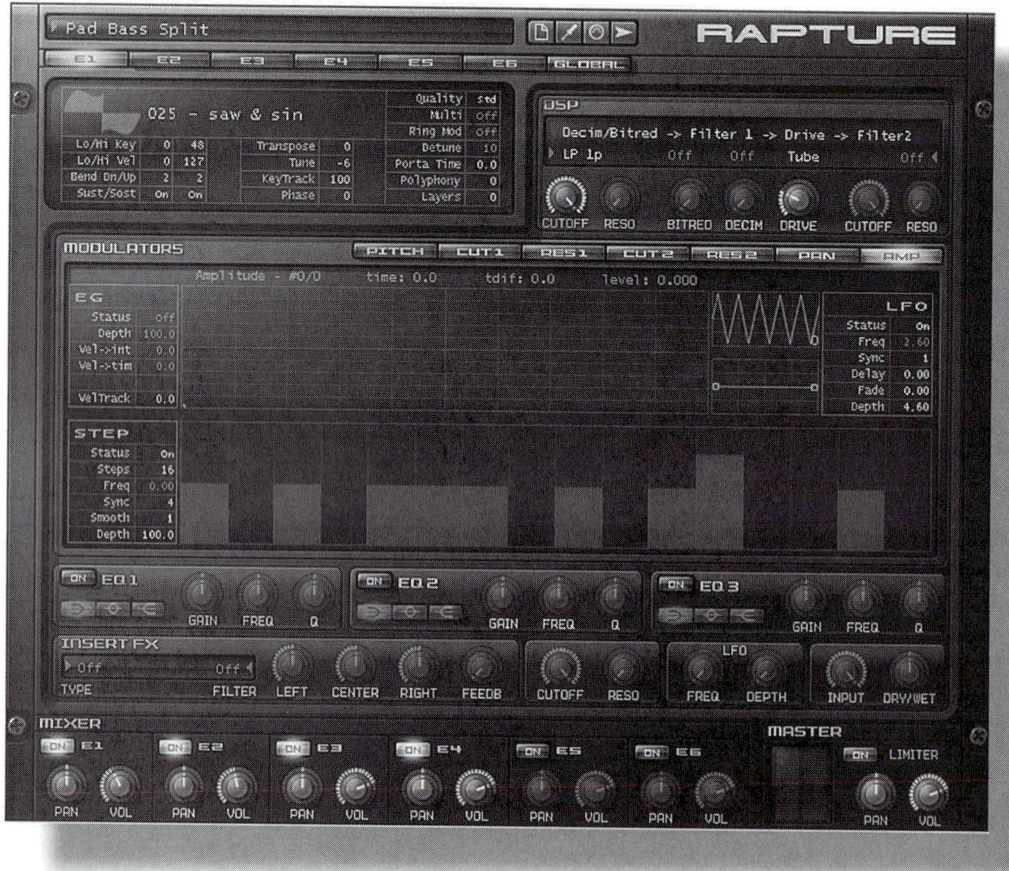

Figure 4.8. *Cakewalk's Rapture synth plug-in. (Courtesy of Twelve Tone Systems, Inc.; www.cakewalk.com.)*

Samplers

A sampler is a device that can convert audio into a digital form that is then imported into internal random access memory (RAM). Once audio has been sampled or loaded into RAM (from disk, disc, or diskette), segments of sampled audio can then be edited, transposed, processed, and played in a polyphonic, musical fashion. In short, a sampler can be thought of as a digital audio memory device that lets you record, edit, and reload samples into RAM. Once loaded, these sounds (whose length and complexity are often only limited by memory size and your imagination) can be looped, modulated, filtered, and amplified (according to user or factory setup parameters) in a way that allows the waveshapes and envelopes to be modified. Signal processing capabilities, such as basic editing, looping, gain changing, reverse, sample-rate conversion, pitch

Figure 4.9. Akai MPC1000 Music Production Center. (Courtesy of Akai Professional, www.akaipro.com.)

change, and digital mixing can also be easily applied to change the sounds in an almost infinite number of ways.

A sampler's design can include a music keyboard or set of trigger pads (Figure 4.9) that let you polyphonically play samples as musical chords, sustain pads, triggered percussion sounds, or sound-effect events, directly from its surface. Or, it might be designed as a sample module that integrates all of their necessary signal processing, programming, and digital control structures into a single, 19-inch rack-mountable unit. Of course, because modules don't have a keyboard, they must be controlled from an external controller (such as an external keyboard controller, drum pads, sequencer, or other controller type).

These samples can be played according to the standard Western musical scale (or any other scale, for that matter) by altering the playback sample rate over the controller's note range. For example, pressing a low-pitched key on the keyboard will cause the sample to be played back at a lower sample rate, while pressing a high-pitched one will cause the sample to be played back at rates that would put Mickey Mouse to shame By choosing the proper sample–rate ratios, these sounds can be polyphonically played (whereby multiple notes are sounded at once) at pitches that correspond to standard musical chords and intervals.

A sampler (or synth) with a specific number of voices—for example, 64 voices—simply means that up to 64 notes can be simultaneously played on a keyboard at any one time. Each sample in a multiple-voice system can be assigned across a performance keyboard, using a process known as *splitting* or *mapping*. In this way, a sound can be assigned to play across the performance surface of a controller over a range of notes, known as a zone (Figure 4.10). In addition to grouping samples into various zones, velocity can enter into the equation by allowing multiple samples to be layered across the same keys of a controller, according to how soft or hard they are played. For example, a single key might be layered so that pressing the key lightly would reproduce a softly recorded sample, while pressing it harder would produce a louder sample with a sharp, percussive

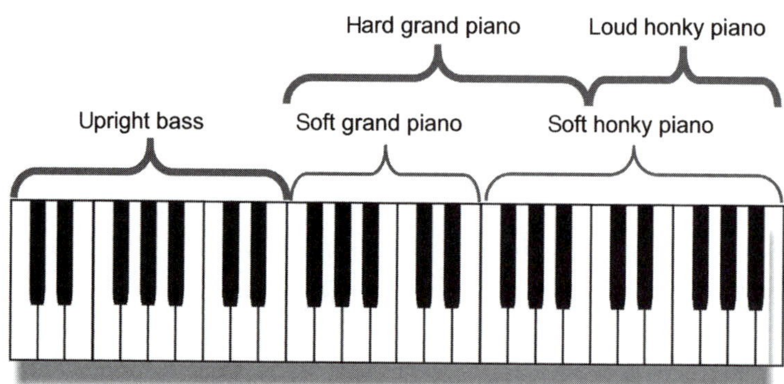

Figure 4.10. *Example of a sampler's keyboard layout that has been programmed to include zones. Notice that the upper register has been split into several zones that are triggered by varying velocities.*

attack. In this way, mapping can be used to create a more realistic instrument or wild set of soundscapes that change not only with the played keys but with different velocities as well.

Most samplers have extensive edit capabilities that allow the sounds to be modified in much the same way as a synthesizer, using such modifiers as:

- Velocity
- Panning
- Expression (modulation and user control variations)
- Low-frequency oscillation (LFO)
- Attack, delay, sustain, and release (ADSR) and other envelope processing parameters
- Keyboard scaling
- Aftertouch

Many sampling systems will often include such features as integrated signal processing, multiple outputs (offering isolated channel outputs for added live mixing and signal processing power or for recording individual voices to a multitrack recording system), and integrated MIDI sequencing capabilities.

Software Samplers

In addition to hardware sampling systems, a fast-growing number of virtual or software samplers are coming on the market that use the computer's existing memory, processing, and signal routing capabilities in order to polyphonically process and reproduce samples in real time. Offering much of the same functionality as their hardware counterparts, these software-based systems (Figures 4.11 and 4.12) are capable of editing, mapping, and splitting sounds across a MIDI keyboard using on-screen graphic controls and digital audio workstation (DAW) integration and have improved to the point of equaling or surpassing their hardware counterparts in cost-effectiveness, power, and ease of use.

Figure 4.11. *HALion 3 software sampling plug-in. (Courtesy of Steinberg Media Technologies GmbH, A Division of Yamaha Corporation; www.steinberg.net.)*

As with a hardware synth, software samplers derive their sounds from recorded and/or imported audio data stored as digital audio data within a personal computer. Using the digital signal processing (DSP) capabilities of today's computers (as well as the recording, sequencing, processing, mixing, and signal-routing capabilities of most DAWs), most software samplers are able to store and access samples within the internal memory of a laptop or desktop computer. Using a graphic interface, these sampling systems often allow the user to:

- Import previously recorded soundfiles (generally in .wav and .aif formats).
- Edit and loop sounds into a usable form.
- Vary envelope parameters (*e.g.*, dynamics) over time.
- Vary processing parameters.
- Save the edited sample setup parameters as a file for later recall.

Figure 4.12. *GigaStudio3 software sampler. (Courtesy of Tascam, www.tascam.com.)*

Another class of software sampler involves the playback and manipulation of prerecorded samples from samplers that are integrated into a unified music production package. Propellerhead's Reason and Abelton's Live software programs come to mind here. Within Reason, .wav or .aif samples can be imported into either the NN-19 or NN-XT sample module (Figure 4.13), allowing them to be edited, processed, and played in much the same way that one would use a hardware or plug-in sampler. Reason's Redrum drum machine module is an example of software module that allows recorded drum samples (or any other type of sounds) to be imported into a drum machine, making it possible for these samples to be integrated into pattern-based samples that can be performed in new and invigorating ways (Figure 4.14).

Sample CDs, DVDs, and the Web

Just as patch data in the form of SysEx dump files can have the effect of breathing new life into your synth, a wide range of free or commercially available samples is commonly available off the web, from a website, or from a business entity that lets you experiment with loading new and fresh sounds into your production system. These files can exist as unedited soundfile data (which can be imported into any sample system or DAW track), or as data that has been specifically programmed

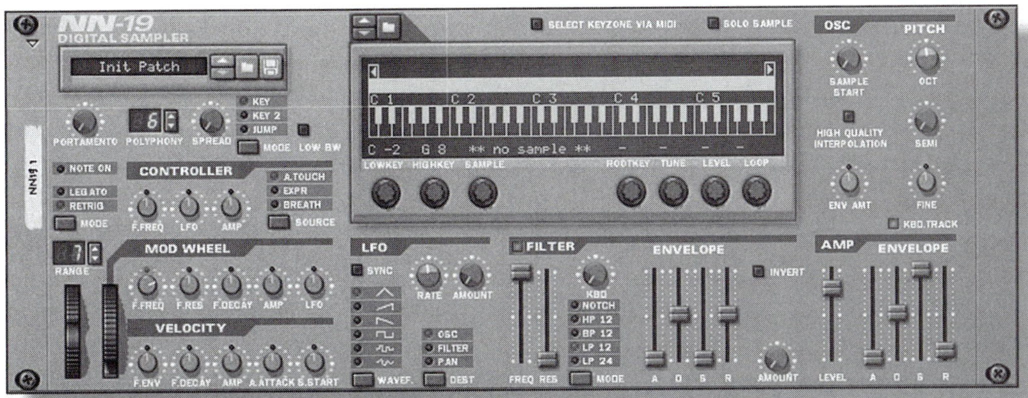

Figure 4.13. *Reason's NN-19 and NN-XT sample modules. (Courtesy of Propellerhead Software, www.propellerheads.se)*

Figure 4.14. *Reason's Redrum drum machine module. (Courtesy of Propellerhead Software, www.propellerheads.se)*

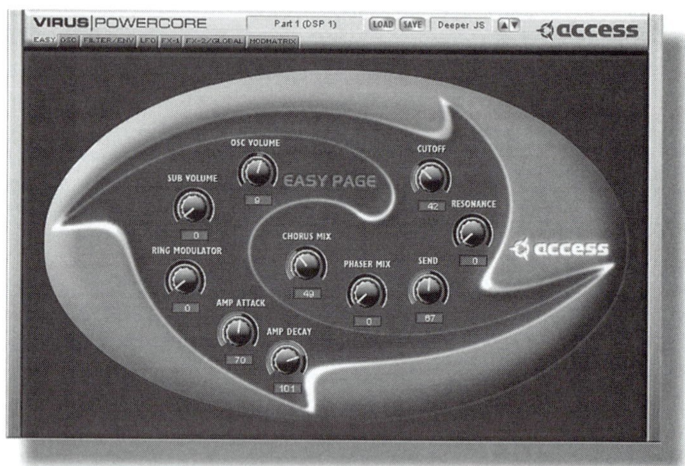

Figure 4.15. *Virus synth plug-in for the TC Electronic PowerCore. (Courtesy of TC Electronic, Inc, www.tcelectronic.com.)*

by a professional musician/programmer to contain all the necessary loops, system commands, and sound-generator parameters so all you ideally need to do is load the sample and begin having fun.

The mind boggles at the range of styles and production quality that has gone into producing samples that are just ready and waiting to give your project a boost. The general production level literally runs the entire amateur-to-pro gamut—meaning that, whenever possible, it's wise to listen to examples to determine their quality and to hear how they might fit into your own personal or project style before you buy. As a final caveat … by now, you've probably heard of the legal battles that have been raging over sampled passages that have been "ripped" from recordings of established artists. In the fall of 2004 decision regarding Bridgeport Music et al. *vs.* Dimension Films, the 6th Circuit U.S. Court of Appeals ruled that the digital sampling of a recording without a license is a violation of copyright, regardless of size or significance. This points to the need for tender loving care when lifting samples off of a CD or off of the web.

Instrument Plug-Ins

In recent years, an almost staggering range of software instruments has come onto the market as *instrument plug-ins.* These systems, which include all known types of synths, samplers, and pitch- and sound-altering devices, are able to communicate MIDI, audio, timing sync, and control data between the software instrument (or effect plug-in) and a host DAW program/CPU processor (Figures 4.15 and 4.16).

Using an established plug-in communications protocol, it's possible for most or all of the audio and timing data to be routed through the host audio application, allowing the instrument or application I/O, timing, and control parameters to be seamlessly integrated into the DAW or application. A few of these protocols include:

- Steinberg's VST
- Digidesign's RTAS
- MOTU's MAS

Figure 4.16. Bass Station virtual bass synth. (Courtesy of Novation Digital Music Systems, Ltd.; www.novationmusic.com.)

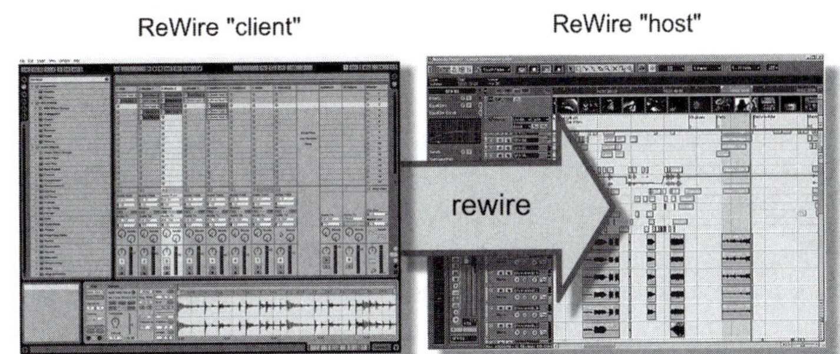

Figure 4.17. ReWire allows a client program to be inserted into a host program (often a DAW) so they can run simultaneously in tandem.

Propellerhead's ReWire is another type of protocol that allows audio, performance, and control data of an independent audio program to be wired into a host program (usually a DAW) such that the audio routing and sync timing of the slave program is locked to the host DAW, effectively allowing them to work in tandem as a single production environment (Figure 4.17). Further reading on virtual and plug-in instruments, plug-in protocols, and applications can be found within Chapter 6.

The MIDI Keyboard Controller

As computers, sound modules, virtual software instruments, and other types of digital devices have come onto the production scene, it's been interesting to note that fewer and fewer devices are being made that include a music keyboard in their design. As a result, the MIDI keyboard controller has gained in popularity as a device that might include a:

- Music keyboard surface
- Variable parameter controls
- Fader, mixing, and transport controls
- Switching controls
- Tactile trigger and control surfaces

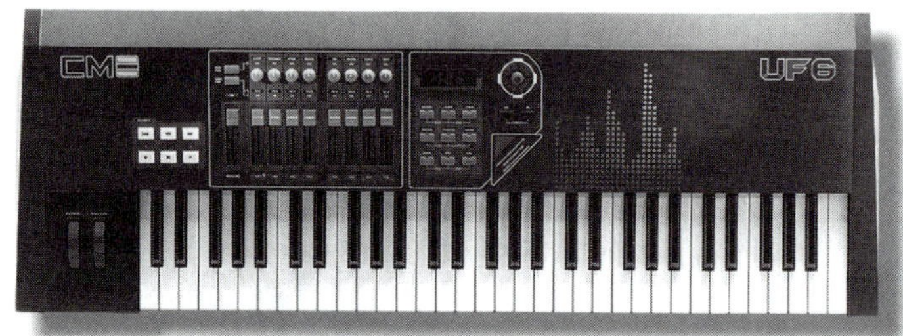

Figure 4.18. CME UF6 controller. (Courtesy of Central Music Company, www.cme-pro.com; distributed by Yamaha Corporation of America, www.yamaha.com.)

A Word About Controllers

A MIDI controller is a device that's expressly designed to control other devices (be they for sound, light, or mechanical control) within a connected MIDI system. As was previously said, these devices contain no internal tone generators or sound-producing elements but often include a high-quality control surface and a wide range of controls for handling control, trigger, and device-switching events. Since controllers have become an integral part of music production and are available in many incarnations to control and emulate many types of musical instrument types, don't be surprised to find controllers of various incarnations popping up all over this book and within electronic music production.

As was stated, these devices contain no internal tone generators or sound-producing elements. Instead they can be used in the studio or on the road as a simple and straightforward surface for handling MIDI performance, control, and device-switching events in real time (Figure 4.18).

As you might imagine, controllers vary widely in the number of features that are offered. For starters, the number of keys can vary from the sporty, portable 25-key models to those having 49 and 61 keys and all the way to the full 88-key models that can play the entire range of a full-size grand piano. The keys may be fully or partially weighted (in the case of the M-Audio O^2 model, a nifty half key height was used to lower the controller's overall height), and in a number of models the keys might be much smaller than the full piano key size—often making a performance a bit difficult.

Beyond the standard pitch and modulation wheels (or similar-type controller), the number of options and general features is up to the manufacturers. With the increased need for control over electronic instruments and music production systems, many models types offer up a wide range of physical controllers for varying an ever-widening range of expressive parameters (Figure 4.19).

The daddy of all keyboard controllers is the MIDI grand piano. Such an instrument may be a fully functional acoustic grand that can be performed in a normal fashion, can output MIDI performance messages, and can respond to MIDI by playing in an automated player-piano style (Figure 4.20). Some electronic pianos offer up a fully weighted keyboard and might be made in a furniture style that mimics a baby mini or small grand but will contain circuitry for playing back synthesized piano, percussion, and other instrument types. In addition to a traditional, weighted keyboard, sustain and una corda pedals on most MIDI grands often offer excellent control over external keyboard instruments.

Figure 4.19. *Novation 25SL USB MIDI Controller. (Courtesy of Novation Digital Music Systems, Ltd.; www.novationmusic.com.)*

Figure 4.20. *Yamaha DS6IV PRO Disclavier MIDI piano. (Courtesy of Yamaha Corporation of America, www.yamaha.com.)*

Not to be outdone, the church organ has eclipsed all others in recent years by becoming the mother of all controllers. These mega keyboard devices have introduced MIDI into houses of worship, allowing recitals and special events to be recorded and then played back at a later time. Imagine walking down the aisle to a "Here Comes the Bride" sequence. (Don't laugh—it's probably happening in Vegas as you read this.) Certain church organ systems have also joined the digital age by offering advanced sample playback engines and controller systems that can mimic the entire organ setup from a DAW environment—and do it in full surround.

Alternative Controllers

On certain occasions, more nontraditional controller types can be used in place of a keyboard or to augment a keyboard's functionality. For example, drum machines are able to transmit MIDI note on/off, velocity and aftertouch messages. In a pinch, these can be programmed to control any number of MIDI device types. Alternatively, control commands, program changes, and other switching functions can also be done on the fly by assigning program-change values to devices such as a MIDI-fied guitar effects foot pedal and any number of the MIDI tools and toys that are on the market. In short, never forget that necessity is often the mother of invention. Problems relating to MIDI control can often be dealt with in interesting and cost-effective ways (Figures 4.21 and 4.22).

The Drum Machine

In its most basic form, the drum machine (Figure 4.23) uses ROM-based, prerecorded waveform samples to reproduce high-quality drum sounds from its internal memory. These factory-loaded sounds often include a wide assortment of drum sets, percussion sets, rare and wacky percussion

Figure 4.21. *The RagsPole MIDI controller. (Courtesy of Rags Tuttle Productions, www.ragstuttle.com.)*

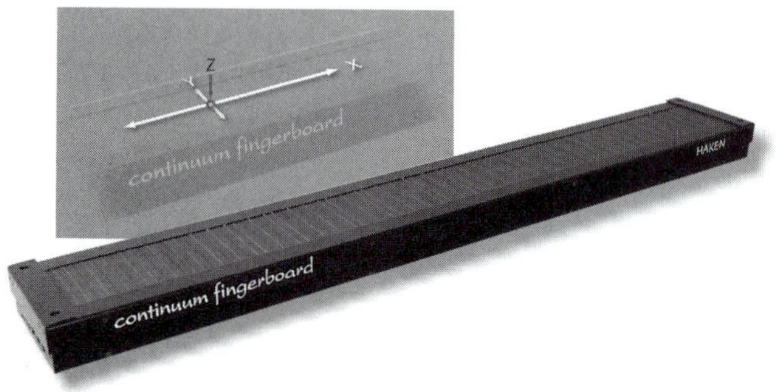

Figure 4.22. *The Continuum Fingerboard tracks and updates MIDI values for the x, y, and z positions of one to sixteen fingers every 1.33 milliseconds; these values appear at the MIDI Out as well as the FireWire® port. (Courtesy of Haken Audio, www.hakenaudio.com.)*

hits, and effected drum sets (*e.g.*, reverberated, gated). Who knows—you might even encounter scream hits by the venerable King of Soul, James Brown. These prerecorded samples can be assigned to a series of playable keypads that are generally located on the machine's top face, providing a straightforward controller surface that often sports velocity and aftertouch dynamics. Sampled voices can be assigned to each pad and edited using control parameters such as tuning, level, output assignment, and panning position.

Because of new cost-effective technology, many drum machines now include basic sampling technology, which allows sounds to be imported, edited, and triggered directly from the box (Figure 4.24). As with the traditional "beat box," these samples can be easily mapped and played from the traditionally styled surface trigger pads. Of course, virtual software drum and groove machines (Figures 4.25 and 4.26) are part of the present-day landscape and can be used in a stand-alone, plugged-in, or rewired production environment.

Figure 4.23. *Alesis SR-16 stereo drum machine. (Courtesy of Alesis, www.alesis.com.)*

Figure 4.24. *Akai MPC2500 Music Production Center. (Courtesy of Akai Professional, www.akaipro.com.)*

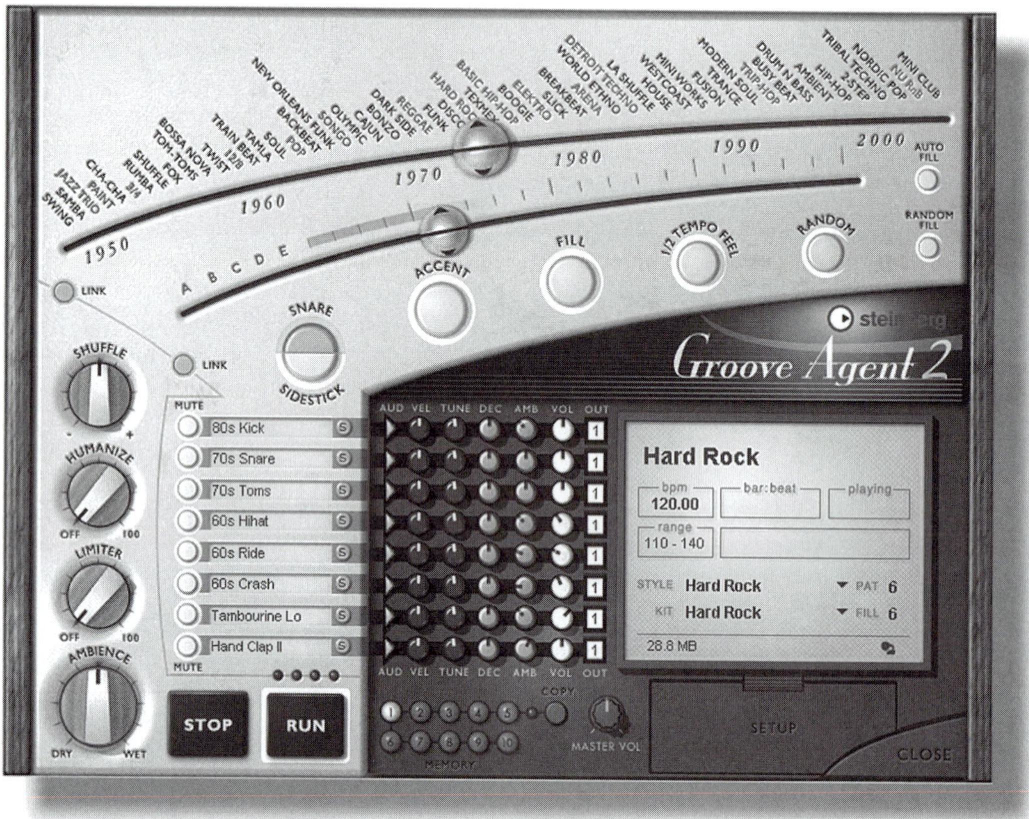

Figure 4.25. *Steinberg's GrooveAgent 2 drum plug-in. (Courtesy of Steinberg Media Technologies GmbH, A Division of Yamaha Corporation; www.steinberg.net.)*

Regardless of their type, drum machines will often have a built-in sequencer that's been specifically designed to arrange drum/percussion sounds into a rhythmic sequence (known as a drum pattern). These patterns often consist of basic variations on a rhythmic groove, or they can be built from patterns taken from an existing library of playing styles (such as rock, country, or jazz). Drum machines that have this feature will often let you chain these patterns together into a continuous song. Once assembled, the song can be played back using an internal MIDI clock source, or it can be synchronously driven from an external clock source (such as a DAW's internal timing source).

Although their design might include a built-in pattern sequencer, it's equally as likely that these studio workhorses will be triggered from a sequencer or DAW MIDI track. This lets us take full advantage of all the real-time performance and editing capabilities that a MIDI pattern track has to offer. For example, sequenced patterns can be easily created in step time, allowing notes to be

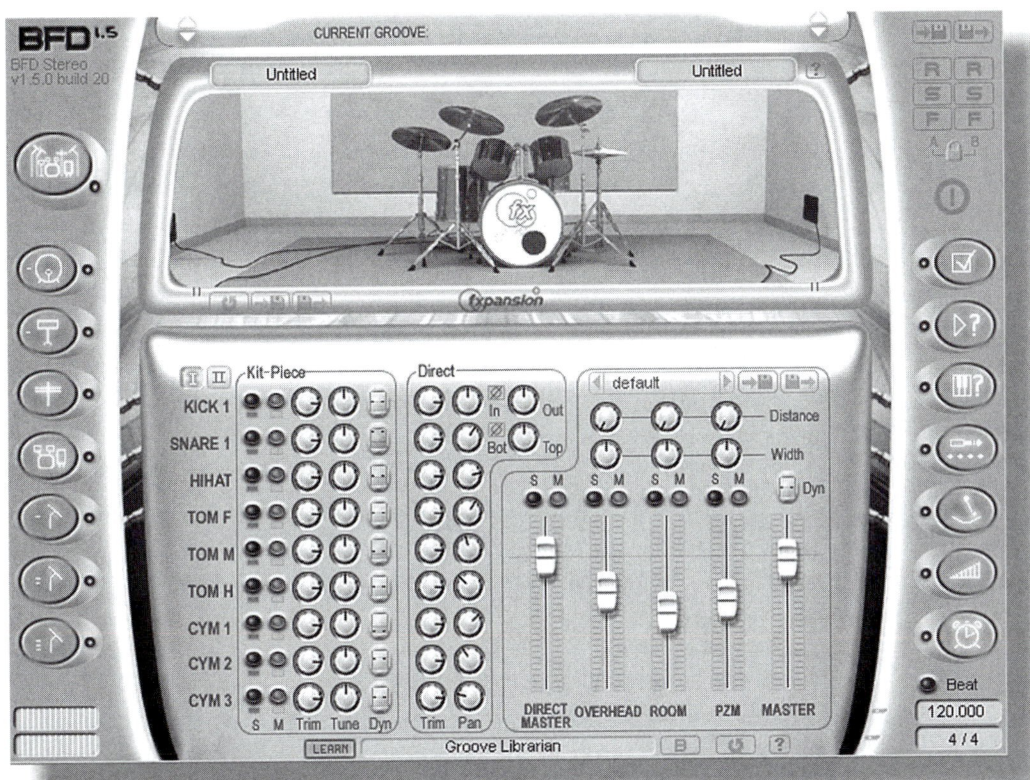

Figure 4.26. *BFD acoustic drum library module. (Courtesy of FXpansion, www.fxpansioncom.)*

entered and assembled into a rhythmic pattern, one note at a time (Figure 4.27), and then be linked into a song that's composed of several rhythmic variations. Alternatively, drum tracks can be played into a sequencer on the fly, creating a live feel, or you can merge step- and real-time tracks together to create a more human-sounding composite rhythm track. In the final analysis, the style and approach to a composition are entirely up to you.

Most drum machine designs include multiple outputs (or virtual signal paths) that let you route individual or groups of voices to a specific mixer input or DAW input. This makes it possible for the isolated voices to be individually mixed, panned, or processed (using equalization, effects, etc.) or to be recorded onto the separate tracks of a recording system.

Alternative Percussion Voices

In addition to the numerous sounds that can be found in a drum machine, a virtually unlimited number of percussion sounds can be obtained and placed into a project from other sources. As was mentioned earlier, a synthesizer will often include several drum and/or percussion setups that are mapped over the entire keyboard surface. Sampler libraries will almost always include

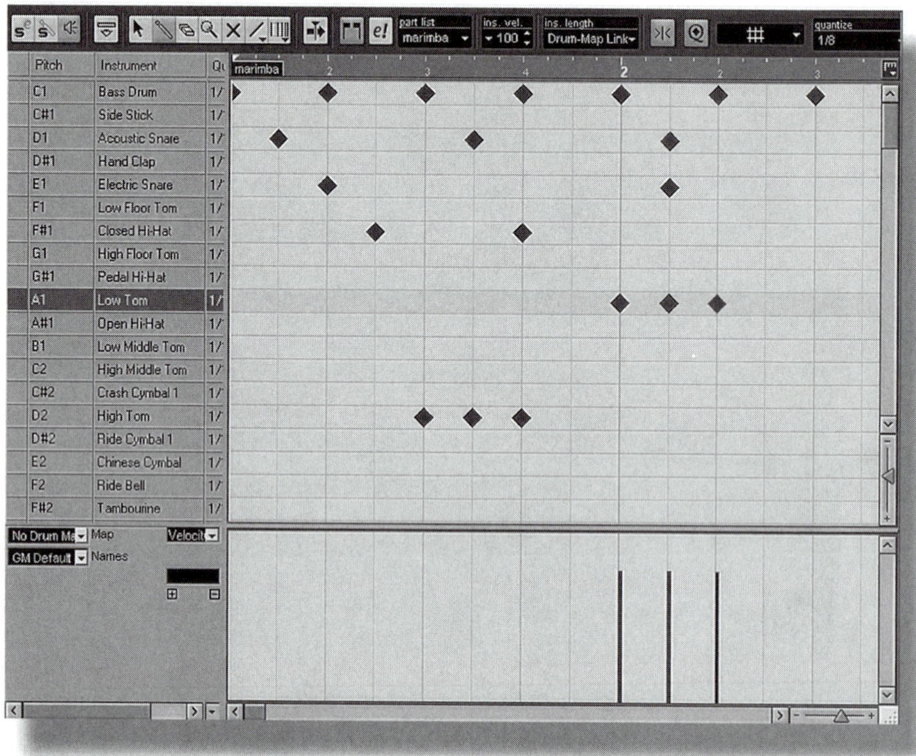

Figure 4.27. Cubase/Nuendo drum pattern track. (Courtesy of Steinberg Media Technologies GmbH, A Division of Yamaha Corporation; www.steinberg.net.)

a never-ending number of percussion instruments and drum sets. Sound files can be loaded into a DAW to build up rhythm tracks, or you can lift percussion loops from loop sources and libraries that are available on CD and DVD. Again, the sky's only limited by your imagination. I'd also like to take this time to remind you of the power that can be harnessed by recording your own samples. If you own a sampling system and can't find the sounds you want, then create your own! It'll give you a great chance to personalize your music and/or effects—in a way that will often cause everyone's ears to perk up.

MIDI Drum Controllers

MIDI drum controllers are used to translate the voicings and expressiveness of a percussion performance into MIDI data. These devices are great for capturing the feel of a live performance, while giving you the flexibility of automating or sequencing a live event. These devices range from having larger pads and trigger points on a larger performance surface to drum-machine-type pads/buttons (Figure 4.28). Coming under the "Don't try this at home" category, these controller pads are generally too small and not durable enough to withstand drumsticks or mallets. For this

Figure 4.28. *Trigger Finger 16-pad MIDI drum control surface. (Courtesy of M-Audio, A Division of Avid Technology, Inc.; www.m-audio.com.)*

reason, they're generally played with the fingers. It's long been a popular misconception that MIDI drum controllers have to be expensive. This simply isn't true. There are quite a few instruments that are perceived by many to be toys but, in fact, are fully implemented with MIDI and can be easily used as a controller. The next few controller sections outline a few of the more common ways of playing, triggering, and creating percussion patterns in the studio.

Drum Machine Button Pads

As was mentioned, one of the most straightforward of all drum controllers is the drum button pad design that's built into most drum machines, portable percussion controllers, and certain keyboard controller styles. By calling up the desired setup and voice parameters, these small footprint triggers let you go about the business of using your fingers to do the walking through a performance or sequenced track. It's also a simple matter to trigger other devices from these keypads. For example, you could assign the pads to a channel that's being responded to by your favorite synth, sampler, or groove machine and then trigger these sounds directly.

The Keyboard as a Percussion Controller

Since drum machines respond to external MIDI data, probably the most commonly used device for triggering percussion and drum voices is a standard MIDI keyboard controller. One advantage of playing percussion sounds from a keyboard is that sounds can be triggered more quickly because the playing surface is designed for fast finger movements and doesn't require full hand/wrist motions. Another advantage is its ability to express velocity over the entire range of possible values (0–127), instead of the limited number of velocity steps that are available on certain drum pad models. Most drum machines allow drum and percussion voices be manually assigned to a particular MIDI note value. As the percussion sounds might not be related to any musical interval, you're free to assign a drum voices to any keyboard note and range that you'd like. In addition, percussion sounds can be assigned to a particular range of notes over a split keyboard arrangement, allowing other sound patches to be simultaneously played on the same surface.

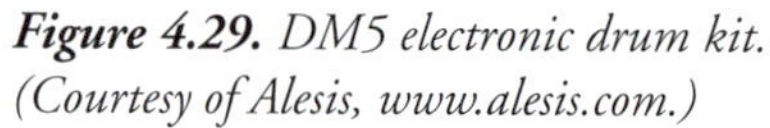
Figure 4.29. *DM5 electronic drum kit. (Courtesy of Alesis, www.alesis.com.)*

Drum Pad Controllers

In more advanced MIDI project studios or live stage rigs, it's often necessary for a percussionist to have access to a playing surface that can be played like a real instrument. In these situations, a dedicated drum pad controller (Figure 4.29) would be better for the job. Drum controllers vary widely in design. They can be built into a single, semiportable case, often having between six and eight playing pads, or the trigger pads can be individual pads that can be fitted onto a special rack, traditional drum floor stand, or a drumset. Of course, these pads can often be triggered over the full velocity range (allowing for sample splits to be built up), and their playing surface can be played with the fingers, hands, percussion mallets, or drumsticks. Besides giving us a more realistic performance interface and full MIDI capabilities, one of the biggest advantages of a drum pad setup is silence You could bang away at a full drum setup in a New York high-rise at 3 a.m., and nobody'd ever know. Often, they're small enough that they can be tucked away in a corner of a bedroom or mid-sized project studio.

MIDI Drums

Taking realism even further, real drums can be MIDI-fied in a number of interesting ways without too much fuss. One of the simpler of these is to incorporate MIDI drum pads into a standard acoustic drum set (Figure 4.30). Such a setup gives us the power and sound of a traditional kit, with the versatility, unique sounds, and sequence capabilities that MIDI has to offer.

Another way to MIDI-fy an acoustic drum is through the use of *trigger* technology. Put simply, triggering is carried out by using a transducer pickup (such as a mic or contract pickup) to change the acoustic energy of a percussion or drum instrument into an electrical voltage. Using a MIDI trigger device (Figure 4.31), a number of pickup inputs can be translated into MIDI so as to trigger programmed sounds or samples from an instrument for use on stage or in the studio.

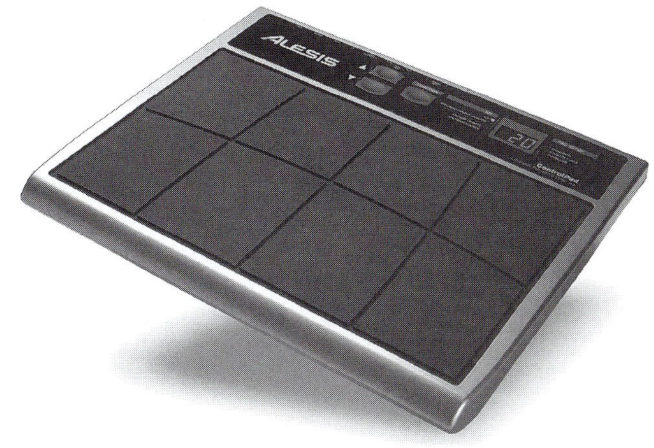

Figure 4.30. *A pad such as the ControlPad USB/MIDI percussion pad controller could be incorporated into a drum set for added versatility. (Courtesy of Alesis, www.alesis.com.)*

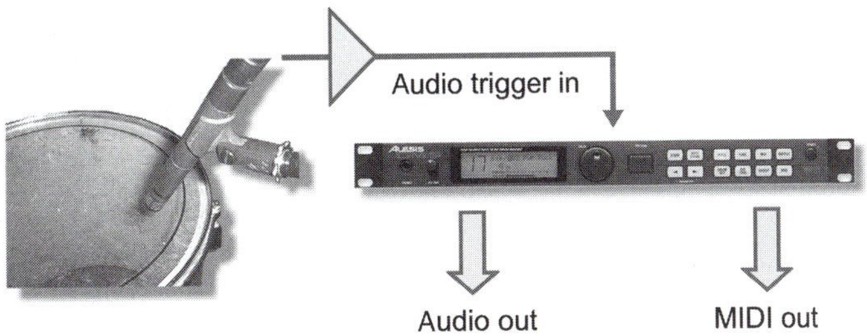

Figure 4.31. *By using a MIDI trigger device (such as an Alesis DM5), a drum's original pickup can be either directly replaced or sent via MIDI to another device or sequenced track. (Courtesy of Alesis, www.alesis.com.)*

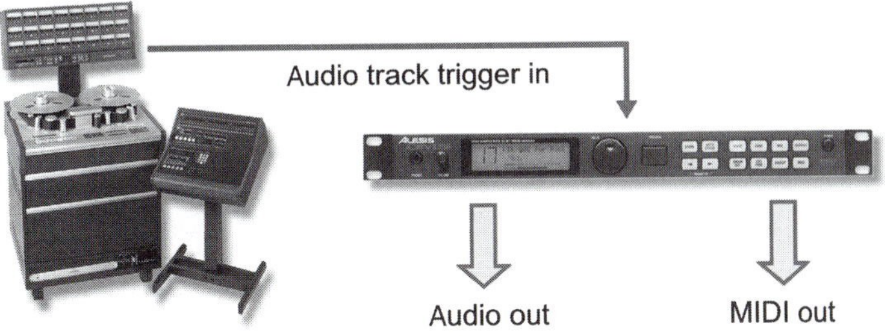

Figure 4.32. *By using a MIDI trigger device (such as an Alesis DM5), a recorded drum track can be either directly replaced or sent via MIDI to another device or sequenced track. (Courtesy of Alesis, www.alesis.com.)*

Because the trigger source is an electrical signal, the original audio source can be almost anything. For example, you could mike a snare drum in the studio and use that signal to trigger a monster snare. Alternatively, you could use the already-recorded snare track as a trigger source, so as to replace the "sucko" track with one that perfectly fits the bill (Figure 4.32). This type of production work definitely comes under the "use your imagination" category.

For those who are using your DAW to record, augment, and even replace recorded drum tracks, software exists that allows you to replace an already recorded drum track or sample with other

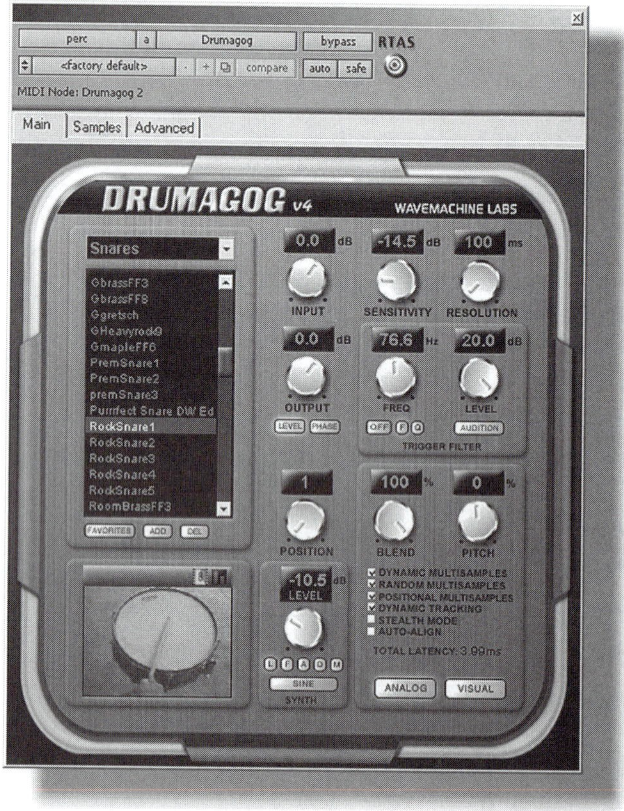

Figure 4.33. *Drumagog professional drum replacer plug-in. (Courtesy of Wavemachine Labs, Inc.; www.drumagog.com.)*

preset or custom sounds … all without changing the original feel or dynamics of the originally recorded track (Figure 4.33).

Other MIDI Instrument and Controller Types

There are literally tons of instruments and controller types out there that are capable of translating a performance or general body movements into MIDI. The following are just a smattering of the more traditional instrument types. You'd be surprised what you'll find searching the web for wild and wacky controllers—both those that are commercially available and those that are made by soldering-iron junkies.

MIDI Vibraphone

Unlike drum machine pads or keyboard controllers, the MIDI vibraphone is generally used by professional percussionists who want a traditional playing surface while making use of the power of MIDI. These vibes are commonly designed with a playing surface that can be fully configured

Figure 4.34. *i2000 Series guitar showing optional connections to USB and to the Roland 13-pin RMC system for fast-tracking access to Roland guitar processors and digital products. (Courtesy of Brian Moore Guitars, www.iguitar.com.)*

using its internal setup memory to provide for user-defined program changes, playing-surface splits, velocity, after touch, modulation, etc.

MIDI Guitars and Basses

Guitar players often work at stretching the vocabulary of their instruments beyond the norm. They love doing nontraditional gymnastics using such tools of the trade as distortion, phasing, echo, feedback, etc. Due to advances in guitar pickup and microprocessor technology, it's now possible for the notes and minute inflections of guitar strings to be accurately translated into MIDI data (Figure 4.34). With this innovation, many of the capabilities that MIDI has to offer are now available to the electric (and electronic) guitarist. For example, a guitar's natural sound can be layered with a synth pad that's been transposed down to give a rich, thick sound that'll shake your boots. Alternatively, recording a sequenced guitar track in sync with the audio tracks would give a producer the option of changing and shaping the sound in mixdown. On-stage program changes are also a big plus for the MIDI guitar. These let the player radically switch between guitar voices from the synth by simply stomping on a MIDI foot controller.

Other guitar-like incarnations replace the traditional strings with a tactile surface and a neck that's chock full o' trigger buttons. To some this is sacrilege; to others it's a whole new way of expressing themselves (truth be known, my MIDI guitar frees me up to loop and groove my way through performances using Abelton's Live in ways that have to be seen and heard to be believed).

MIDI Wind Controllers

MIDI wind controllers (Figure 4.35) differ from keyboard and drum controllers because they're expressly designed to bring the breath and key articulation of a woodwind or brass instrument

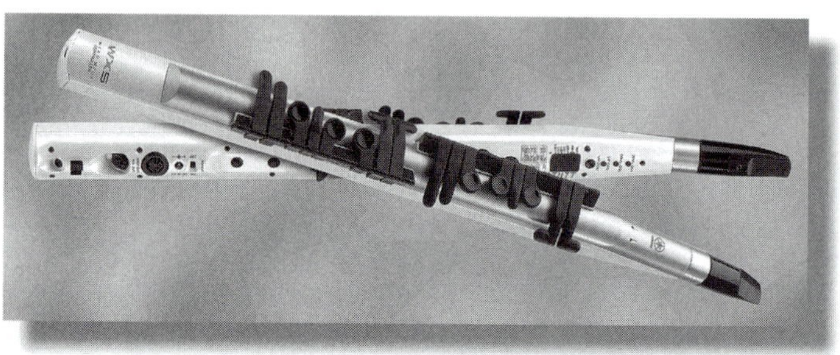

Figure 4.35. *Yamaha WX5 Wind MIDI Controller. (Courtesy of Yamaha Corporation of America, www.yamaha.com.)*

into the world of the MIDI performance. These controller types are used because many of the dynamic- and pitch-related expressions (such as breath and controlled pitch glide) simply can't be communicated from a standard music keyboard. In these situations, wind controllers can often help create a dynamic feel that's more in keeping with their acoustic counterparts by using an interface that provides special touch-sensitive keys, glide- and pitch-slider controls, and sensors for outputting real-time breath control over dynamics.

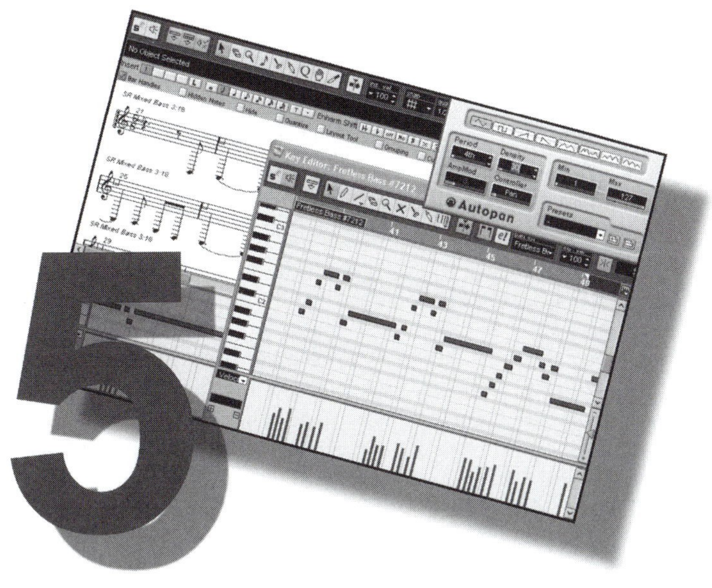

Sequencing

Apart from electronic musical instruments, one of the most important tools that can be found in the modern-day project studio is the MIDI sequencer. Basically, a *sequencer* is a digital device or software application that's used to record, edit, and output MIDI messages in a sequential fashion. These messages are generally arranged in a track-based format that follows the modern production concept of having instruments (and/or instrument voices) located on separate tracks. This traditional interface makes it easy for us humans to view MIDI data as tracks on a digital audio workstation (DAW) or analog tape recorder that follow along a straightforward linear time line.

These tracks contain MIDI-related performance and control events that are made up of such channel and system messages as Note-On, Note-Off, Velocity, Modulation, Aftertouch, and Program/Continuous Controller messages. Once a performance has been recorded into a sequencer's memory, these events can be graphically arranged and edited into a musical performance. The data can then be saved as a file or within a DAW session and recalled at any time, allowing the data to be played back in its originally recorded or edited order.

As was mentioned previously, most sequencers are designed to function in a way that's similar to their distant cousin, the multitrack tape recorder. This gives us a familiar operating environment in which each instrument, set of layered instruments, or controller data can be recorded onto separate, synchronously arranged tracks. Like its multitrack cousin, each track can be re-recorded, erased, copied, and varied in level during playback. However, since the recorded data is digital in nature, a MIDI sequencer is far more flexible in its editing speed and control in that it offers all the cut-and-paste, signal-processing, and channel-routing features that we've come to expect from a digital production application.

Hardware Sequencers

Hardware sequencers (Figure 5.1) are stand-alone devices that are designed for the sole purpose of sequencing MIDI data. These systems are more or less a thing of the past, as most hardware systems are now generally integrated into hardware MIDI production systems called *hardware workstations.*

Like their software counterparts, hardware sequencers are designed to emulate the basic functions of a tape transport (record, play, start/stop, fast forward, and rewind) in a basic and straightforward fashion. Generally, these systems offer only a moderate number of editing features, such as velocity and other controller messages, program change, cut-and-paste, track merging capabilities, tempo changes, and the ability to change note values.

These devices incorporate a specially designed CPU and operating system, memory, MIDI ports, and integrated controls for performing sequence-specific functions. An LCD is commonly used to display programming, tracking, and editing information. Of course, these displays are often small in size and resolution and are limited to information that relates to one parameter or track at a time.

Figure 5.1. *Yamaha QY700 Palmtop Music Sequencer. (Courtesy of Yamaha Corporation of America, www.yamaha.com.)*

Integrated Workstation Sequencers

A type of keyboard synth and sampler system known as a *keyboard workstation* will include much of the necessary production hardware that's required for music production, including effects and an integrated hardware sequencer. These systems have the advantage of letting you take your instrument and sequencer on the road without having to drag your whole system along. Similar to the stand-alone hardware sequencer, a number of these sequencer systems have the disadvantage of offering few editing tools beyond transport functions, punch-in/out commands, and other basic edit functions. With the advent of more powerful keyboard systems that include a larger, integrated LDC display, the integrated sequencers within these systems are becoming more powerful, resembling their software sequencing counterparts (Figure 5.2). In addition, other types of palm-sized microworkstations (Figure 5.3) offer such features as polyphonic synth voices, drum machine kits, effects, MIDI sequencing, and, in certain cases, facilities for recording multitrack digital audio in an all-in-one package that literally fits in the palm of your hand!

Software Sequencers

By far, the most common sequencer type is the *software sequencing program.* These programs run on all types of personal and laptop computers and take advantage of the hardware and software

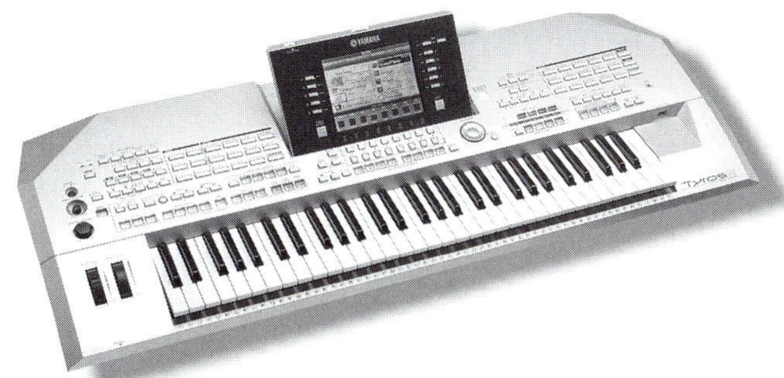

Figure 5.2. *Tyros2 61-key keyboard workstation. (Courtesy of Yamaha Corporation of America, www.yamaha.com.)*

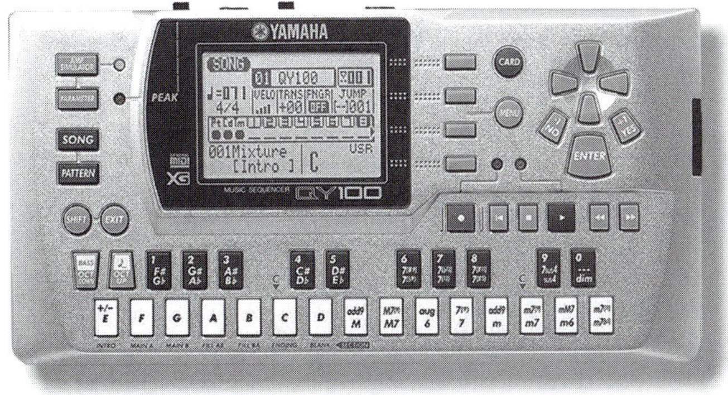

Figure 5.3. *Yamaha QY100 Palmtop Music Sequencer. (Courtesy of Yamaha Corporation of America, www.yamaha.com.)*

versatility that only a computer can offer in the way of speed, hardware flexibility, memory management, signal routing, and digital signal processing. Software sequencers offer several advantages over their hardware counterparts. Here are just a few highlights:

- Increased graphics capabilities (giving us direct control over track and transport-related record, playback, mix and processing functions)
- Standard computer cut-and-paste edit capabilities
- Ability to easily change note and controller values, one note at a time or over a defined range
- A window-based graphic environment (allowing easy manipulation of program and edit-related data)
- Easy adjustment of performance timing and tempo changes within a session
- Powerful MIDI routing to multiple ports within a connected system
- Graphic assignment of instrument voices via Program Change messages
- Ability to save and recall files using standard computer memory media

Basic Introduction to Sequencing

When dealing with any type of sequencer, one of the most important concepts to grasp is that these devices don't store sound directly; instead, they encode MIDI messages that instruct instruments as to what note is to be played, over what channel, at what velocity, and at what, if any, optional controller values. In other words, a sequencer simply stores command instructions that follow in a sequential order. These instructions tell instruments and/or devices how their voices are to be played or controlled. This means that the amount of encoded data is a great deal less memory intensive than its digital audio or digital video recording counterparts. Because of this, the data overhead that's required by MIDI is very small, allowing a computer-based sequencer to work simultaneously with the playback of digital audio tracks, video images, Internet browsing, etc., all without unduly slowing down the computer's CPU. For this reason, MIDI and the MIDI sequencer provide a media environment that plays well with other computer-based production media (Figure 5.4).

As you might expect, scores of keyboard and computer-based DAWs exist on the market that include integrated MIDI sequencing capabilities. Each type and model offers a unique set of advantages and disadvantages. Obviously, it's also true that each production software or device will have its own basic operating feel; as a result, choosing one over another is totally a matter of personal preference. As with anything, care should be taken when buying a DAW or hardware system:

- Shop carefully.
- Keep your personal working habits and future growth needs in mind.
- Most of all—choose the system or version that's right for you.

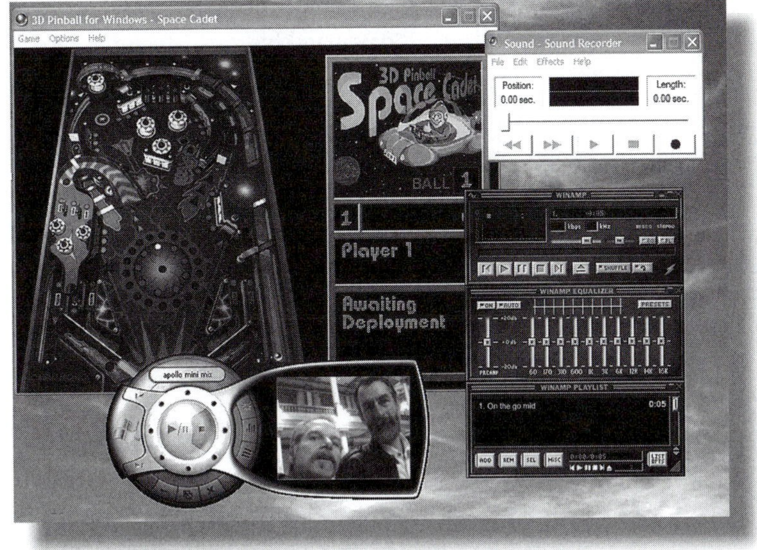

Figure 5.4. *MIDI is definitely part of the multimedia environment.*

Often this will mean spending research time on the web, talking with friends, spending some time at your favorite music store, and checking out the latest online software demos. Now that we've had a basic introduction to the power and capabilities of MIDI sequencing, let's take a closer look at some of the functions that computer-based MIDI sequencing tools have to offer, using screenshots and general terminology from such popular DAW programs as Steinberg's Cubase/Nuendo and Digidesign's Pro Tools.

> In this chapter, we'll be using a number of DYIs (do-it-yourself instruction tutorials). I definitely recommend that you go to the website of your favorite DAW manufacturer, download a demo of your choice, and follow along. As the DYIs are generic in nature, feel free to download a demo of the DAW of your choice; cross-check any setups, procedures, and processing functions with the company's manual; and follow along with the chapter.
>
> *Note:* Although this chapter covers the basics of traditional sequencing, one other program that often breaks these rules, and then creates its own new ones is Reason from the folks at Propellerheads (http://www.propellerheads.se) Although this program is definitely based on sequencing principles, I've decided to cover this program in Chapter 7 ("Groove Tools and Techniques"). I definitely recommend that you download the program's demo and take its quirky, but powerful, sequencer for a spin.

Recording

Commonly, a MIDI sequencer is an application within a digital production workspace for creating personal compositions in environments that range from the bedroom to more elaborate professional and project studios. Whether hardware or software based, most sequencers use a working interface that's roughly designed to emulate a traditional multi-track-based environment. A tape-like set of transport controls lets you move from one location to the next using standard play, stop, fast forward, rewind, and record command buttons Beyond using the traditional record-enable button to select the track or tracks that you want to record onto, all you need to do is select the MIDI input (source) port and output (destination) port, MIDI channel, instrument patch, and other setup requirements. Then press the record button and begin laying down your track.

Once you've finished laying down a track, you can jump back to the beginning of the recorded passage and listen to it. From this point, you could then "arm" (a term used to denote placing a track into the record-ready mode) the next track and go about the process of laying down additional tracks until a song begins to form.

Now that we've gone over the basics of laying down a simple MIDI passage, let's step back a bit and run through a few of the checkpoints for setting up a successful session and or MIDI track.

Setting the Session Tempo

When beginning a MIDI session, one of the first aspects to consider is the *tempo and time signature*. The beats per minute (bpm) value will set the general tempo speed for the overall session. This is important to set at the beginning of the session, so as to lock the overall "bars and beats" timing elements to this initial speed that's often essential in electronic music production. Here are but a few examples:

- Set the initial tempo, which allows a click track to be used as a tempo guide throughout the session (Figure 5.5).
- This tempo/click element can then be used to lock the timing elements of other instruments and/or rhythm machines to the session (*e.g.*, a drum machine plug-in can be pulled into the session that'll automatically be locked to the session's speed and timing).

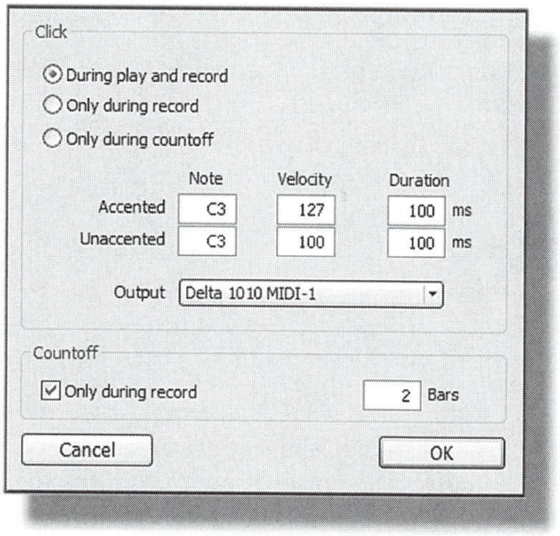

Figure 5.5. *ProTools click setup window. (Courtesy of Digidesign, A Division of Avid Technology, Inc.; www.digidesign.com.)*

- When effects plug-ins are used, their delay and timing elements will be locked to a session.
- The session can be timed to visual elements (such as video and film), allowing sound cues and tempos to match the picture.

Failure to set a tempo will make it difficult or impossible for these advantages to be put to good use, as the tempo of a freewheeling performance will almost certainly not match or can drift from the undefined tempo of a recorded MIDI session. This unfortunate event will severely limit the above options.

Once a track or set of tracks have been laid down, changing the tempo to a new value will change the overall length and notation timing values. For example, let's say that we forgot to set the session tempo and left it at the sequencer's default tempo (often 120 bmp). If no click track is used, we might start laying down tracks at the song's native tempo (oh, say 90 bpm). During production it's decided that we need a click track. Upon setting the tempo to 90 bpm, we immediately notice that the song's tempo slows down considerably, the recorded notes lengthen, and the click bears no relationship to the song whatsoever! Now we're stuck! Of course, there are ways to bring the timing into a range that can closely or at least reasonably match the desired tempo; however, these actions can be time consuming and might only be a Band-Aid, requiring that you start the session over. You might ask: What about playing the song at 120 bpm, while recording it to another sequencer that's set to 90 bpm? Well, since MIDI has a clocking element that's inherently built into the MIDI message structure, the cloned sequence will still be timed to 120 bpm—the wrong tempo. Again, we're stuck!

In short, in order to avoid any number of unforeseen obstacles to a straightforward production, it's often wise to set your session tempo *before* pressing any record buttons. Of course, the tempo of a session can be changed over its duration by creating a tempo map that can cause the speed to vary by defined amounts at specific points within a song.

Care should also be taken in setting the proper time signature at the session's outset. Listening to a 4/4 click can be disconcerting, when the song's being played in 3/4 time.

Changing Tempo

If synchronization to an outside media isn't a concern, the tempo of a MIDI production can be easily changed without worrying about changing the program's pitch or real-time control parameters. In short, once you know how to avoid potential conflicts and pitfalls, tempo variations can be made after the fact with relative ease. Basically, all you need to do is set the tempo of a sequence (or part of a sequence) to best match the feel of the overall feel of the song.

Click Track

When musical timing is important (as is often the case in modern music and visual media production), a click track can be used as a tempo guide for keeping the performance as accurately on the beat as possible (Figure 5.6). A click track can be set to make a distinctive sound on the measure boundary or (for a more accurate timing guide) on the first beat boundary and on subsequent meter divisions (*e.g.*, tock, tick, tick, tick, tock, tick, tick, tick, ...). Most sequencers can output a click track by either using a dedicated beep sound (often outputting from the device or main speakers) or by

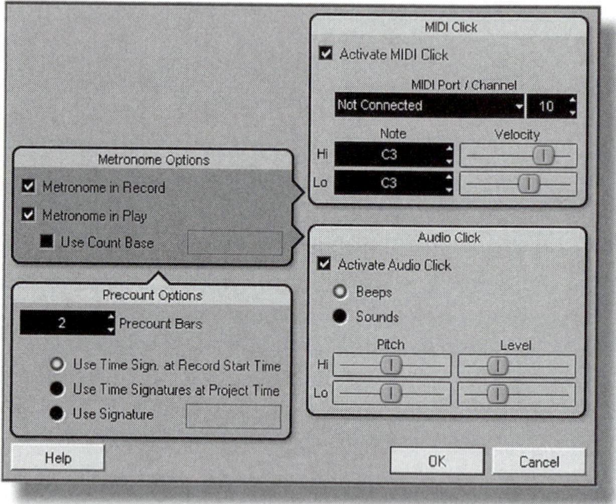

Figure 5.6. *Steinberg Cubase/Nuendo click setup window. (Courtesy of Steinberg Media Technologies GmbH, A Division of Yamaha Corporation; www.steinberg.net.)*

sending Note-On messages to a connected instrument in the MIDI chain. The latter lets you use any sound you want and often at definable velocity levels. For example, a kick could sound on the beat, while a snare sounds out the measure divisions. A strong reason for using a click track (at least initially) is that it serves as a rhythmic guide that can improve the timing accuracy of a performance.

The use of a click track is by no means a rule—in certain instances, it can lead to a performance that sounds stiff. For compositions that loosely flow and are legato in nature, a click track can stifle the passage's overall feel and flow. As an alternative, you could turn the metronome down, have it sound only on the measure boundary, and then listen through one headphone. As with most creative decisions, the choice is up to you and your current circumstance.

Setting Up a Track

In keeping with the concept that "MIDI is not audio," the first steps toward setting up the system for recording a MIDI track to a sequencer is to make sure that both the MIDI *and* audio connections are properly made. If the MIDI connections between the instruments and interface are to be made via standard MIDI cables, you'll probably want to make sure that the MIDI In and MIDI Out cables are properly routed to and from the device ports. No matter what your setup, the audio pathways will also need to be connected. This might mean:

- Routing the audio outputs to an external audio mixer.
- Routing the audio outputs to the inputs on a virtual DAW mixer (via the audio interface).
- Routing the virtual outputs of an instrument plug-in to the virtual inputs of the DAW.

Once the connections are made, the next step is to make sure that the MIDI Ins and Outs are properly assigned on the MIDI track that's to be recorded on. Once done, simply arm the track and check for MIDI activity by playing the source MIDI controller and listening for sound. If there's no sound, check your connections and general settings (including your channel volume [controller 7] and instrument Local On/Off settings). If at first you don't succeed, be patient, trace through your system and settings, and try, try again.

Setting Up a Session and Laying Down a MIDI Track

- Pull out a favorite MIDI instrument or call up a favorite plug-in instrument.
- Route the instrument's MIDI and audio cables to their appropriate workstation and audio mixer devices according to the situation at hand.
- Call up a MIDI sequencer (either by beginning a DAW session or by using a hardware sequencer) and create a MIDI track that can be recorded to.
- Set the session to a tempo that feels right for the song.
- Assign the track's MIDI input to the port that's receiving the incoming MIDI data.
- Assign the track's MIDI output to the port and proper MIDI channel that'll be receiving the outgoing MIDI data during playback.
- If a click track is desired, turn it on.
- Name the track. This will make it easier to identify the MIDI passage in the future.
- Place the track into the Record Ready mode.
- Play the instrument or controller. Can you hear the instrument? Do the MIDI activity indicators light up on the sequencer, track, and MIDI interface? If not, check your cables and run through the checklist again. If so, press Record and start laying down your first track.

Multiple Track Recording

Although only one MIDI track is commonly recorded at a time, most mid- and professional-level sequencers allow us to record multiple tracks at one time. This feature makes it possible for a multi-trigger instrument or for several performers to record to a sequence in one, live pass. For example, such an arrangement would allow for each trigger pad of a MIDI drum controller to be recorded to its own track (with each track being assigned to a different MIDI channel on a single port. Alternatively, several instruments of an on-stage electronic band could be captured to a sequence during a live performance and then laid into a DAW session for the making of an album project.

Punching In and Out

Almost all sequencers are capable of punching in and out of record while playing a sequence (Figure 5.7). This commonly used function lets you drop in and out of record on a selected track (or series of tracks) in real time, in a way that mimics the traditional multitrack overdubbing process. Although punch-in and punch-out points can often be manually performed on the fly, most sequencers can automatically perform a punch by graphically or numerically entering in the measure/beat points that mark the punch-in and punch-out location points. Once done, the sequence can be rolled back to a point a few measures before the punch-in point and the artist can play along while the sequencer automatically performs the necessary switching functions.

Figure 5.7. *Steinberg Cubase/Nuendo punch-in and punch-out controls. (Courtesy of Steinberg Media Technologies GmbH, A Division of Yamaha Corporation; www.steinberg.net.)*

Punching During a Take

- Set up a track and record a musical passage. Save the session to disk.
- Roll back to the beginning of the take and play along. Manually punch in and out during a few bars. Was that easy? Difficult? Now, undo or revert back to your originally saved session.
- Read your DAW/sequencer manual and learn how to perform an automated punch (placing your punch-in and punch-out points at the same place as before).
- Roll back to a point a few bars before the punch and go into the Record mode. Did the track automatically place itself into record? Was that easier than doing it manually?
- Feel free to try other features, such as record looping or stacking.

Figure 5.8. *ProTools with record loop enabled. (Courtesy of Digidesign, A Division of Avid Technology, Inc.; www.digidesign.com.)*

PRE-COUNT

A pre-count or pre-roll feature will sound a metronome count before placing itself into the playback or record mode, giving the performer time to prepare for laying down his or her track. It's important that the proper tempo and time signature be entered into the session at the outset in order to get the right feel for the pre-count.

Record Loop

As the name suggests, the loop function in a DAW or sequencer lets you cycle between a predetermined in-point and out-point in a continuous fashion. By placing the transport into a loop and then entering into record, it's possible to continuously cycle through a defined section in the Record mode, allowing the section to be repeated until the phrase has been recorded to your liking (Figure 5.8). A pre-count can be placed at the loop's beginning, so as to give a preparatory break between takes.

Stacking

Certain DAW programs allow a record loop to be set up in such a way that each subsequent pass will be recorded to disk (Figure 5.9). This multiple take feature can be extremely useful in a number of ways; for example:

- When in the midst of laying down overdubs in the record-loop mode, it's often difficult to be objective about which take was the best. Stacking lets you decide at a later time.
- Multiple stacks can be combined to create a rich, layered effect on vocals, guitar—you name it.
- The best segments from various passes can be edited into a single, composite, winning take.

Step Time Entry

In addition to laying down a performance track in real time, most sequencers will allow us to enter note values into a sequence one note at a time. This feature (known as step time, step input, or pattern sequencing) makes it possible for notes to be entered into a sequence without having to worry about the exact timing. Upon playback, the sequenced pattern will play back at the session's original tempo. Fact is, step time entry can be an amazing tool, allowing a difficult or a blazingly fast passage to be meticulously entered into a pattern and then be played out or looped with a degree of technical accuracy that would otherwise be impossible for most of us to play. Quite often, this data entry style is used with fast, high-tech musical styles where real time entry just isn't possible or accurate enough for the song.

The ways that a passage can be entered into a sequence in step time varies between devices and software passages. Upon entering the Step Time mode and selecting an initial note length setting, you can go about the business of manually entering the notes directly into the pattern window screen from a keyboard or other controller, continuing to enter notes of a specified duration into the sequence until a rest value or new note of a different duration value is selected.

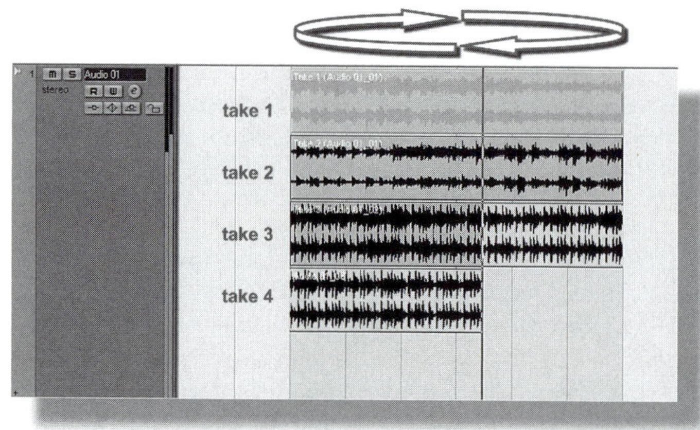

Figure 5.9. *Multiple audio and/or MIDI takes can be created in a looping fashion using the stack function. (Courtesy of Steinberg Media Technologies GmbH, A Division of Yamaha Corporation; www.steinberg.net.)*

Saving Your MIDI Files

Just as it's crucial that we carefully and methodically back up our program and production media, it's important that we save our MIDI and session files while we're in production. This can be done in two ways:

- Update your files by periodically saving files over the course of a production. A number of programs can be set up to automatically perform this function at regular intervals.
- At important points throughout a production, you might choose to save your session files under new and unique names, thereby making it possible to easily revert back to a specific point in the production. This can be an important recovery tool should the session take a wrong turn or for recreating a specific effect and/or mix.

> When working with MIDI within a DAW session, it's *always* a good idea to save the original MIDI tracks within the session. This makes it easy to go back and change a note, musical key, or sounding voice or to make any other alterations you want. Not saving these files could lead to some major headaches or worse.

MIDI files can also be converted to and saved as a *standard MIDI file* for use in exporting to and importing from another MIDI program or for distributing MIDI data for use on the web, in cell phones, etc. These files can be saved in either of two formats:

- Type 0—Saves all of the MIDI data within a session as a single MIDI track. The original MID channel numbers will be retained. The imported data will simply exist on a single track
- Type 1—Saves all of the MIDI data within a session onto separate MIDI tracks that can be easily imported into a sequencer in a multitrack fashion

Editing

One of the more important features a sequencer has to offer is its ability to edit sequenced tracks. Editing functions and capabilities often vary from one sequencer to the next, with certain hardware systems offering only the most basic cut-and-paste, signal processing, and data routing capabilities. Most software sequencers, on the other hand, generally offer a wide range of editing functions that can manipulate MIDI data in a simple, easy-to-use graphic environment.

As was mentioned, hardware sequencers will often display entry- and editing-related information using an LCD screen that's often 2 inches by 5 inches or smaller. As a result, the way in which parameters are edited can be somewhat limited by the display type. To some, this simplifies the editing process by requiring that you do much of your editing by ear, with entry and edit resolutions being entered at the measure level, instead of at the note and clock resolution level.

Software workstations and sequencers, on the other hand, display their edit, processing, and transport controls over a much larger surface of your computer monitor (or preferably multiple monitors). As software can be easily reconfigured to best deal with the task at hand, these sequencing applications can graphically tailor their display and control surface in a manner that's intuitive and easy to use (at least in theory).

The remainder of this section outlines the various control and editing features that can be found in current sequencing and DAW production packages. For practicality's sake, we'll be looking at examples and general functionality from a software point of view, using screenshots from Steinberg's Cubase/Nuendo and Digidesign's Pro Tools as a visual guide. Although hardware sequencers often vary in their overall capability, many of the following examples can be applied to these devices as well.

Main Edit Screen

The main screen of a sequencer application (Figure 5.10) will generally include a transport control surface, track lanes (within a DAW, these tracks can contain audio, MIDI, video, automation, etc.), and parameter/mixer functions. The track *edit window* is used to display such information

Figure 5.10. *ProTools showing the main edit screen. (Courtesy of Digidesign, A Division of Avid Technology, Inc.; www.digidesign.com.)*

as track-related data, track names, MIDI port assignments for each track, program change assignments, volume controller values, mix and control automation, and other production-related commands. An elapsed time and/or measure/beat display bar is placed at the top of the edit window, allowing the user to easily zoom to a specific point within the session and to have a reliable time and measure reference. Depending on the display type, the existence of MIDI data on a track is generally indicated by small bars that appear in the track in a piano roll fashion.

Port assignments are usually made from the track window, allowing port and MIDI channels to be assigned for each MIDI track. The number of ports that are available will depend on the number of hardware interface and software instrument/applications that are available to the system. For example, Figure 5.11 shows the MIDI assignment dialog boxes that can access a wide range of ports and software instrument plug-ins.

MIDI Grouping

In addition to dealing with MIDI tracks in a linear fashion that continuously flow from the beginning to the end of a song, most sequencers allow musical phrases to be grouped into defined chunks that can be looped, moved, and processed as a single entity with relative ease and in a building block fashion. Other programs also allow these groups to be named and then arranged into a consecutive, playlist style session that can be arranged and rearranged at will.

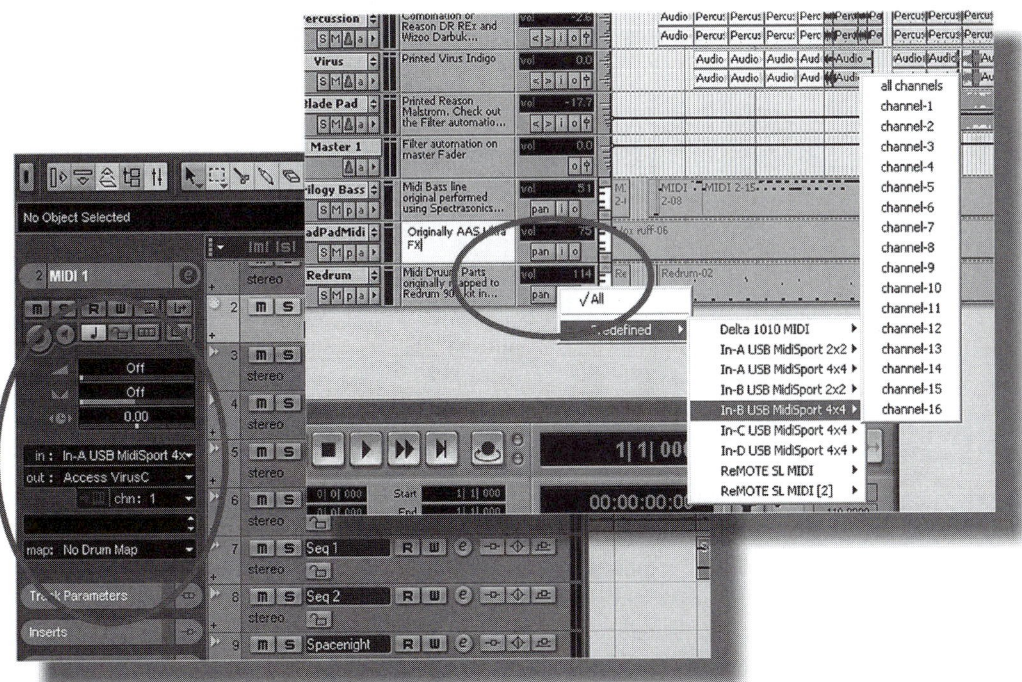

Figure 5.11. *MIDI assignment dialog boxes for both Cubase/Nuendo and ProTools.*

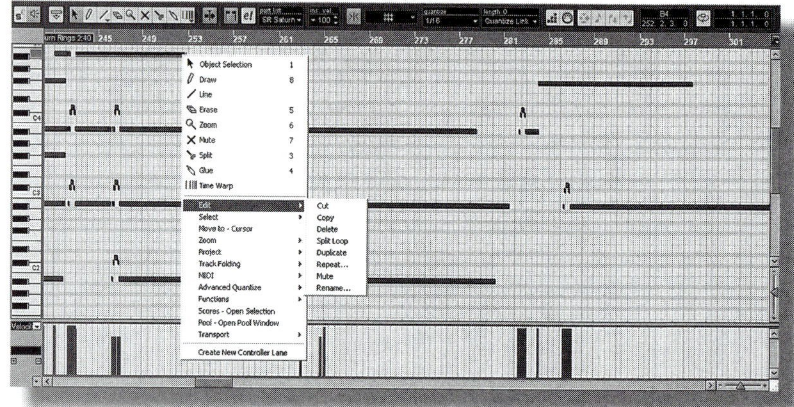

Figure 5.12. Steinberg Cubase/ Nuendo piano roll edit window. (Courtesy of Steinberg Media Technologies GmbH, A Division of Yamaha Corporation; www. steinberg.net.)

Piano Roll Editor

One of the easiest ways to see and edit MIDI track note values is through the use of the sequencer's piano roll edit window (Figure 5.12). This intuitive window lets you graphically edit MIDI data at the individual note level by displaying values on a continuous piano roll grid that's broken down into time (measures) and pitch (notes on a keyboard). Depending on where on the note's bar you click, you can:

- Change its beginning start time by clicking on the bar's front-end and dragging to the proper point.
- Change it to a new note number by vertically dragging it up or down the musical scale. Often, a keyboard is graphically represented at the window's left-hand side as a guide.
- Change its duration by clicking near the end of the bar and dragging to the proper point.

As you might expect, the way in which notes can be edited might change from one sequencer to the next. The operation examples listed above are fairly common; however, you should consult your program's manual for further insights and details.

Snap to ...

Before continuing on, let's take a quick look at the concept of snapping an event to a defined boundary. In short, the snapping feature on a DAW or any type of sequencing system allows a MIDI, audio, or other media type to magnetically jump to the closest defined time or measure boundary (Figure 5.13). For instance, in the above example, we said that we could click on the beginning of a note event within a piano roll display and move its beginning point in time. By turning on the "snap to measure" control, the beginning note would magnetically jump to the precise start point of the nearest measure. Of course, these snap boundaries can be defined in measures, measure subdivisions, or time-related events. Again, now would be a good time to get out your trusty manual and read up on the snap feature.

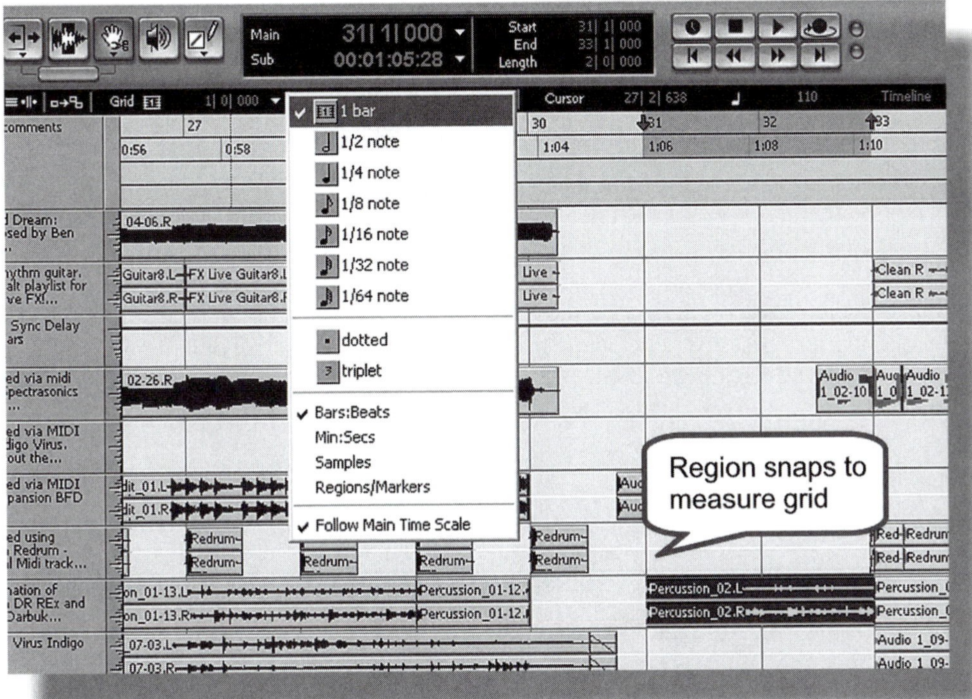

Figure 5.13. *ProTools showing Snap and Grid Value pop-up menu. (Courtesy of Digidesign, A Division of Avid Technology, Inc.; www.digidesign.com.)*

Snapping to ...

- Record, open, or import a MIDI track into your sequencing app.
- Open the edit wind for that track (consult the manual; however, this is often done by double-clicking on the MIDI data track).
- Zoom into the sequence so the event bars can be easily dragged in time.
- Turn off any snap functions.
- Click in the beginning point of an event and drag it in time.
- Turn on the "snap to measure" feature.
- Click in the beginning point of an event and drag it in time. Does it snap to the measure boundary?
- Change the various snap settings and experiment with this useful feature.

Notation Editor

Many sequencing packages allow notes to be entered and edited into a sequence using standard musical notation (Figure 5.14). This window works in much the same way as the piano roll editor, except that pitch, duration, and other values are displayed and edited directly into measures as musical notes on any clef. One really nice by-product of this feature is that most programs also let us print a track's musical part in standard notation. This can be really useful for creating basic lead sheets (lyrics can even be entered into some sequences) for a particular instrument or part. Certain software packages are also able to print all of the tracks and parts of a sequence in a score form (where all the parts are arranged and printed onto the pages that can be reviewed and/or handed out to musicians during a session). Because the layout and printing demands are greater when laying out a printed score, many manufacturers offer a separate program for carrying out this fairly specific production task.

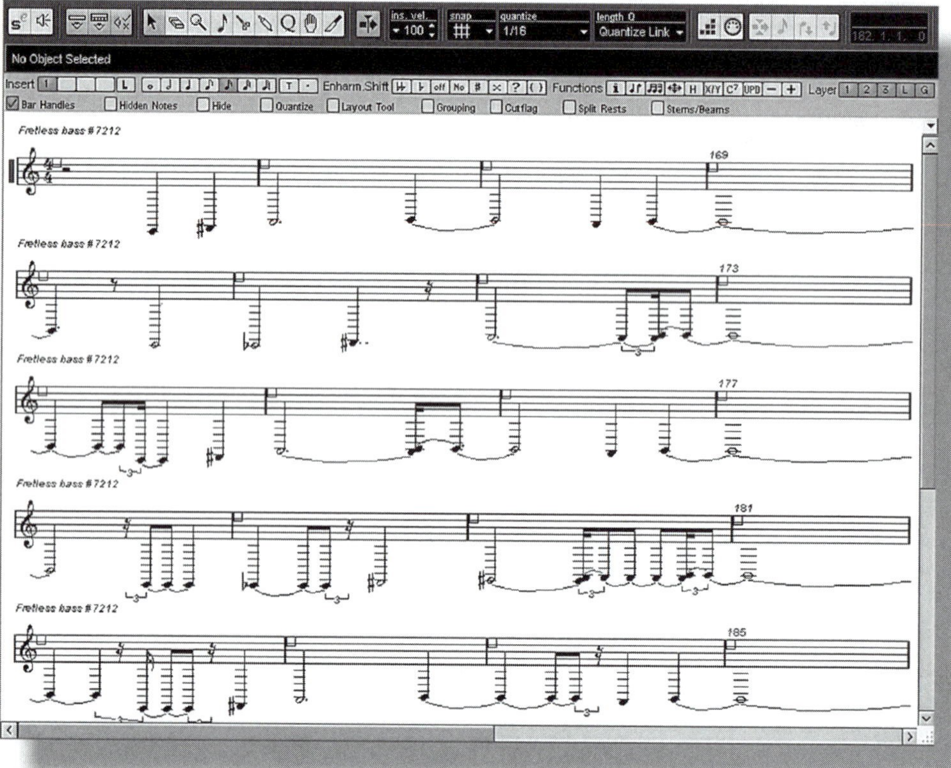

Figure 5.14. *Steinberg Cubase/Nuendo notation edit window. (Courtesy of Steinberg Media Technologies GmbH, A Division of Yamaha Corporation; www.steinberg.net.)*

Drum Editor

One of the more useful edit windows for creating drum- and percussion-based patterns and grooves is the drum edit window (Figure 5.15). This graphic interface makes use of a grid-like pattern display that vertically lists the various instrument parts of a kit on the right-hand side of the grid, while the pattern notes are placed horizontally along the horizontal axis, allowing patterns to be built up in a way that offers in-depth edit control over quantization (timing), velocity, and most of the standard controller options. Using this grid, patterns of any complexity can be built up, looped, and chained in combinations with other varying patterns in a way that can contribute to a song that's creative and rhythmically diverse. In addition to constructing patterns, a drum edit window (or the occasional stand-alone drum editing program) can be used to edit and finesse patterns that have been performed into the sequence. Of course, the web is a continual source of drum patterns that can be imported into a session for further editing and combining in new and unique ways.

Creating a Drum Pattern

- Read your DAW/sequencer manual and learn how to open and use the basic functions of yours system's drum editor.
- Open a new session, create a MIDI track and assign its MIDI outs to a drum machine or percussion section of an instrument/plug-in.
- Open the Drum Edit window for that track (let's restrict our practice to a 4-bar pattern).
- Create a kick pattern by clicking on the first beat of every bar. Did it sound when you clicked on the grid? When it played back?
- If so, click the snare part on the first and third beats.
- Hopefully, you've created a simple pattern that can be trimmed and looped within your basic rhythmic session.
- Now, copy that pattern and make a few variations to the beat
- Continue practicing and have fun!

Event Editor

Although they're used less often than either the piano roll or notation edit windows, an event editor displays all of the MIDI messages that exist on a track (or in an entire sequence) as a sequential list of events. For example, Figure 5.16 shows a list of all MIDI events that occur in a MIDI track over time. In it, we can see that near the fourth beat of measure 167 a Note-On event is being transmitted for note D2 with a velocity of 111 on channel 12. Although this type of edit interface isn't often used for musical editing, it can be useful for instructing a sequencer to perform a task-based event (*i.e.*, do something at such-and-such a time). For example, you could instruct the sequencer to trigger a specific sample at a specific time or insert a program change on a certain channel that tells an instrument to switch an instrument voice on cue.

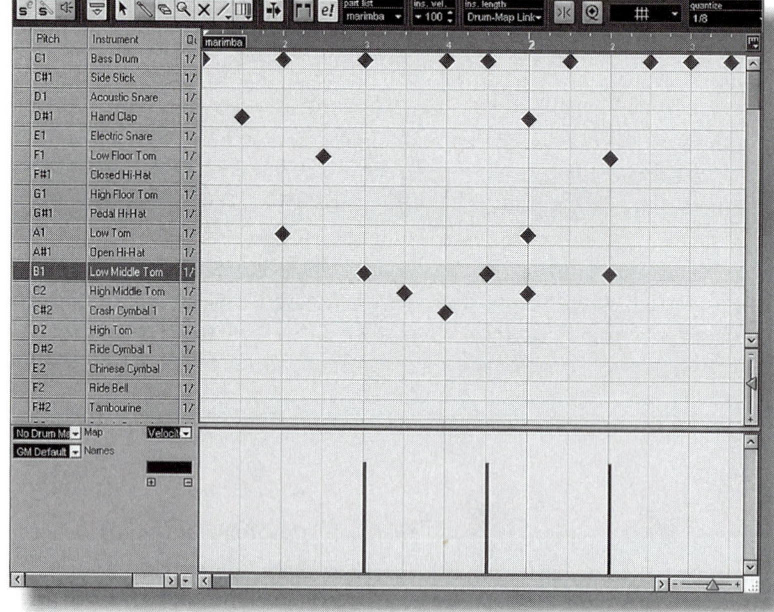

Figure 5.15. *Steinberg Cubase/ Nuendo drum edit window. (Courtesy of Steinberg Media Technologies GmbH, A Division of Yamaha Corporation; www.steinberg.net.)*

Practical Editing Techniques

When it comes to learning the ins, outs, and thrus of basic sequencing, absolutely nothing can take the place of diving in and experimenting with your setup. Here, I'd like to paraphrase Craig Anderton, who said: "Read the manual through once when you get the program (or device), then play with the software and get to know it before you need it. Afterwards, reread the manual to pick up the system's finer operating points." Wise words! In this section, we'll be covering some of the basic techniques that'll speed you on your way to sequencing your own music. Note that there are no rules to sequencing MIDI. As with all of music production (and the arts, for that matter), there are as many right ways to perform and sequence music as there are musicians. Just remember that these are no hard and steadfast rules to music production—but there are definitely guidelines and tips that can speed and improve the process.

Transposition

As was mentioned earlier, a sequencer app is capable of altering individual notes in a number of ways including pitch, start time, length, and controller values. In addition, it's generally a simple matter for a defined range of notes in a passage to be altered as a single entity in ways that could alter the overall key, timing, and controller processes. Changing the pitch of a note or the entire key of a song is extremely easy to do with a sequencer. Depending on the system, a song can be transposed up or down in pitch at the global level, thereby affecting the musical key of a song. Likewise, a segment can be shifted in pitch from the main edit, piano roll, or notation edit windows by simply highlighting the bars and tracks that are to be changed and then calling up the transpose function from the program menu.

Trilogy Bass | Options | Insert | 196 Events

<table>
<tr><th>Start</th><th colspan="4">Event</th><th>length/info</th></tr>
<tr><td>1| 1| 000</td><td></td><td>7</td><td>51</td><td></td><td>volume</td></tr>
<tr><td>5| 4| 015</td><td>♩</td><td>F-1</td><td>94</td><td>64</td><td>0| 2| 000</td></tr>
<tr><td>6| 1| 000</td><td>♩</td><td>D1</td><td>104</td><td>64</td><td>0| 1| 000</td></tr>
<tr><td>6| 2| 000</td><td>♩</td><td>D1</td><td>82</td><td>64</td><td>0| 0| 480</td></tr>
<tr><td>6| 2| 480</td><td>♩</td><td>Eb2</td><td>96</td><td>64</td><td>0| 1| 000</td></tr>
<tr><td>6| 3| 480</td><td>♩</td><td>D2</td><td>95</td><td>64</td><td>0| 0| 541</td></tr>
<tr><td>6| 4| 000</td><td>♩</td><td>C2</td><td>91</td><td>64</td><td>0| 0| 562</td></tr>
<tr><td>6| 4| 480</td><td>♩</td><td>D2</td><td>91</td><td>64</td><td>0| 0| 480</td></tr>
<tr><td>7| 1| 000</td><td>♩</td><td>D1</td><td>108</td><td>64</td><td>0| 1| 000</td></tr>
<tr><td>7| 2| 000</td><td>♩</td><td>D1</td><td>108</td><td>64</td><td>0| 0| 480</td></tr>
<tr><td>7| 2| 480</td><td>♩</td><td>G#1</td><td>98</td><td>64</td><td>0| 1| 000</td></tr>
<tr><td>7| 3| 480</td><td>♩</td><td>G1</td><td>91</td><td>64</td><td>0| 0| 480</td></tr>
<tr><td>7| 4| 000</td><td>♩</td><td>F1</td><td>96</td><td>64</td><td>0| 0| 480</td></tr>
<tr><td>7| 4| 480</td><td>♩</td><td>G1</td><td>100</td><td>64</td><td>0| 0| 480</td></tr>
<tr><td>8| 1| 000</td><td>♩</td><td>D1</td><td>108</td><td>64</td><td>0| 1| 000</td></tr>
<tr><td>8| 2| 000</td><td>♩</td><td>D1</td><td>96</td><td>64</td><td>0| 0| 480</td></tr>
<tr><td>8| 2| 480</td><td>♩</td><td>Eb2</td><td>98</td><td>64</td><td>0| 1| 000</td></tr>
<tr><td>8| 3| 480</td><td>♩</td><td>D2</td><td>83</td><td>64</td><td>0| 0| 551</td></tr>
<tr><td>8| 4| 000</td><td>♩</td><td>C2</td><td>74</td><td>64</td><td>0| 0| 480</td></tr>
<tr><td>8| 4| 480</td><td>♩</td><td>D2</td><td>84</td><td>64</td><td>0| 0| 480</td></tr>
<tr><td>9| 1| 000</td><td>♩</td><td>D1</td><td>108</td><td>64</td><td>0| 1| 000</td></tr>
<tr><td>9| 2| 000</td><td>♩</td><td>D1</td><td>108</td><td>64</td><td>0| 0| 480</td></tr>
<tr><td>9| 2| 480</td><td>♩</td><td>G#1</td><td>98</td><td>64</td><td>0| 1| 000</td></tr>
<tr><td>9| 3| 480</td><td>♩</td><td>G1</td><td>91</td><td>64</td><td>0| 0| 480</td></tr>
<tr><td>9| 4| 000</td><td>♩</td><td>F1</td><td>96</td><td>64</td><td>0| 1| 000</td></tr>
<tr><td>10| 1| 000</td><td>♩</td><td>D1</td><td>96</td><td>64</td><td>0| 1| 000</td></tr>
<tr><td>10| 2| 000</td><td>♩</td><td>D1</td><td>98</td><td>64</td><td>0| 0| 480</td></tr>
<tr><td>10| 2| 480</td><td>♩</td><td>Eb2</td><td>96</td><td>64</td><td>0| 1| 053</td></tr>
</table>

Figure 5.16. *ProTools showing the MIDI Event List. (Courtesy of Digidesign, A Division of Avid Technology, Inc.; www.digidesign.com.)*

Quantization

By far, most common timing errors begin with the performer. Fortunately, "to err is human," and standard performance timing errors often give a piece a live and natural feel. However, for those times when timing goes beyond the bounds of nature, an important sequencing feature known as *quantization* can help correct these timing errors. Quantization allows timing inaccuracies to be adjusted to the nearest desired musical time division (such as a quarter, eighth, or sixteenth note). For example, when performing a passage where all involved notes must fall exactly on the quarter-note beat, it's often easy to make timing mistakes (even on a good day). Once the track has been recorded, the problematic passage can be highlighted and the sequencer can recalculate each note's start and stop times so they fall precisely on the boundary of the closest time division (Figure 5.17). Such quantization resolutions often range from full whole-note to sixty-fourth-note values and can also include triplet values.

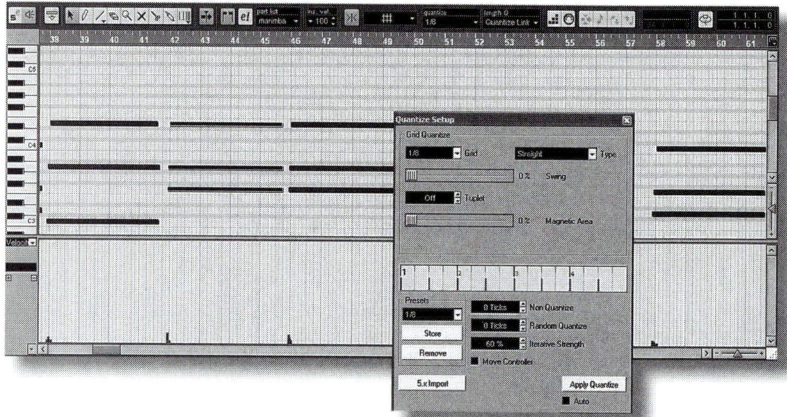

Figure 5.17. Steinberg Cubase/ Nuendo MIDI edit and quantize window. (Courtesy of Steinberg Media Technologies GmbH, A Division of Yamaha Corporation; www.steinberg.net.)

Because quantization is used to "round off" the timing elements of a range of notes to the nearest selected beat resolution, it might be advisable to try to lock your playing in time with the sequencer's own timing resolution. This is done simply by selecting the tempo and time signature that you want and turning on the click track. By playing along with a click track, you're basically using the sequencer as a musical metronome. Once the track or song has been sequenced, the quantization function can be called up, which will further correct the selected note timing values.

In another attempt to extol the virtues of setting your initial tempo and click track settings, suppose you were to record the same sequence without a timing reference (most often a click track). In this case, the notes would be quantized to a timing benchmark that doesn't exist. That's not to say that it's impossible to quantize a sequence that wasn't played to a click—it's just trickier business. In short, whenever you feel you might need to quantize a segment, always give special consideration to a sequence's initial timing elements (*i.e.*, using a click track).

Quantization

- Read your DAW/sequencer manual and learn its quantization features.
- Open a new session, create a MIDI track, and assign its MIDI outs to an instrument or plug-in.
- Set the session to a tempo that feels right for the song and turn on the click track.
- Record a small musical passage and save the file.
- Zoom into a few of the notes and notice that they probably don't start exactly on the measure subdivisions.
- Highlight the notes in the passage and set the quantize feature to a value that best matches the correct note length values (*e.g.*, whole note, quarter note, one-sixteenth note).
- Undo the change and try quantizing to various subdivision settings (*e.g.*, whole note, half note, quarter note). Save the results to several session files.

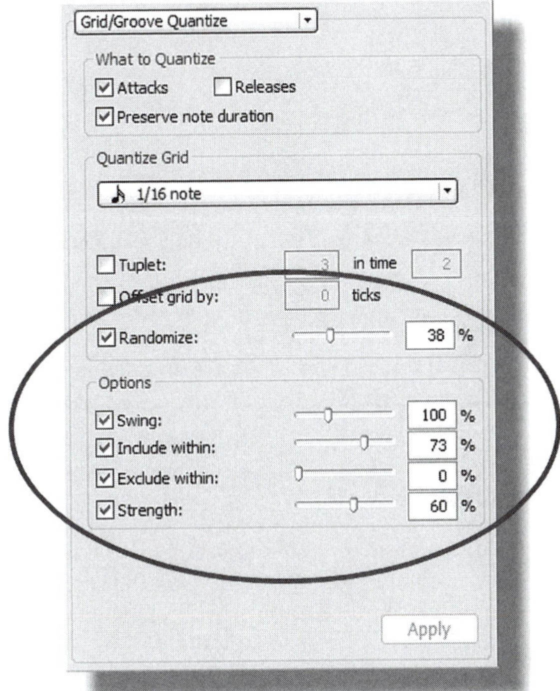

Figure 5.18. *The randomize (humanize) function within Pro Tools. (Courtesy of Digidesign, A Division of Avid Technology, Inc.; www. digidesign.com.)*

Humanizing

When you get right down to it, one of the most magical aspects of music is its ability to express emotion. A major factor used for conveying expression is the minute timing variations that are introduced during a performance. Whenever these variations become so large that they become sloppy, the first task at hand is to tighten up the section's timing through quantization. The downfall of quantization, however, is that it can introduce a robotic accuracy that can take away from the basic human variations in the music, making it sound rigid and machine-like. One of the ways to reintroduce these variations in timing back into a quantized segment is through a process known as *humanization.*

The humanization process is used to randomly alter all of the notes in a selected segment according to such parameters as timing, velocity, and note duration (Figure 5.18). The amount of randomization can often be limited to a user-specified value or percentage range, and parameters and can be individually selected or fine-tuned for greater control. Beyond the obvious advantages of reintroducing human-like timing variations back into a track, this process can help add expression by randomizing the velocity values of a track or selected tracks. For example, humanizing the velocity values of a percussion track that has a limited dynamic range can help bring it to life. The same type of life and human swing can be effective on almost any type of instrument. Let's give it a try.

Humanizing

- Read your DAW/sequencer manual and learn its basic humanization features.
- Open a quantized file from the above DYI on quantization.
- Select a range of measures that have been quantized (preferably a series of fast, staccato notes).
- Call up the humanize function and experiment with various time, velocity, and length variables. (You can almost always undo your last move before trying a new set of values.)

Swing

Another form of humanization, called swing, serves to introduce timing variables into the rhythmic feel of a passage by changing the timing of every other note in a musical part (or every second position in a pattern grid). This swing effect, which is generally expressed as a percentage value, gives a loose, shuffle-like feel to the pattern—often adding a degree of humanity to the track (if it's not overdone). You might want to swing out by trying the above DYI with the swing function.

Slipping in Time

Another timing variable that can be introduced into a sequence to help change the overall feel of a track is the slip time feature (Figure 5.19). Slip time is used to move a selected range of notes either forward or backward in time by a defined number of clock pulses. This has the obvious effect of changing the start times for these notes, relative to the other notes or timing elements in a sequence.

This function can be used to microtune the start times of a track so as to give them a distinctive feel. For example, nudging the notes of a sequenced percussion track forward in time by only a small number of clock pulses will effectively make the percussion track rush the beat, giving it a heightened sense of urgency or expectation. Likewise, retarding a track by any factor will give it a slower, backbeat kind of feel.

Slipping can also be used to move a segment manually into relative sync with other events in a sequence. For example, let's say that we've created a song that grooves along at 96 bpm; we've searched our personal archives and found a bridge (a musical break motif) that would work great in the piece. After inserting the required number of empty measures into the sequence and pasting he break into it, we found that the break comes in 96 clocks too late. No problem! We can simply highlight the break and slip it forward in time by 96 clocks. Often this process will take several tries and some manual nudging to find the timing that feels right, but the persistence could definitely pay off.

Figure 5.19. *The slip feature can be used to move a note (or range of notes) in time. (Courtesy of Steinberg Media Technologies GmbH, A Division of Yamaha Corporation; www.steinberg.net.)*

Editing Controller Values

Almost every sequencer package allows controller message values to be edited or changed, and they often provide a simple, graphic window whereby a line or freeform curve can be drawn that graphically represents the effect that relevant controller messages will have on an instrument or voice (Figure 5.20). By using a mouse or other input device, it becomes a simple matter to draw a continuous stream of controller values that correspondingly change such variables as velocity, modulation, pan, etc.

With the advancement of software workstations, music production software, and plug-in applications, control over MIDI controller messages has become fully integrated into the basic mixing and sequencing applications, such that all you have to do is assign the control to a specific

Some of the more common controller values that can effect a sequence and/or MIDI mix values include the following (for the full listing of controller ID numbers, see Table 2.2 in Chapter 2):

Control Number	Parameter
1	Modulation Wheel
2	Breath Controller
4	Foot Controller
7	Channel Volume (formerly Main Volume)
8	Balance
10	Pan
11	Expression Controller
64	Damper Pedal On/Off (Sustain) (values 0–63 off, 64–127 on)

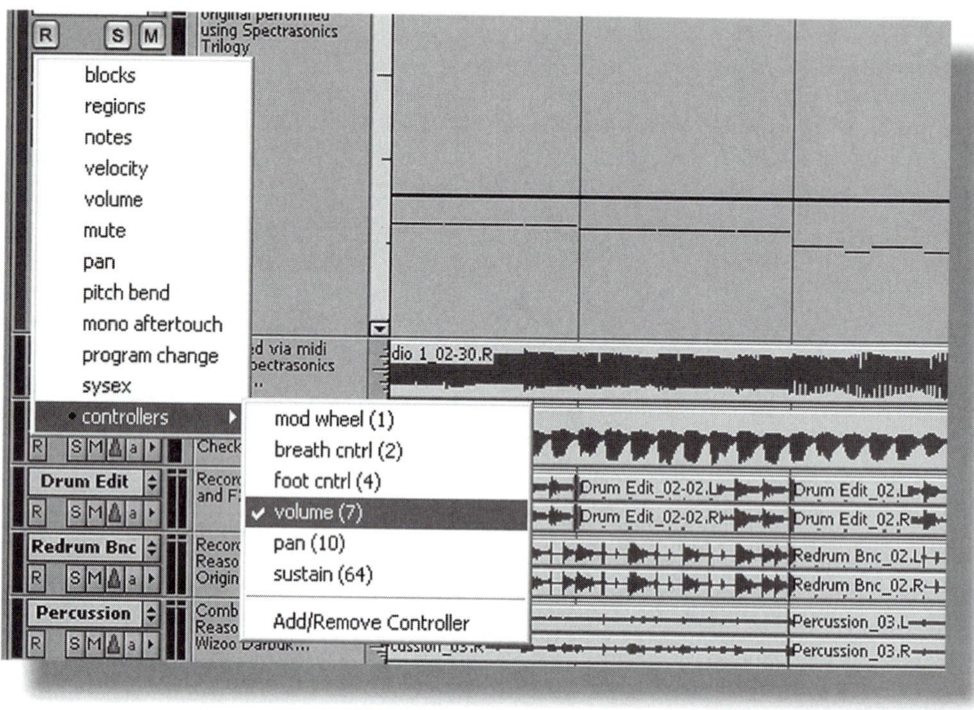

Figure 5.20. *Window from ProTools showing a MIDI track's controller parameter selector. (Courtesy of Digidesign, A Division of Avid Technology, Inc.; www.digidesign.com.)*

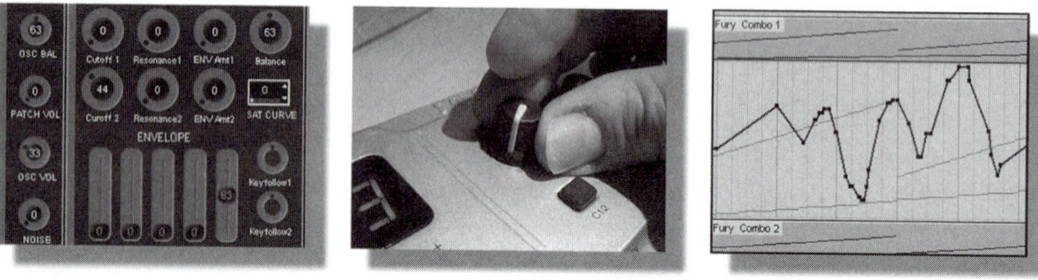

Figure 5.21. *Examples of how a controller value can be varied as parameter faders, physical hands-on controls, or drawn parameter curves.*

parameter and then twiddle the knob, move the fader, or graphically draw the variables on-screen in a WYSIWYG ("what you see is what you get") fashion (Figures 5.21 and 5.22).

It almost goes without saying that a range of controller events can be altered on one or more tracks by allowing a range of MIDI events to be highlighted and then altered by entering in a parameter or processing function from an edit dialog box. This ability to define a range of events often comes in handy for making changes in pitch/key, velocity, main volume, and modulation (to name a few).

Figure 5.22. Mackie C4 plug-in and virtual instrument controller. (Courtesy of Loud Technologies, Inc.; www.mackie.com.)

Changing Controller Values

- Read your DAW/sequencer manual and learn its basic controller editing features.
- Open or create a MIDI track.
- Select a range of measures and change their Channel Volume settings (controller 7) over time. Does the output level change over the segment?
- Highlight the segment and reduce the overall Channel Volume levels by 25% (a setting of about −30).
- Take the segment's varying controller settings and set them to an absolute value of 95. Did that eliminate the level differences?
- Now, refer to your DAW/sequencer manual for how to scale MIDI controller events over time.
- Again, rescale the Channel Volume settings so they vary widely over the course of the segment.
- Highlight the segment and scale the velocity values so they have a minimum value of 64 and a maximum of 96. Could you see and hear the changes?
- Again, highlight the segment and instruct the software to fade it from its current value to an ending value of 0. Hopefully, you've just created a Channel Volume fade. Did you see the MIDI channel fader move?
- Undo and start the fade with an initial value of 0 and a current value of 100% ending. Did the segment fade in?

Defining a range of note events can also be useful for scaling controller values (upward, downward, or over time as a faded event change). For example, you could define a segment and place a minimum and maximum limit on the velocity values, effectively limiting the track's overall dynamic range (without using a compressor). You can also process a range of velocity or main volume messages so they ramp up or down over time, effectively creating a smooth fade-in or fade-out in the MIDI domain (not always a wise thing to do if you intend to transfer the MIDI track to an audio soundfile, in which case it's generally best to make your changes in the audio domain).

Thinning Controller Values

Often, you'll find that these physically controlled or drawn curves will have their resolution set so high that literally hundreds of controller changes can be introduced into the sequence over just a few measures. In most cases, this resolution simply isn't necessary and might even create a data bottleneck when playing back a complex sequence. To filter out the gazillions of unnecessary messages that could be placed into a track, you might want to lower the control change resolution (if your sequencer and or controller hardware has such a feature) or you might want to thin the controller data down to a resolution that's more reasonable for the task at hand.

Thinning Controller Values

- Read your DAW/sequencer manual and learn its basic controller thinning features.
- Open or create a MIDI track.
- Select a range of measures, choose an obvious controller (*e.g.*, 7 [volume] or 1 [modulation]) and draw some complex and wacky curve.
- Highlight the drawn curve, select the "thin controller data" function, and thin the data by about half. Did it remove tons of redundant controller messages?
- Listen to the thinned track. Could you tell any difference in the sound? Or visually in the controller window?

Filtering MIDI Data

Most sequencing applications are capable of filtering MIDI in a way that allows specific message or controller types within a datastream to be either recognized or ignored (Figure 5.23). This feature can come in handy in a studio or on-stage setting in that it can be used to block unwanted commands or controller types that might accidentally pass through. Depending on the program or hardware sequencer, a MIDI filter might be inserted into the entire sequencing system or on a track-by-track basis. In the latter instance, a filter could be inserted on one channel so as to block

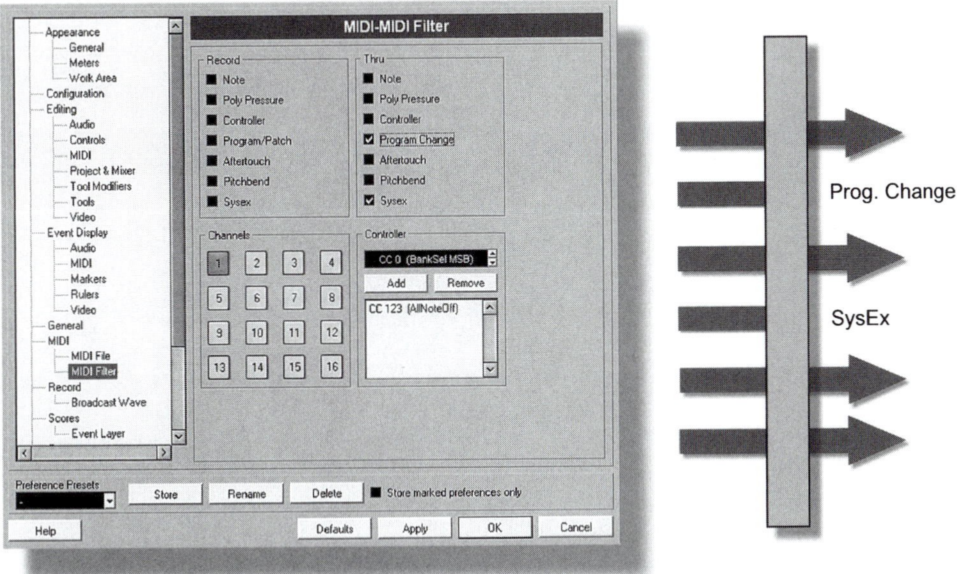

Figure 5.23. *Most sequencing systems are capable of filtering out specific types of MIDI data. (Courtesy of Steinberg Media Technologies GmbH, A Division of Yamaha Corporation; www.steinberg.net.)*

Program Change messages that might inadvertently be passed to an important on-stage instrument, while SysEx messages could be blocked on another data port/line so as to keep intense data bursts from clogging up a sensitive MIDI line.

Mapping MIDI Data

MIDI mapping is a process by which the value of a data byte or range of data bytes can be reassigned to another value or range of values. This function lets you change one or more of the parameters in an existing MIDI datastream to an alternate value or set of values. This process can be applied to a MIDI datastream to reassign channel numbers, transpose notes, change controller numbers and values, and limit controller values to a specific range. As with MIDI filtering, it's often possible to map data on a single channel so specific information can be mapped without affecting data on other channels. In this way, only a single device, chain of devices in a data line, or specific instrument voice will be affected.

Program Change Messages

Another type of automation that's supported by most sequencing packages involves the use of Program Change messages. As we saw in Chapter 2, the transmission of a Program Change message (ranging from 0 to 127) on a specific MIDI channel can be used to change the program or patch a preset number of an instrument or voice. By assigning a program change number to a specific sequence track, it becomes possible for an instrument (or a single part within a polyphonic instrument) to be automatically recalled to the desired patch setting In short, an instrument or device patch can be automatically assigned to each track in a sequence, allowing the patch to be automatically recalled upon opening the file. Ideally, all you need to do is press the play button. Program changes can also be inserted in the middle of a sequence. Such a message can be used to instruct a synth voice to change from one patch to another in the middle of a song. For example, a synth could be used to play a B3 organ for the majority of a song; however, during a break, a Program Change message could instruct it to switch to an acoustic piano voice and then back again to finish the song. Patch changes such as these can also be used to change the settings for effects devices that can respond to MIDI program changes (*i.e.*, changing a processor's settings from a rich room plate to a wacky echo setting in the middle of a sequence).

System Exclusive: The Musician's Pal

As you may have guessed from comments in previous chapters, I'm a big fan of the power and versatility that System Exclusive can bring to almost every electronic musician's project studio. Most sequencers are able to read and transmit this instrument- and device-exclusive data, allowing you to create your own sounds, grab sound patches from the Internet, swap patches with your friends, or buy patch data disks from commercial vendors.

As we saw in Chapter 2, the System Exclusive message (or SysEx for short) makes it possible for MIDI manufacturers, programmers, and designers to communicate customized MIDI messages between devices that talk MIDI. The general idea behind SysEx is that it uses MIDI to transmit and receive device-specific program, and patch data or real-time parameter information from one device to another. Basically, it's capable of turning an instrument or device into a digital chameleon. One moment an instrument can be configured with a certain set of sound patches and/or setup parameters and then, after having received a new SysEx data dump, you could easily end up with a whole new setup that's literally full of new and exciting (or not-so-exciting) sounds and settings. OK, let's take a look at a few examples of how SysEx can be put to good use.

Let's say that you have a Brand X Model Z synthesizer and it turns out that you have a buddy across town who also has a Brand X Model Z. That's cool, except your buddy's synth has a completely different set of sound patches—and you want them! SysEx to the rescue! All you need to do is go over and transfer your buddy's patch data into your synth or into a MIDI sequencer as a SysEx data dump. To make life easier, make sure you take your instruction manual along (just in case you run into a snag) and follow these simple guidelines. I'll caution you that you're taking on these tasks at your own risk. Take your time—be patient and be careful during these procedures:

1. Back up your current patch data! This can be done by transmitting a SysEx dump of your synthesizer's entire patch and setup data to your sequencer's SysEx dump utility or SysEx track on your sequencer (of course, you should get out both the device's manual and your sequencer's manual and follow their SysEx dump instructions very carefully during the process). This is so important that I'll say it again: Back up your current patch data before attempting a SysEx dump! If you forget and download a new SysEx dump, your previous settings could easily be lost.
2. Save the data, according to your sequencer's manual.

3. Check that the dump was successful by reloading it back into the device in question. Did it reload properly? If so, your current patch data is now saved.
4. Next, connect your buddy's device to your sequencer. Dump this data to your sequencer. Save the new patch data (using a new and easily identifiable filename) according to your sequencer's manual and then safely back this data up.
5. Reconnect the sequencer to your synth and load the new data dump into it. Does your synth have a bunch of new sounds? Now reload your original SysEx dump back into your device. Are the original sounds restored?

If you were successful, you'll effectively have access to lots more sounds, the number of which will be limited solely by the number of SysEx dumps that you have hoarded for that particular device.

Another use for SysEx that many folks overlook is its ability to act as a backup medium for storing your patch and setup data just in case your system's memory gets corrupted or your instrument's RAM memory battery decides to go belly up. Here's an example of how a SysEx backup can save the day. A few years ago, I crashed the data in my WaveStation SR's memory by loading sounds from a WaveStation EX that wasn't totally compatible. Everything worked until I got to a certain patch and then the box went into a continual reboot loop—major freak-out time! No matter what I did, the system was totally locked up! The solution? I opened up the box and took out the backup memory battery (which in turn cleared out the error as well as the box's RAM patch memory). All I had to do was reload the factory SysEx patch settings back into the box and I was back in business in no time.

Work Those Tracks!

There are no rules, only guidelines! Of course, this wise adage strongly applies to the creative process of manipulating, combining, and playing with MIDI tracks to create an amazing musical experience. A whole host of editing and creative tools are at your disposal to breathe more life into a sequenced track or set of tracks; a few of these are track splitting, track merging, voice layering, and echoing.

Splitting Tracks

Although a MIDI track is encoded as a single entity that includes musical notes and other performance-related events, the data that's contained in a track can often be separated out in creative ways and then split into separate tracks for transposition, further processing, or routing to another instrument. As an example, let's say that we're working on a sequence that includes a synth part that was played live on stage. We, as co-producers, have decided that we'd like to have a greater degree of level mixing control over the left-hand part while leaving the bass part alone. This isn't really a problem. The best solution would be to split the track into two sequenced tracks by making an exact copy of the synth track and then pasting it onto a separate track that's assigned to a different MIDI channel or cable. Once done, we can call up the piano roll window for the original track (allowing the note ranges to be easily seen), highlight the bass notes, and

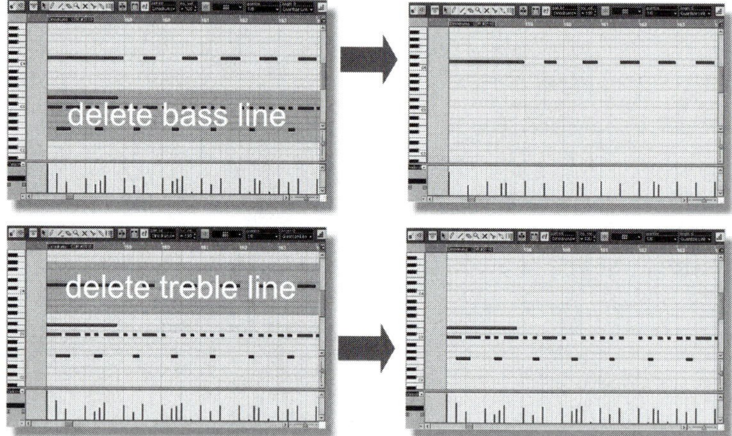

Figure 5.24. *By copying a MIDI passage, it's relatively simple to delete parts of the passage on each track copy, thus allowing control over volume, velocity, and controllers and for effects to be altered after the fact. Truly, the ability to edit is one of MIDI's strongest assets!*

delete them. We can then repeat the process on the copied track and delete the upper notes (Figure 5.24). Now that the part has been split into two tracks, we can go about doing any number of things to them—after the fact during postproduction. For example, we could:

- Change the channel volumes on each track to change their relative levels.
- Assign each part to a different instrument voice.
- Transpose the upper notes or tag on a transposed harmony.
- Quantize the bass notes while leaving the upper notes untouched.

Are you beginning to get the idea that there are very few limits to this useful technique?

Merging Tracks

Just as a track can be split into two or more tracks, multiple tracks can be merged into a single track. On most sequencers, this can be done easily by highlighting the entire track or segment that you want to merge and copying it into memory. Once done, you can select the destination track and measure into which the copied data is to be merged and then invoke the sequencer's merge command. When merging MIDI data, you should keep in mind that it might be difficult to unmerge data that has been combined into a single track. For this reason, it's generally wise to keep the tracks separate and then simply assign them to the same MIDI channel and port or to save the original unmerged tracks as an archived backup.

Another use of track merging involves the playing back of an instrument track while recording real-time controller data from a pitch, modulation, or other type of controller onto a separate track. This controller overdub process can help in keeping the original track intact, while real-time controller performances are recorded to a new track that's been assigned to the same port and channel. I often find that this process improves a performance, because it frees us up to concentrate on the performance. Once the controller tracks have been made to your liking, you can merge them with the original track (or simply keep them separate for future editing or archive purposes).

Layering Tracks

One of the more common and powerful tricks of the trade is the ability for instrument voices to be layered together to create a new, composite sound. Although digital samples and modern-day synthesis techniques have improved over the years, you're probably aware that many of these sounds will often have a character that can be easily distinguished from their acoustic counterparts. One of the best ways to fill out a sound and make it richer and more realistic is to layer multiple tracks together and then to assign each track to a different voice; when combined, these tracks will make a single, more complex and interesting voice. One example of layering would be to take the sound of a piano from synthesizer A and combine it with the sounds of a sampled Steinway from sampler B.

The process of layering can be carried out in several ways; for example:

- A layered track could be manually overdubbed onto another track and then mixed together as a combined voice.
- A single take could be duplicated to another track, assigned to another instrument voice, and then mixed as a combined voice.
- A single take could be duplicated onto another track. This new track could then be assigned to another instrument or part and changes could be made (such as humanization and changes to note lengths). And, finally, combined sounds could be mixed together in a way that would create a richer, fuller sound.
- Layers can also be made by simply by assigning the output of a MIDI track to two or more instruments by transmitting the same data over multiple MIDI ports, channels, and cables.

Of course, each layered part can be individually exported to an audio track or simply combined in the mix in ways that will achieve the best or most convincing results. For example, when working in surround sound, a full soundscape can be made by steering one stereo voice to the front speakers … and then steering a layered track to the rear.

MIDI Echo, Echo, Echo, …

They say that you can never have too many effects boxes in your toy chest. Well, MIDI can also come to the rescue to help you set up effects in more ways than you might expect. For example, a part can be easily repeated in a digital delay fashion by copying a track to another track and then slipping that track (either forward or backward) in time. By assigning these tracks to the same destination (or by merging these tracks into one), you'll be setting up a fast and free MIDI echo effect. Simply repeat the process if you want to add more echoes.

MIDI Processing Effects

Just as signal processors can be used in an audio chain to create an effect by changing or augmenting an existing sound, a MIDI processor can often be inserted into a DAW MIDI track to alter its performance and processing functions, strictly in the MIDI domain. The following MIDI effects plug-ins are but a few that are included within Steinberg's Cubase SX/Nuendo DAW program (of course, you should check your DAW's own manual for MIDI plug-in applications that might be included in its processing toolbox):

- *Arpache 5*—A typical arpeggiator accepts a chord (organized group of simultaneous MIDI notes) as input and plays back each note in the chord separately in the playback order and speed set by the user (Figure 5.25).
- *Autopan*—This plug-in works a bit like an LFO (low frequency oscillator) in a synthesizer, allowing you to send out continuously changing and evolving MIDI controller messages (Figure 5.26). One typical use for this is automatic MIDI panning (hence, the name), but you can select any MIDI continuous controller event type.
- *Chorder*—The Chorder is a MIDI chord processor that allows complete chords to be assigned to single keys in a multitude of variations. There are three main modes of operation: Normal, Octave, and Global (Figure 5.27).
- *Compress*—The MIDI compressor plug-in is used for evening out or expanding differences in velocity. Though the result is similar to what you get with the velocity compression track parameter, the Compress plug-in presents the controls in a manner more like regular audio compressors (Figure 5.28).

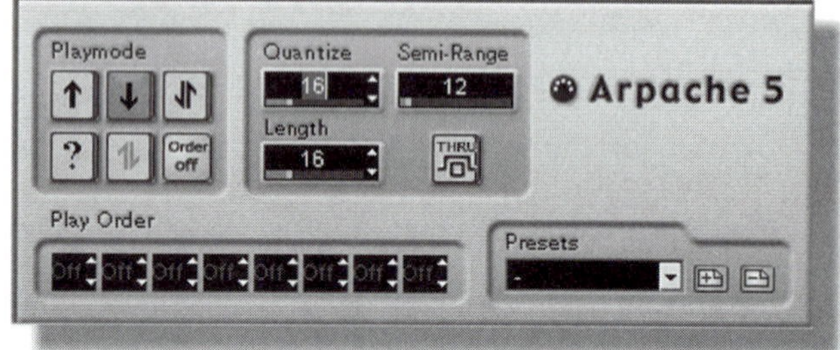

Figure 5.25. *Steinberg Cubase/Nuendo Arpache 5 MIDI arpeggiator. (Courtesy of Steinberg Media Technologies GmbH, A Division of Yamaha Corporation; www.steinberg.net.)*

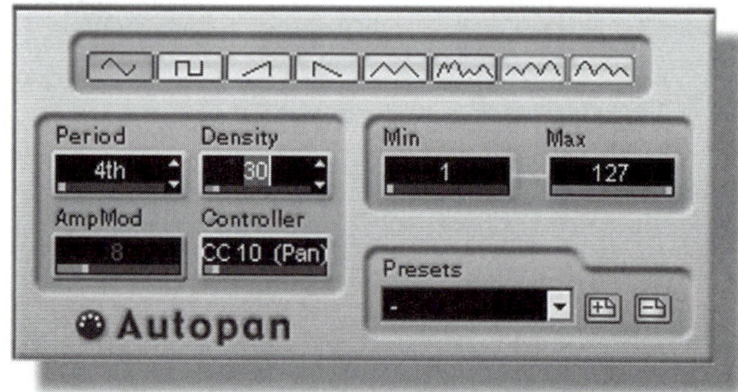

Figure 5.26. *Steinberg Cubase/Nuendo Autopan control changer. (Courtesy of Steinberg Media Technologies GmbH, A Division of Yamaha Corporation; www.steinberg.net.)*

- *MIDI Echo*—This is an advanced MIDI Echo that will generate additional echoing notes based on the MIDI notes it receives (Figure 5.29). It creates effects similar to a digital delay but also features MIDI pitch shifting and much more. As always it is important to remember that the effect doesn't echo the actual audio—it echoes the MIDI notes that will eventually produce the sound in the synthesizer.
- *Step Designer*—The Step Designer is a MIDI pattern sequencer that sends out MIDI notes and additional controller data according to the pattern you set up (Figure 5.30). It does not make use of the incoming MIDI, other than automation data (such as recorded pattern changes).

Figure 5.27. *Steinberg Cubase/Nuendo Chorder MIDI chord processor. (Courtesy of Steinberg Media Technologies GmbH, A Division of Yamaha Corporation; www.steinberg.net.)*

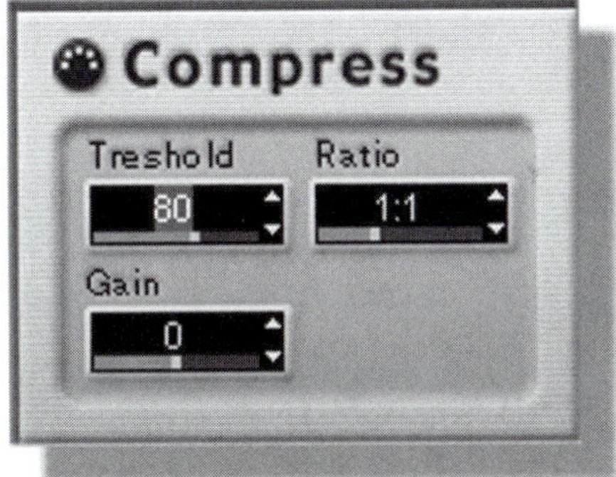

Figure 5.28. *Steinberg Cubase/Nuendo Compress MIDI compressor plug-in. (Courtesy of Steinberg Media Technologies GmbH, A Division of Yamaha Corporation; www.steinberg.net.)*

Figure 5.29. *Steinberg Cubase/Nuendo MIDI Echo plug-in. (Courtesy of Steinberg Media Technologies GmbH, A Division of Yamaha Corporation; www.steinberg.net.)*

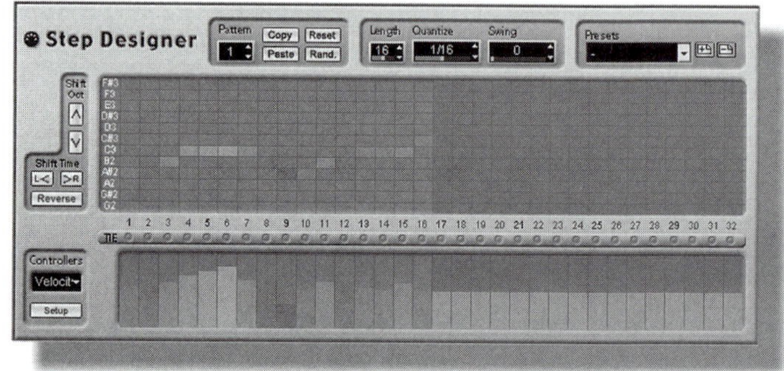

Figure 5.30. *Steinberg Cubase/Nuendo Step Designer MIDI pattern sequencer. (Courtesy of Steinberg Media Technologies GmbH, A Division of Yamaha Corporation; www.steinberg. net.)*

Audio-to-MIDI Interpretation

A number of stand-alone software and sequencer applications are able to accept an audio file (often in a wide range of formats), interpret the data, and then convert the passage into a MIDI track or standard MIDI file. Not surprisingly, this interpretive art isn't always precise but can often take simple and even polyphonic sound and create a MIDI file that can be used for effect, for layering sounds over an existing audio track, or for notation purposes.

Replacing Audio Tracks via MIDI

In the world of recording production, it's widely known that one of the most difficult instruments to properly record is the drum set. "Thuddy" kicks, "ringy" snares, phase leakage—any and more of these problems can render a recorded drum track bad or even useless. Fortunately, MIDI can again come to the rescue by helping to resurrect a truly horrific drum track through the refined art of track replacement. Let's take a look at a few tools and tricks that can help save a project's butt:

- *Acoustic triggers*—For starters, acoustic or electro-acoustic pickups can be attached to a drum kit (that will also probably be acoustically miked for a session or on-stage event). These pickups can then be plugged into a drum module unit that's equipped with individual analog trigger inputs which, in turn, are able to take an input signal from each drum and instrument in the kit and convert this into MIDI messages that can be synchronously recorded to a MIDI track or tracks (either to MIDI tracks on a DAW or to a synchronized sequence track). Once the data has been encoded to MIDI, each drum sound can be altered, slipped in time, effected, and used to replace the original sound.
- *Triggering from track*—Just as the trigger inputs can be derived from a pickup, the source can also come from a previously recorded track. This means that a pitiful drum sound can be resurrected after the fact, allowing the offending track or tracks to be converted to MIDI, where the desired sounds can be triggered from a synth, sampler, or plug-in (Figure 5.31).
- *Drum replacement software*—In addition to the above methods, a more automated system exists in the form of a software plug-in that can actually extract the originally recorded drum information and convert it into replacement tracks that can easily be placed into a DAW session (Figure 5.32). As is often the case with plug-in technology, a wide range of replacement sounds and trigger timing options are available.

Figure 5.31. *DM-5 18-bit drum module. (Courtesy of Alesis, www.alesis.com.)*

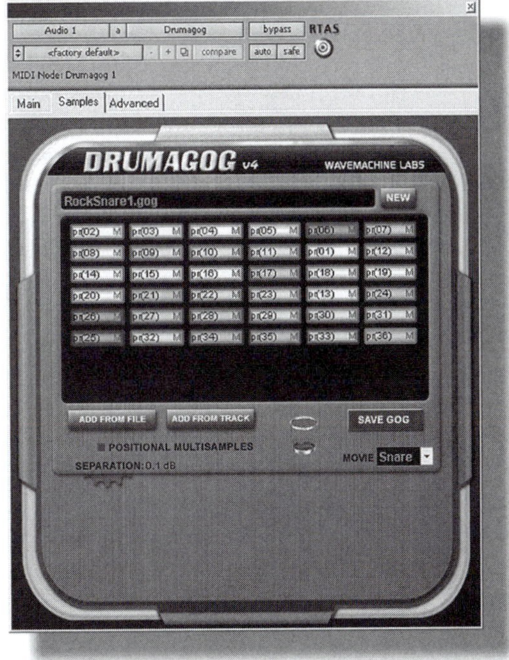

Figure 5.32. *Drumagog drum replacer plug-in for the VST, RTAS, and AU formats. (Courtesy of WaveMachine Labs, Inc.; www.drumagog.com.)*

Playback

Once a sequence is composed and saved to disk, all of the sequence tracks can be transmitted through the various MIDI ports and channels to the instruments or devices to make music, create sound effects for film tracks, or control device parameters in real time. Because MIDI data exists as encoded real-time control commands and not as audio, you can listen to the sequence and make changes at any time. You could change the patch voices, alter the final mix, or change and experiment with such controllers as pitch bend or modulation—even change the tempo and key signature. In short, this medium is infinitely flexible in the number of versions that can be created, saved, folded, spindled, and mutilated until you've arrived at the overall sound and feel you want. Once done, you'll have the option of using the data for live performance or mixing the tracks down to a final recorded media, either in the studio or at home.

During the summer, in a wonderful small-town tavern in the city where I live, there's a frequent performer who'll wail the night away with his voice, trusty guitar, and a backup band that consists of several electronic synth modules and a laptop PC/sequencer that's just chock-full of country-'n'-western sequences. His set of songs for the night is loaded into a song playlist feature that's programmed into his sequencer. Using this playlist, he queues his sequences so that when he's finished one song, taken his bow, introduced the next song, and complimented the lady in the red dress, all he needs to do is press the space bar and begin playing the next song. Such is the life of an on-the-road sequencer.

In the studio, it's become more the rule than the exception that MIDI tracks will be recorded and played back in sync with a DAW, analog multitrack machine, or digital multitrack. As you're aware, using a DAW-based sequencer automatically integrates the world of audio and video with the MIDI production environment. Whenever more than one playback medium is involved in a production, a process known as synchronization is required to make sure that events in the MIDI, analog, digital, and video media occur at the same point in time. Locking events together in time can be accomplished in various ways (depending on the equipment and media type used). Further and more in-depth reading on this subject can be found in Chapter 11.

Mixing a Sequence Using Continuous Controllers

Almost all DAW and sequencer types will let you mix a sequence in the MIDI domain using various controller message types. This is usually done by creating a physical or software interface that incorporates these controls into a virtual on-screen mixer environment (Figure 5.33). Instead of directly mixing the audio signals that make up a sequence, these controls are able to directly access such track controllers as Main Volume (controller 7), Pan (controller 10), and Balance (controller 8), most often in an environment that completely integrates into the workstation's overall mix controls. Since the mix-related data is simply MIDI controller messages, an entire mix can be easily stored within a sequence file. Therefore, even with the most basic sequencer, you'll be able to mix and remix your sequences with complete automation and total settings recall whenever a new sequence is opened. As is almost always the case with a DAW's audio and MIDI graphical user interface (GUI), the controller and mix interface will most always have moving faders and pots—a feature that's not only useful for letting you know where your settings are but also fun!

Tips and Tricks

In addition to all of the above information, here are a few hints that can help the process go more smoothly. Hopefully, you can create a diary or document that can add to the list.

1 Exporting MIDI Tracks to Audio

Although it's often a matter of personal preference, many production musicians and producers will transfer their MIDI tracks into the audio domain through the use of a DAW's export or bounce-to-disk function. This process allows the audio to be edited, copied, and processed using the same edit and plug-in tools that are used for the audio tracks. In short, the MIDI instruments don't need to be recalled and used, and the basic process becomes simpler in the long run. When exporting

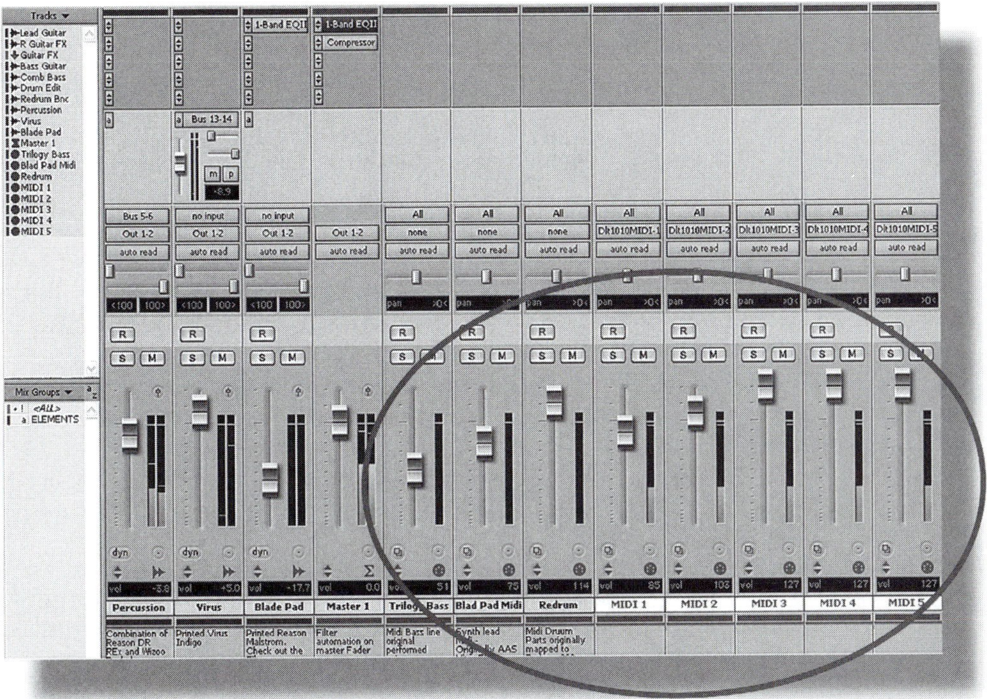

Figure 5.33. MIDI tracks can be added into the mix for real-time parameter control over hardware and/or software devices. (Courtesy of Digidesign, A Division of Avid Technology, Inc.; www.digidesign.com.)

MIDI instrument or plug-in tracks to an audio track, a number of considerations should be kept in mind so as not to make the process more difficult during mixdown; for example:

- Set the Channel Volume to a value that allows the audio track to be recorded at a reasonably high gain without distorting the signal.
- If the tracks are stereo, make sure that the Pan or Balance controllers are set to the 64 center position, thereby allowing the balance to be accurately set in the audio domain.
- For the same reasons of control during mixdown, it might be wise to strip all gain changes (including fades) from the MIDI track, allowing these changes to be carried out in the audio domain.
- Depending on the style and general makeup of the track, you might or might not want to print the instrument sounds to a track that includes digital signal processing effects. For example, a synth might have a really bad reverb effect on a track that might have to be turned off (often requiring that you get out the device's handy-dandy manual). In another case, the instrument effect might be integral to the track's overall sound, in

which case you might want that sound (print the track twice, both with and without effects).

- It's generally wise to begin the export from the beginning of the song. This sets the track's beginning time to the obvious beginning point in the song, should the project be exported to another DAW program or for some other unforeseen reason.

2 Track Considerations

Just as a professional recording engineer will often create and assign tracks according to his or her own habits and logistic considerations, it's often a good idea to create a standard MIDI setup that makes the most sense for you. Most DAW programs allow for a basic setup template to be created and recalled at will. One of the most important aspects of successful sequencing is the creation of a system that's as simple and straightforward to operate as possible. I know it's not always easy, but keeping your studio setup as simple as possible can go a long way towards battling tangled wires, confused connections, and unintuitive session setups.

3 Document, Document, Document!

It's generally a good idea to document your session, writing down any equipment and special settings that were used during a project. This is especially helpful with the passage of time. We all know that it's all too easy to forget special equipment or techniques that were used several years ago. This also holds true if another person is helping with the project's production or mixdown. Writing down important settings, tools, and general stuff can help keep confusion to a minimum.

4 PDF Manuals to the Rescue!

During this entire chapter (and book, for that matter), you've been asked to refer back to a device's or software package's manual. This process can be made much easier by creating a directory on your hard disk that's strictly devoted to storing and archiving manuals. Going to the web or your program discs, grabbing their manuals in PDF form, and then archiving them in one convenient place can come in mighty handy in a pinch. Having a copy of this directory on your laptop can also make for some interesting reading on your next cross-country trip. Happy reading!

Digital Audio Production

Over the years, digital audio technology has grown to play a strong role in MIDI production. This merging factor is largely due to the fact that MIDI is a digital medium and as such can easily be interfaced with devices that output or control digital audio. Devices such as samplers, digital audio workstations, hard disk recorders, and digital audio recorders (of the DAT, MINI Disc, and modular digital multitrack varieties) are commonly used to record, reproduce, and transfer sound within such an environment.

In recent years, the way that electronic musicians store, manipulate, and transmit digital audio has changed dramatically. As with most other media, these changes have been brought about by the integration of the personal computer into the modern-day project studio environment. In addition to sequencing MIDI data and controlling production-related devices in a MIDI system, newer generations of computers and their hardware peripherals have been integrated into the MIDI environment to receive, edit, manipulate, and reproduce digital audio with astonishing ease. This chapter is dedicated to the various digital system types and to the details of how they relate to the modern-day project studio.

The Digital Recording/Reproduction Process

The following sections provide a basic overview of the various stages that are involved in the encoding of analog signals into equivalent digital data and then converting this data back into its original analog form.

The encoding and decoding phases of the digitization process center around two processes:

- Sampling
- Quantization

In a nutshell, *sampling* is a process that effects the overall bandwidth (frequency range) that can be encoded within a sound file, while *quantization* refers to the resolution (overall quality and distortion characteristics) of an encoded signal compared to the original analog signal at its input.

Sampling

In the world of analog audio, signals are passed, recorded, stored, and reproduced as changes in voltage levels that continuously change over time (Figure 6.1). The digital recording process, on the other hand, doesn't operate in such a continuous manner; rather, digital recording takes periodic samples of a changing audio waveform (Figure 6.2) and transforms these sampled signal levels into a representative stream of binary words that can be manipulated or stored for later processing and/or reproduction.

Within a digital audio system, the sampling rate is defined as the number of measurements (samples) that are taken of an analog signal over the course of a second. Its reciprocal (sampling time) is the elapsed time that occurs between each sampling period. For example, a sample rate of

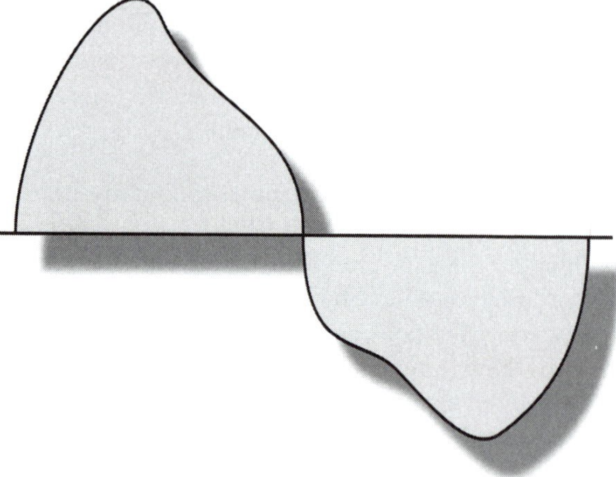

Figure 6.1. *Analog signals are continuous in nature.*

44.1 kHz corresponds to a sample time of 1/44,100 of a second. Because sampling is tied directly to the component of time, the sampling rate of a system determines its overall bandwidth (Figure 6.3), meaning that a recording made at a higher sample rate will be capable of storing a wider range of frequencies (effectively increasing the signal's bandwidth at its upper limit).

The sampling process can be likened to a photographer who takes a series of action sequence shots. As the number of pictures taken in a second increases, the accuracy of the captured event will likewise increase until the resolution is so great that you can't tell that the successive, discrete pictures have turned into a continuous and (hopefully) compelling movie.

Quantization

Quantization represents the amplitude component of the digital sampling process. It is used to translate the voltage levels of a continuous analog signal (at discrete sample points over time) into binary digits (bits) for the purpose of manipulating or storing audio data in the digital domain. By sampling the amplitude of an analog signal at precise intervals over time, the converter's determines the exact voltage level of the signal (during a sample interval, when the voltage level

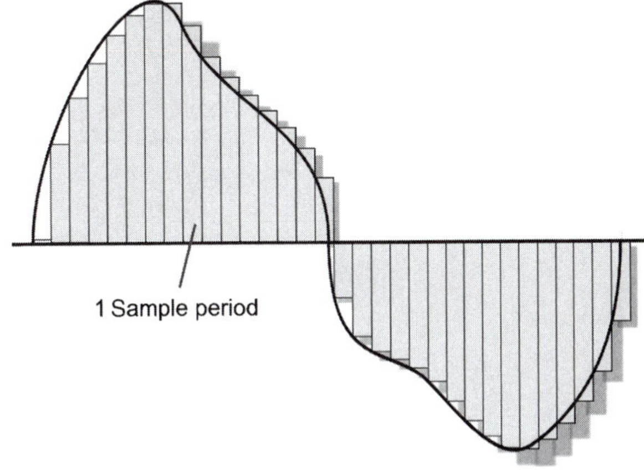

Figure 6.2. *Digital signals make use of periodic sampling to encode information.*

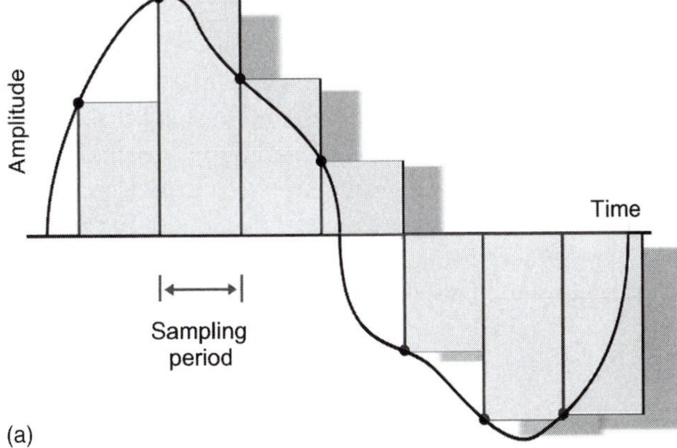

Figure 6.3. *Discrete time sampling. (a) Whenever the sample rate is set too low, important data between sample periods will be lost. (b) As the rate is increased, more frequency-related data can be encoded. (c) Increasing the sampling frequency further can encode the recorded signal with an even higher bandwidth range.*

Figure 6.3. Continued.

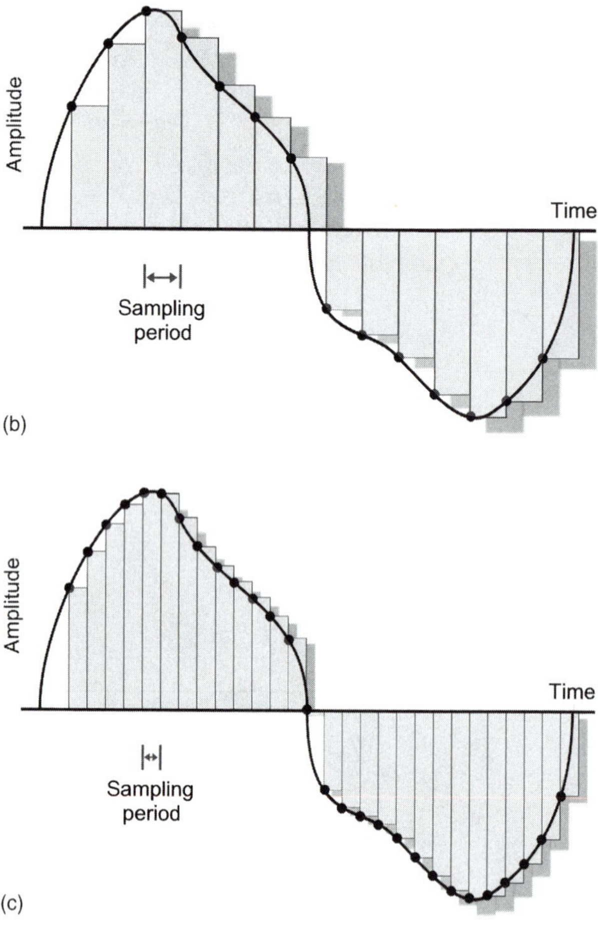

is momentarily held) and then outputs the signal level as an equivalent set of binary numbers (as a grouped word of *n*-bits length) that represents the originally sampled voltage level (Figure 6.4). The resulting word is used to encode the original voltage level with as high a degree of accuracy as can be permitted by the word's bit length and the system's overall design.

Currently, the most common binary word length for audio is 16-bit (for example, [0110010100101101]); however, bit-depths having 20- and 24-bit resolution are also in common use. In addition, computers and signal-processing devices are capable of performing calculations internally at the 32- and 64-bit resolution level. This added internal headroom at the bit level helps reduce errors in level and performance at low-level resolutions whenever multiple audio datastreams are mixed or processed within a digital signal processing (DSP) system. This greater internal bit resolution is used to reduce errors that might accumulate within the least significant bit (LSB; the final and smallest numeric value within a digital word). As multiple signals are mixed together and multiplied (a regular occurrence in gain change and processing functions), lower-bit-resolution numbers play a more important role in determining the signal's overall accuracy and distortion. Since the internal bit depth is higher, these resolutions can be preserved

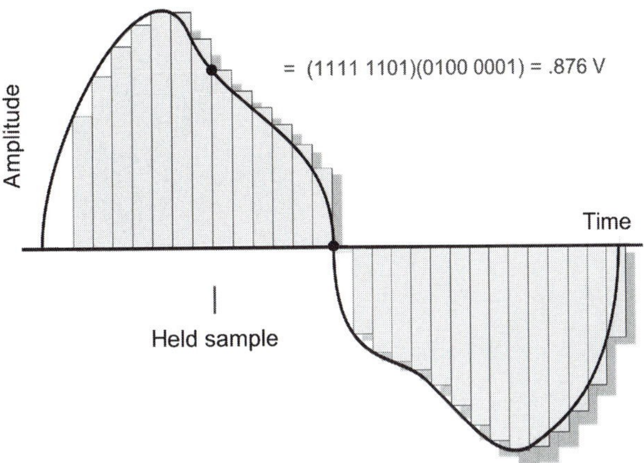

Figure 6.4. *The instantaneous amplitude of the incoming analog signal is broken down into a series of discrete voltage steps, which are then converted into an equivalent binary-encoded word.*

(instead of being dropped by the system's hard- or software processing functions), with a final result being an *n*-bit datastream that's relatively free of errors.

This leads us to the conclusion that greater word lengths will often directly translate into an increased resolution (and thus higher quality) due to the added number of finite steps into which a signal can be digitally encoded. The following box details the number of encoding steps that are encountered for the most commonly used bit lengths:

8-bit word = (*nnnnnnnn*) = 256 steps
16-bit word = (*nnnnnnnn nnnnnnnn*) = 65,536 steps
20-bit word = (*nnnnnnnn nnnnnnnn nnnn*) = 1,048,576 steps
24-bit word = (*nnnnnnnn nnnnnnnn nnnnnnnn*) = 16,777,216 steps
32-bit word = (*nnnnnnnn nnnnnnnn nnnnnnnn nnnnnnnn*) = 4,294,967,296 steps

Although the details of the digital audio record/playback process can get quite detailed and complicated, the basic steps that must be taken include sampling (in the truest sense of the word) an analog voltage signal at precise intervals in time, converting these samples into a digital word value that most accurately represents these voltage levels, and then storing the numeric values within a digital memory device. Upon playback, these digital words are then converted back into discrete voltages (again, at precise intervals in time), allowing the originally recorded audio signal to be recreated, processed, or played back.

Samplers

Historically, one of the first production applications in digital audio gave us the ability to record and play back drum and percussion sounds. This made it possible for electronic musicians

Figure 6.5. *Akai MPC2000XL Music Production Center. (Courtesy of Akai Professional, www.akaipro.com.)*

(mostly keyboard players) to add percussion samples to their own compositions. Out of this sprang a major class of sample and synthesis technology that has fueled electronic production technology over its history.

As we saw in Chapter 4, a *sampler* (Figure 6.5) is a hardware or software device that's capable of recording, musically transposing, processing, and reproducing segments of digitized audio directly from RAM memory. Because this memory is often limited in size (relative to digital audio's memory-intensive nature), the segments are generally limited in length and range from only a few seconds to one or more minutes.

Assuming that sufficient memory is available, any number of audio samples can be loaded into a system in such a way that their playback can be transposed in real time (either up or down) over a number of octave ranges in a musical fashion. Quite simply, this musical transposition occurs by reproducing the recorded digital audio segments at sample rates that correspond to established musical intervals. These samples can then be assigned to a specific note on a MIDI controller or mapped across a keyboard range in a multiple-voice fashion.

Sample Editing

Whenever a recorded sound is transferred into a hardware or software sampler, the original source material may contain extraneous noises, breathing/fidget sounds, or other music that occurs both before and after the desired sample (Figure 6.6a). Using the sampler's edit function, these unwanted sounds can be deleted by trimming the in and out points to include only the desired sounds. Trimming is accomplished by instructing the system's microprocessor to ignore (not access or reproduce) the samples that exist before a user-defined in point or following a desired out point (Figure 6.6b). Once done, the final sample can be played, looped, dynamically processed (if a fade or other function needs to be performed), and then saved for later recall.

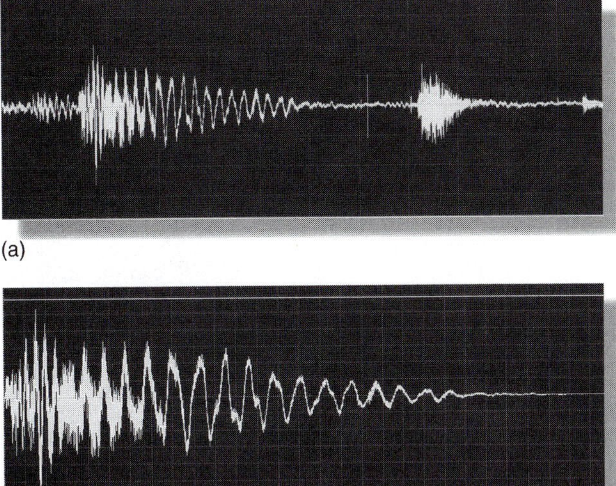

Figure 6.6. Sample editing: (a) unedited sample; (b) sample that has been trimmed, normalized, and faded at its end.

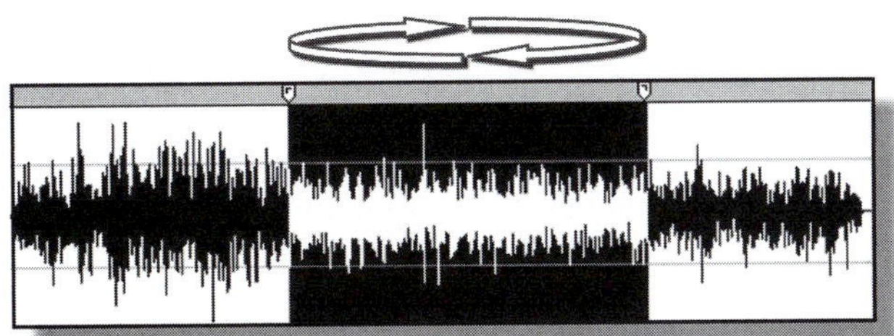

Figure 6.7. Example of a sound file with a highlighted sustain loop.

Sample Looping

A common editing technique that's regularly used to maximize the system's available RAM and disk-based memory is a process known as looping. Using this technique, a sample that occupies a finite memory space in RAM can be played in a repeated fashion. In this way, a carefully edited sample can be sustained for long periods of time (well past the length of the original sample) by continuing to hold down the note key. Once the key is released, the note can be programmed to fade out over time in a natural fashion. Such a loop can be created by finding a sustaining segment of sound within a sample that doesn't significantly change in amplitude and timbre over time. By setting the in and out loop points to repeat within this sustained segment, the sampler can repeatedly access the loop over time within RAM (Figure 6.7).

The most challenging part of creating a realistically convincing loop is making sure that the loop splice point is carefully matched in level and frequency balance. This process can be simplified by following this simple rule: Be sure to carefully match the waveform shape and amplitude at the beginning of the loop with the waveform shape and amplitude at the end of the loop. This simply means that the beginning and end amplitudes must match (Figure 6.8). If they don't, the

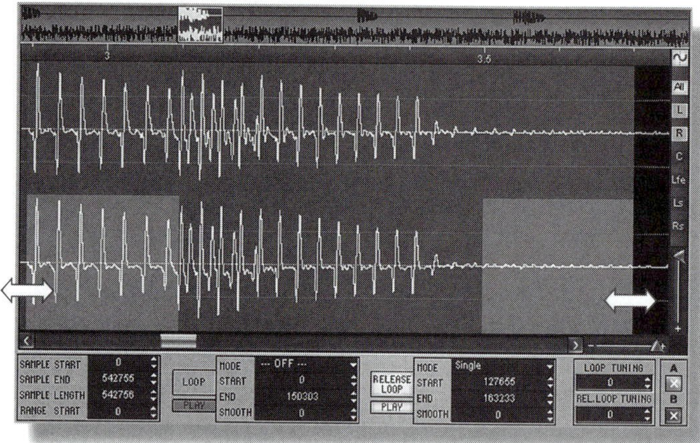

Figure 6.8. *A loop waveform window allows the beginning and end levels of a loop to be manually matched. (Courtesy of Steinberg Media Technologies GmbH, A Division of Yamaha Corporation; www.steinberg.net.)*

signal levels will vary and an annoying "pop," audible "tick," or discontinuity in the sample will result. Many samplers and sample-editing programs provide a way to automatically search out the closest level match or display the loop crossover points on a screen; however, more often than not the levels will need to be manually finessed to find their final crossover points.

In addition to allowing multiple, layered samples to be assigned to a single note, many samplers are able to access multiple loop points within a single sample. This has the effect of making the sample sound less repetitive and more natural and adds expressiveness when played on a keyboard. In addition to having multiple sustain loops, a different release loop can be programmed to create a unique decay that occurs whenever the sample note is released.

Distribution of Sampled Audio

Since MIDI's inception, several sample-dump transmission and file formats have been developed that allow samples to be communicated among hardware sampling systems, software samplers, and the computer. Over the years, several file formats have become standardized throughout the electronic music industry. The existence of such standardized file formats makes it possible for entire libraries of pre-edited samples to be commercially or freely distributed between sample-based devices. In addition to the export and import of standardized file formats, the distribution of samples and related information can be communicated between supporting devices through the use of the slower MIDI sample-dump standard, a high-speed small computer system interface (SCSI) port, or the FireWire® transfer protocol. As always, consulting the manuals for each device before attempting a data transfer will most likely reduce stress, frustration, and wasted time.

MIDI Sample-Dump Standard

The MIDI Sample Dump Standard (SDS) was developed and proposed by the MIDI Manufacturers Association as a protocol for transmitting sampled digital audio and sustain-loop information between sampling devices. This data is transmitted over regular MIDI lines as a series of MIDI System Exclusive (SysEx) messages, which are unspecified in length and data structure. Although samplers of different manufacture and model type can be used to perform similar musical functions, the inner electronics and the way that data is internally structured can vary widely from device to device. As a result, most samplers communicate using their own unique SysEx data structure (as identified by a unique manufacturer and device ID number). In order for data to be successfully transmitted between samplers, they must be of the same or compatible manufacture and design. If this isn't the case, a computer-based program (such as a sample editor or import function) must be used to translate from one sampler's data format and structure into one that can be understood by another make or model. It should be noted that the MIDI Sample Dump Standard is rarely used in modern production, as it has the distinct disadvantage of being rather slow, transmitting audio data over standard MIDI lines at a rate of 31.25-k baud. When transmitting anything longer than a short sample, be prepared to take a long coffee break.

SCSI Sample Dump Formats

It should be noted that a number of computer-based digital audio systems and professional samplers are also capable of transmitting and receiving sampled audio via SCSI (a bidirectional communications line that's commonly used by personal computers to exchange digital data at high speeds). This protocol could be used to provide a direct parallel data link for transferring sound files at a rate of 16 MB/sec or higher to and from older sampler devices (literally hundreds of times faster than MIDI but still slow by most modern standards). Although the data format will change from one device to the next (meaning that data can only be transmitted between like devices or via specific device/computer system combinations), SCSI still wins out as a fast and straightforward way to transfer data to and from an editing program, hard disk, or CD-ROM sample library.

The above sound file formats and communications protocols make it possible for sampled audio to be imported into a dedicated sample edit program or the application of a software sampler. This in turn makes it possible for audio and edit loop points to be trimmed and manipulated in an intuitive, on-screen environment that lets us take full advantage of the computer's sample-editing functions, allowing us to:

- Store samples to hard disk, arrange them in a categorized library that best suits the user's needs, and transmit them to any hardware sampling device in the system.
- Edit and arrange samples using standard computer cut and paste edit tools.
- Have greater, more intuitive control over multiple loop and split points within a sample.
- Alter samples using such functions as gain changing, mixing, equalization, inversion, reversal, muting, fading, cross-fading, and time compression.

Hard-Disk Recording

With the introduction of the sample-based drum machine (the first practical digital playback system), advances in digital and semiconductor technology soon opened the doors for recording audio as longer and longer sample files into RAM. Computer technology advanced to the point of being affordable to the average user … and through the use of specialized hardware, software, and I/O interfacing, digital audio could be recorded to, edited on, and played back from a computer's hard disk. Thus, the concept of the hard-disk recorder was born. As most are now fully aware, there are numerous advantages to using a hard-disk recording system in an audio production environment:

- *The ability to handle long sample files*—Hard-disk recording time is often limited only by the size of the disk itself.
- *Random-access editing*—Once audio (or any type of data) is recorded onto a disk, any point within the program can be instantly accessed at any time, regardless of the order in which it was recorded.
- *Nondestructive editing*—This process allows audio segments (often called regions) to be placed back in any context or order within a program without changing or affecting the originally recorded sound file in any way. Once edited, these tracks and segments can be reproduced to create a single, cohesive program.
- *DSP*—Digital signal processing can be performed on a sound file and/or segment in either real time or non-real time (often in a nondestructive fashion).

Add to this the fact that computer-based digital audio devices can integrate many of the tasks that are related to both digital audio and MIDI production in a unified fashion that's often powerful, easy to use, and cost effective … and you have a system that offers the artist and engineer an unprecedented degree of production power.

The Digital Audio Workstation

In recent years, the term digital audio workstation (DAW) has increasingly come to signify an integrated, computer-based, hard-disk recording system that commonly offers such features as:

- Advanced multitrack recording, editing, and mixdown capabilities
- MIDI sequencing, edit, and score capabilities
- Integrated and plug-in signal processing support
- Support for integrating software plug-in instruments (VSTi, AU, and RTAS) and peripheral music programs (ReWire)
- Integration of peripheral hardware devices such as controllers and audio and MIDI interface devices

Truth of the matter is, by offering a staggering amount of production power for the buck these software-based programs (Figure 6.9) and their peripherally connected devices have revolutionized the faces of professional, project, and personal studios in a way that touches almost every life within the audio and music production communities.

Integration Now … Integration Forever!

Throughout the history of music and audio production, we've gotten used to the idea that certain devices were only meant to perform a single task: A recorder records and plays back, a limiter

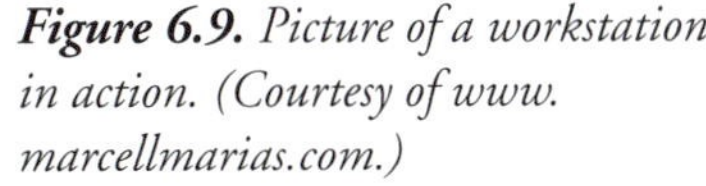

Figure 6.9. *Picture of a workstation in action. (Courtesy of www.marcellmarias.com.)*

limits, and a mixer mixes. Fortunately, the age of the microprocessor has totally broken down these traditional lines in a way that has created a breed of digital chameleons that can change their functional colors as needed to match the task at hand. Along these same lines, the digital audio workstation isn't so much a device as a systems concept that can perform a wide range of audio production tasks with ease and speed. Some of the characteristics that can (or should be) offered by a DAW include:

- *Integration*—One of the major features of a workstation is its ability to provide centralized control over the digital audio recording, editing, processing, and signal-routing functions, as well as to provide transport- or time-based control over MIDI/electronic music systems, external tape machines, and video recorders.
- *Communication*—A DAW should be able to communicate and distribute pertinent audio-, MIDI-, and automation-related data throughout the connected network system. Digital timing (wordclock) and synchronization (SMPTE time code and/or MTC) should also be supported.
- *Speed and flexibility*—These are probably a workstation's greatest assets. After you've become familiar with a particular system, most production tasks can be tackled in far less time than would be required using similar analog equipment. Many of the extensive signal processing, automation, and system's communications features would simply be next to impossible to accomplish in the analog domain.
- *Automation*—Because all of the functions are in the digital domain, the ability to instantly recall a session and to undo a performed action becomes a relatively simple matter.
- *Expandability*—Most DAWs are able to integrate new and important hardware and software components into the system with little or no difficulty.

- *User-friendly operation*—An important element of a digital audio workstation is its ability to communicate with its central interface unit: you! The operation of a workstation should be relatively intuitive and shouldn't obstruct the creative process by speaking "computerese."

I'm sure that you've gathered from the above that a software system (and its associated hardware) that's capable of integrating audio, video, and MIDI under a single, multifunctional umbrella can be a major investment, both in financial terms and in terms of the time that's spent to learn and master the overall program environment. When choosing a system for yourself or your facility, be sure to take the above considerations into account. Each system has its own strengths, weaknesses, and particular ways of working. When in doubt, it's always a good idea to research the system as much as possible before committing to it. Feel free to contact your local dealer for a salesroom test drive. As with a new car, purchasing a DAW can be an expensive proposition that you'll probably have to live with for a while. Once you've taken the time to make the right choice, you can get down to the business of making music.

DAW Hardware

Keeping step with the modern-day truism "technology marches on," the hardware and software specs of a computer and the connected peripherals continue to change at an ever-increasing pace. This is usually reflected as general improvements in such areas as their:

- Need for speed
- Increased computing power
- Increased disk and RAM memory and speed
- Operating system (OS) and peripheral integration
- General connectivity (networking and the web)

In this section, we'll be taking a look at many of the hardware devices and connected peripheral devices that help to make a DAW work.

The Desktop Computer

Desktop computers are often (but not always) too large and cumbersome to lug around. As a result, these systems are most often found as a permanent installation in the professional, project, and home studio (Figures 6.10 and 6.11). One of the most commonly asked questions is "Which one ... Mac® or PC?" The actual answer as to what OS to invest in actually depends upon:

- Your preference
- Your needs
- The kind of software you currently have
- The kind of computer platform and software your working associates have

Figure 6.10. *The MacBook™ Pro with display. (Courtesy of Apple Computers, Inc.; www.apple.com.)*

Figure 6.11. *Event—Digital Media Workstation. (Courtesy of Rain Recording, www.rainrecording.com.)*

Truth of the matter is, beyond these important questions, the choice it strictly up to you. Once you've decided which side of the platform tracks that you'd like to live on, the more important questions that you should be asking are:

- Is my computer fast and powerful enough for the tasks at hand?
- Does it have enough hard disks that are large and fast enough for my needs?
- Is there enough RAM memory?
- Do I have enough monitor space (real estate) to see the important things at a glance?

On the "need for speed" front, it's always a good idea to buy (or build) a computer at the top of its performance range at any given time. Keeping in mind that technology marches on, the last

Figure 6.12. Rain Recording Storm FireWire® and USB2 external drive. (Courtesy of Rain Recording, LLC; www.rainrecording.com.)

thing that you'll want to do is buy a new computer only to soon find out that it's underpowered for the tasks ahead.

There's never been a better time for choices on the hard-disk front. With today's faster and higher capacity IDE, serial ATA, and SCSI drives, it's a simple matter to install cost-effective drives, each with a capacity of hundreds of gigabytes. In addition, portable drive cases (Figure 6.12) can be plugged into either a FireWire® or USB2 port (and in some cases both), making it easy to take your own personal drive with you into the studio. Data transfer speeds can be an issue when buying a hard drive for either audio or video applications.

The speed at which the disc platters turn will often affect a drive's access time. Modern drives that spin at 7200 rpm or higher are often preferable. On-board buffer memory can also be helpful in transferring data and in freeing up the system for other processing functions. It should be mentioned that the portable FireWire® and USB2 drives mentioned above will often have reduced access times (over their internal counterparts), making them a better medium for backing up data—although, with the introduction of FireWire® 800 and the onward march of technology, this disadvantage could change at a moment's notice.

Regarding random access memory, it's always a good idea to use as much (and as fast) RAM as you can muster. If a system doesn't have enough RAM, data will often have to be swapped to the system's hard drive, which can slow things down and affect overall performance. When dealing with video and digital images, having a sufficient amount of RAM becomes even more of an issue.

Just like there never seems to be enough space around the house or apartment, having a single, undersized monitor can leave you feeling cramped for visual "real estate." For starters, a sufficiently large monitor (either LCD or CRT) that's capable of working at higher resolutions will greatly increase the size of your visual desktop; however, if one is a good thing, two can be better! Both Windows® XP and newer OS versions for the Mac® offer support for dual monitors (Figure 6.13). By adding a dual-head video card or by simply adding another video card, these systems can be easily configured so that the two monitors will literally double your working space for less

Figure 6.13. *You can never have enough "real-estate"!*

Figure 6.14. *Rain Recording LiveBook™ laptop. (Courtesy of Rain Recording, LLC; www.rainrecording.com.)*

bucks than you might think. I've found that it's truly a joy to have your edit window, mixer, effects sections, and transport controls in their own places—all in plain and accessible view.

The Laptop Computer

One of the most amazing characteristics of the digital age is miniaturization. At the forefront of the studio-a-go-go movement is the laptop computer (Figures 6.14 and 6.15). With the dawning of smaller, lighter, and more powerful notebooks has come the technological Phoenix of the portable DAW. USB and FireWire® audio interfaces, controllers, and other peripheral devices are now literally capable of handling most (if not all) of the edit and processing functions that can be handled in the studio. In fact, these AC/battery-powered systems have become powerful

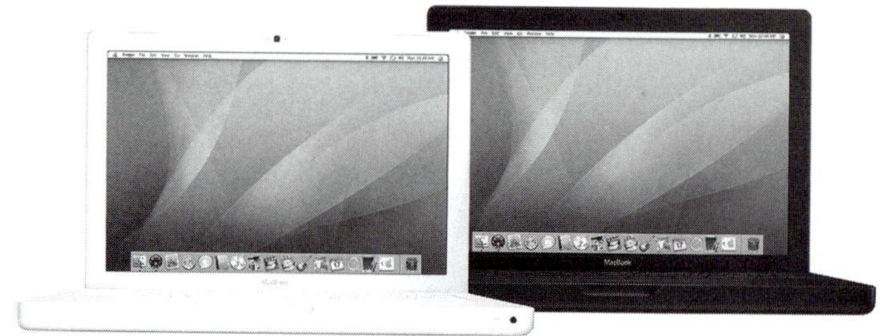

Figure 6.15. *The MacBook™. (Courtesy of Apple Computers, Inc.; www.apple.com.)*

enough to handle advanced DAW edit/mixing functions and can happily handle a wide range of plug-in effects and virtual instruments, all in the comfort of … virtually anywhere!

That's the good news! Now, the downside of all this portability is the fact that, since laptops are optimized to run off of a battery with as little power drain as possible, their:

- Processors will often (but not always) run slower.
- The BIOS (the important subconscious brains of a computer) might be different (especially with regards to battery-saving features).
- Hard drives might not spin as fast.
- Video display capabilities are sometimes limited when compared to a desktop.
- The internal audio interface usually isn't so great.

Although the central processing unit (CPU) will often run slower (usually to reduce power consumption in the form of heat), most modern laptops are more than powerful enough to perform on the road. For this reason, it's always best to get a system with the fastest CPU that you can afford.

Most often, the primary problems with a laptop lie in their basic BIOS and OS battery-saving features. When it comes to making music with a laptop, there actually is a real difference between the Mac® and a PC. Basically, there's little difference between a Mac® laptop and a Mac® tower, as the BIOSs are virtually identical. Conversely, the BIOS of a laptop PC is often limited in power and functional capabilities when compared to a desktop. When shopping for a PC laptop, it's often good to research how a particular BIOS chipset will work for music, particularly with regard to certain audio interface devices (some interfaces won't work well or at all with certain chipsets).

Both PC and Mac® laptops have an automatic power-saving feature (called "speed step" and "processor cycling," respectively) that changes the CPU's speed in much the same way that a vehicle changes gears in order to save energy. Often these gear changes can wreak havoc on many of the DSP functions of an audio workstation. Turning them off will greatly improve performance, at the expense of having a reduced battery life.

Hard-drive speeds on a laptop are often limited when compared to a desktop computer, resulting in slower access times and fewer track counts on a multitrack DAW. Even though these speeds are often more than adequate for general music applications, speeds can often be improved through the use of an external FireWire® drive.

Again, when it comes to RAM, it's often a good idea to pack as much into the laptop as you can. This will reduce data swapping to disk when using larger audio applications (especially software synths and samplers). Within certain systems, the laptop's video card capabilities will run off of the system's RAM (which can severely limit audio processing functions in a DAW). Memory-related problems can also crop up when using a motherboard to run a dual-monitor (LCD and external monitor) setup.

It almost goes without saying that the internal audio quality of most laptops ranges from being acceptable to abysmal. As a result, about the only true choice is to find an external audio interface that works best for you and your applications. Fortunately, there are a ton of audio interface choices for either FireWire® or USB, ranging from a simple stereo I/O device to those that include multitrack audio, MIDI, and controller capabilities in a small, on-the-go package.

In addition to being a studio-on-the-go, the laptop can act as an expansion module for a desktop setup. You could trigger a loaded software synth or sampler via MIDI or through the use of VST System Link (a Steinberg product that allows multiple computers to act as a single, connected system via a network connection), and any number of effects, instruments, and VST software devices can work in unison with the main DAW.

The Audio Interface

An important device that deserves careful consideration when putting together a DAW-based production system is the digital audio interface. These devices can have a single, dedicated purpose, or they might be multifunctional in nature—in either case, their main purpose in the studio is to act as a connectivity bridge between the outside world of analog audio and the computer's inner world of digital audio (Figures 6.16 through 6.18). Audio interfaces come in all shapes, sizes, and functionalities; for example, an audio interface can be:

- Built into a computer (although, more often than not, these devices are often limited in quality and functionality)
- A simple, two-I/O audio device
- Multichannel, offering eight analog I/Os and numerous I/O expansion options
- Fitted with one or more MIDI I/O ports
- One that offers digital I/O, wordclock, and sync options
- Fitted with a controller surface (with or without motorized faders) that provides for hands-on DAW operation

These devices may be designed as hardware cards that fit directly into the computer, or they might plug into the system via USB or FireWire®. An interface might have as few as two inputs and two outputs, or it might have as many as 24. It might offer a limited number of sample-rate and bit-depth options, or it might be capable of handling rates up to 96 kHz/24 bits or higher. Unless you buy a system that's been designed to operate with a specific piece of hardware (most

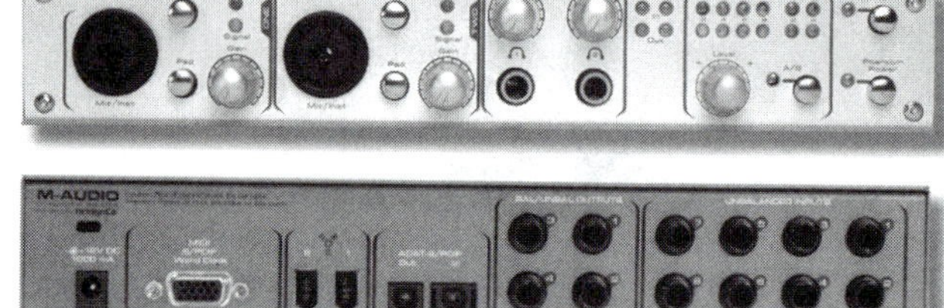

Figure 6.16. FireWire 1884 FireWire® audio interface. (Courtesy of M-Audio, A Division of Avid Technology, Inc.; www.m-audio.com.)

Figure 6.17. Mackie Onyx 1200f studio recording preamp and 192 k FireWire® interface. (Courtesy of Loud Technologies, Inc.; www.mackie.com.)

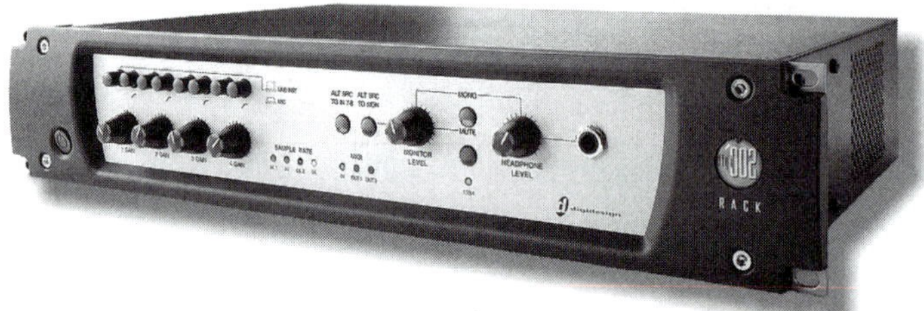

Figure 6.18. Digidesign Digi 002 rack audio interface. (Courtesy of Digidesign, A Division of Avid; www.digidesign.com.)

notably, those offered by Digidesign), you should weigh the vast number of interface options and capabilities with patience and care—the options could easily affect your future expansion choices.

Audio Driver Protocols

Audio driver protocols are software programs that set standards for allowing data to be communicated between the system's software and hardware. A few of the more common protocols are:

- WDM—The Windows® Driver Model is a robust driver implementation that's directly supported by Microsoft® Windows®. Software and hardware that conform to this basic standard can communicate audio to and from the computer's basic audio ports.
- ASIO—The Audio Stream Input/Output architecture forms the backbone of VST. It does this by supporting variable bit depths and sample rates, multichannel operation, and synchronization. This commonly used protocol offers low latency, high performance, easy set-up, and stable audio recording within VST.
- MAS—The MOTU Audio System is a system extension for the Mac® that uses an existing CPU to accomplish multitrack audio recording, mixer, bussing, and real-time effects processing.

- CoreAudio—This driver allows compatible single-client, multichannel applications to record and play back through most audio interfaces using Mac® OS X. It supports full-duplex recording and playback of 16-/24-bit audio at sample rates up to 96 kHz (depending on your hardware and CoreAudio client application).

In most circumstances, it won't be necessary for you to be familiar with the protocols—you just need to be sure that your software and hardware are compatible for use with a driver protocol that works best for you. Of course, further reading can always be found at the respective company's website.

Latency

When discussing the audio interface as a production tool, it's important that we touch on the issue of *latency*. Quite literally, latency refers to the buildup of delays (measured in milliseconds) in audio signals as they pass through the audio circuitry of the audio interface, CPU, internal mixing structure, and I/O routing chains. When monitoring a signal directly through a computer's signal path, latency can be experienced as short delays between the input and monitored signal. If the delays are excessive, they can be unsettling enough to throw a performer off time. For example, when recording a synth track, you might actually hear the delayed monitor sound shortly after hitting the keys (not a happy prospect). Because we now have faster computers, improved audio drivers, and better programming, latency has been reduced to such low levels that it's not even unnoticeable. For example, the latency of a standard Windows® audio driver can be truly pitiful (upward to 500 ms). By switching to a supported ASIO driver and by optimizing the interface/DAW buffer settings to their lowest operating size (without causing the audio to stutter), these delay values can be reduced down to an unnoticeable range.

Wordclock

One aspect of digital audio recording that never seems to get enough attention is the need for synchronization at the sample level within a series of interconnected digital audio devices (such as integrating an audio interface with a digital mixer). In order to reduce such gremlins as clicks, pops, and jitter (oh my!), it's often necessary to lock the overall sample rate timing of the devices to a single master clock signal (so the conversion sample and hold states for all digital audio channels and devices will occur at exactly the same point in time) through the use of a single timing reference known as *wordclock*.

As an example, let's assume that we're in a room that has four or five clocks, and none of them shows the same time! In places like this, you never quite know what the time really is—the clocks could be running at different speeds or at the same speed but are set to different times. Basically, trying to accurately keep track of the time while simultaneously looking at all of the clocks would end up being a jumbled nightmare. On the other hand, if all of these clocks were locked to a single, master clock (remember those self-correcting clocks that are installed in most schools?) keeping track of the time (even when moving from room to room) would be much simpler. In effect, wordclock operates in a similar fashion. If the sample clock (the timing reference that determines the sample rate and DSP traffic control) for each device were set to operate in a freewheeling, internal fashion, the timing references of each device within the connected digital audio chain wouldn't accurately match up. Even though the devices are all running at the

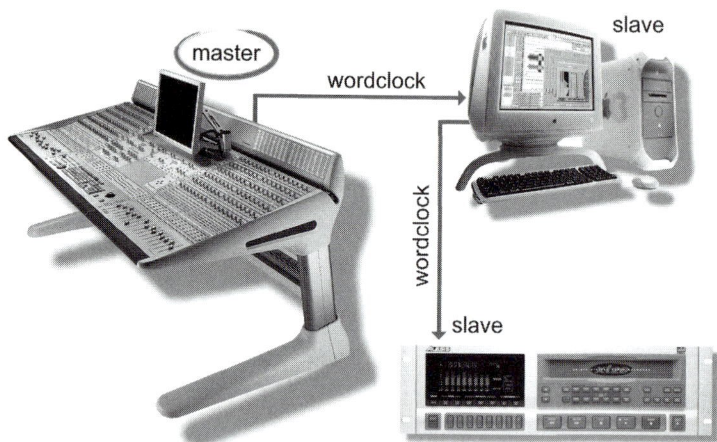

Figure 6.19. Example of a wordclock distribution chain showing that there can only be one master clock.

same sample rate, these resulting mismatches in time will often result in clicks, ticks, excessive jitter, and other unwanted distortions. In order to correct this, the internal clocks of all the digital devices within a connected chain must then be referenced to a single master wordclock timing element (Figure 6.19).

Similar to the distribution of time code, there can only be one master wordclock reference within a connected digital distribution network. This reference source can be derived from a digital mixer, soundcard, or any desired source that can transmit wordclock. This reference pulse can be chained between the involved devices through the use of BNC and/or RCA connectors and low-capacitance cables (often 75-ohm, video-grade coax cable is used, although this cable grade isn't always necessary when shorter cable runs are used). It's interesting to note that wordclock isn't generally needed when making a digital copy from one device to another (via such protocols as AES, S/PDIF, MADI, or TDIF2), as the timing information is actually embedded within the data bitstream itself. Only when we begin to connect devices that share and communicate digital data will we see the immediate need for wordclock.

There will often be differences in connections and parameter setups from one system to the next. In addition to the need for proper cabling and impedance termination considerations throughout the network, specific hardware and software setups may be required to get all of the device blocks to communicate. In order to better understand your particular system's setup (and to keep frustration to a minimum), it's always a good idea to keep your device manuals close at hand.

DAW Controllers

One of the more common complaints against the digital audio editor and workstation environment (particularly when relating to the use of on-screen mixers) is the lack of a hardware controller that gives the user access to hands-on controls. In recent years, this has been addressed by major manufacturers and third-party companies in the form of a hardware DAW controller interface (Figures 6.20 through 6.22).

These controllers generally mimic the design of an audio mixer in that they offer slide or rotary gain faders, pan pots, solo/mute, and channel select buttons—with the added bonus of a full

Figure 6.20. *Mackie Control Universal DAW controller. (Courtesy of Loud Technologies, Inc.; www.mackie.com.)*

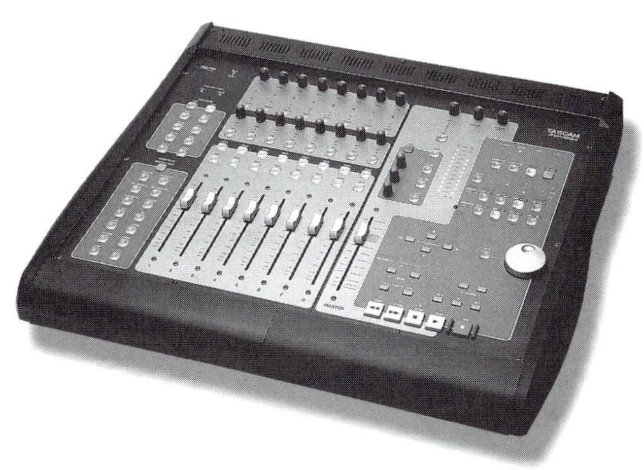

Figure 6.21. *Tascam FW-1884 FireWire® audio/MIDI interface and control surface. (Courtesy of Tascam, www.tascam.com.)*

Figure 6.22. *Digidesign ICON integrated controller/console. (Courtesy of Digidesign, A Division of Avid Technology, Inc.; www.digidesign.com.)*

transport remote. A channel select button is used to actively assign a specific channel to a section that contains a series of grouped pots and switches that relate to EQ, effects, and dynamic functions. These controllers offer direct mixing control over eight input strips at a time. By switching between the banks in groups of 8 (1–8, 9–16, 17–24, ...), any number of the grouped inputs can be accessed by the virtual mixer. These devices will also often include software function keys that can be programmed to give quick and easy access to the DAW's more commonly used program keys.

Controller commands are most commonly transmitted between the controller and audio editor via device-specific MIDI SysEx messages. Therefore, in order to be able to integrate a controller into your system, the DAW's current version must be specifically programmed to accept the control codes from a particular controller, unless the DAW and controller make use of a new plug-in architecture that allows compatible devices to be freely connected. Most controller surfaces communicate these messages to the DAW host via the easy-to-use USB or FireWire® protocols.

Certain controllers also offer all-in-one capabilities that can be straightforward and cost-effective devices for first-time buyers. Often, these devices include a multichannel audio interface, MIDI interface ports, monitor capabilities, and full controller functions. Others may already have an existing digital mixer that can actually be used as a fully functional controller (and in certain circumstances, as a multichannel audio interface) when connected to a DAW host program. For these and other reasons, taking the time to research your needs and current equipment capabilities can save you both time and money.

> It's important to note that there are numerous controllers from which to choose—and just because others feel that the mouse is cumbersome doesn't mean that you have to feel that way; for example, I have several controllers in my own studio, but the mouse is still my favorite tool. As always, the choice is totally up to you.

Sound File Formats

An amazingly varied number of sound file formats exist within audio and multimedia production. Here is a list of the most commonly used audio production formats that don't use data compression:

- Wave (.wav)—The Microsoft Windows format supports both mono and stereo files at a variety of bit and sample rates. WAV files contain PCM coded audio (uncompressed pulse code modulation formatted data) that follows the Resource Information File Format (RIFF) spec, which allows extra user information to be embedded and saved within the file itself.
- Broadcast wave (.wav)—In terms of audio content, broadcast wave files are the same as regular wave files; however, text strings for supplying additional information can be imbedded in the file according to a standardized data format.

- Wave64 (.w64)—This proprietary format was developed by Sonic Foundry, Inc. (now operating under the Sony name). In terms of audio quality, Wave64 files are identical to wave files except that their file headers use 64-bit values whereas wave uses 32-bit values. As a result, Wave64 files can be considerably larger than standard wave files, and this format is a good choice for long recordings (*e.g.*, surround files and file sizes over 2 GB).
- Apple AIFF (.aif or .snd)—This standard sound file format from Apple® supports mono or stereo, 8-bit or 16-bit audio at a wide range of sample rates. Like broadcast wave files, AIFF files can contain embedded text strings.
- Sound Designer I and II (.sd and .sd2)—Digidesign uses Sound Designer as a sound file format. SDI was first released in 1985 and can still be found on many CD-ROM and sound file discs; it was primarily used to store 16-bit, mono samples of short duration (often on the order of seconds). In its latest incarnation, SDII can encode 16- or 24-bit sound files of any practical length at a variety of sample rates.

Format Interchange and Compatibility

At the sound file level, most software editors and DAWs are able to read a wide range of uncompressed and compressed formats, which can then be saved into a new format. At the session level, there are several standards that allow for the exchange of data for an entire session from one platform, OS, or hardware device to another. These include the following:

- The Advanced Authoring Format (AAF; www.aafassociation.org) is a multimedia file format that's used to exchange digital media and metadata between different systems and applications across multiple platforms. Designed by the top media software companies, this format will greatly help media creators of all types by allowing them to exchange projects between applications without losing valuable metadata such as fades, automation, and processing information.
- Open Media Framework Interchange (OMFI) is a platform-independent session file format intended for the transfer of digital media between different DAW applications; it is saved with an .omf file extension. OMF (as it is commonly called) can be saved in either of two ways: (1) "export all to one file," when the OMF file includes all of the sound files and session references that are included in the session (be prepared for this file to be extremely large), or (2) "export media file references," when the OMF file will not contain the sound files themselves but will contain all of the session's region, edit, and mix settings; effects (relating to the receiving DAW's available plug-ins and ability to translate effects routing); and I/O settings. This second type of file will be small by comparison; however, the original sound files must be transferred into the session folders.
- Developed by the Audio Engineering Society, the AES31 standard is an open file interchange format that was designed to overcome format incompatibility issues between different software and hardware systems. Transferred files will retain event positions, mix settings, fades, etc. AES31 makes use of Microsoft's FAT32 file system, with broadcast wave as the default audio file format. This means that an AES31 file can be transferred to

any DAW that supports AES31, regardless of the type of hardware and software used, as long as the workstation can read the FAT32 file system, broadcast wave, or regular wave files.

- OpenTL is a file exchange format that was developed for Tascam hard-disk recording systems. An imported OpenTL project file will contain all audio files and edits that were made within the Tascam system, with all events positioned correctly in the project window. Conversely, a session can be edited and then exported to a disk in the OpenTL format, making it possible to transfer all edits and audio files back to the Tascam hard-disk device.

Sound File Sample Rates

The sample rate of a recorded bitstream directly relates to the resolution at which a recorded sound will be digitally captured. Just as with a moving image, if you take more samples of the image as it moves through time, you'll have a more accurate representation of that recorded image. If the number of samples is too low, the resolution will be substandard and lossy. On the other hand, too high of a rate might result in a recorded bandwidth that's so high that the gain in resolution is lost on the audience's ability to discriminate it. Or, the storage requirements might become so great that the files become inordinately large. Beyond the basic adherence to certain industry sample rate standards, such are the choices and personal decisions that must be made regarding which is the best sample rate to use on a project. Although other sample-rate standards exist, the following are the most commonly used in the professional, project, and general audio production community:

- 32 k—This rate is often used by broadcasters to transmit and receive digital data via satellite. With its overall 15-kHz bandwidth and reduced data requirements, certain devices also use it in order to conserve on memory. Although this rate isn't generally used by the professional community, it's surprising just how good a sound can be captured at 32 k, given a high-quality converter.
- 44.1 k—The long-time standard of consumer and pro audio production, 44.1 is the chosen rate of the CD-audio standard. With its overall 20-kHz bandwidth, the 44.1-k rate is generally considered to be the minimum sample rate for professional audio production. Assuming that high-quality converters are used, this rate is capable of recording lossless audio while conserving on memory storage requirements.
- 48 k—Widely used in post production for TV, this standard was adopted early on as a standard sample rate for professional audio applications (particularly when referring to hardware digital audio devices).
- 96 k—With the onset of 24-bit recording capabilities, higher-rate and bit-rate recordings have made it feasible for recordings to be encoded at 96 kHz and higher rates (e.g., 24/96). 96 kHz is also the accepted rate for DVD-audio production.
- 192 k—This is also an accepted rate for DVD-audio production.

Sound File Bit Rates

The bit rate of a digitally recorded sound file directly relates to the number of quantization steps that are encoded into the bitstream. As a result, the bit rate (or bit depth) is directly correlated to the:

- Accuracy at which a sampled level (at one point in time) is to be encoded
- Signal-to-error figure and thus the overall dynamic range of the recorded signal

Although other bit-rate standards exist, the following are the most commonly used in the professional, project, and general audio production community:

- 16 bits—Like the 44.1-k sound file sample rate, the 16-bit depth is the consumer and professional standard and the chosen bit depth of the CD-audio standard (offering a theoretical dynamic range of 97.8 dB). It is generally considered to be the minimum depth for high-quality professional audio production. Again, when high-quality converters are used, this rate is capable of recording lossless audio while conserving on memory storage requirements.
- 20 bits—Before the 24-bit rate came onto the scene, 20 bits was considered to be the standard for high-bit-depth resolution. Although it is used less commonly, it can still be found in high-definition audio recordings (offering a theoretical dynamic range of 121.8 dB).
- 24 bits—Offering a theoretical dynamic range of 145.8 dB, this standard bit rate is often used in professional audio, high-definition, and DVD audio applications.

Further reading on sound files and compression codecs can be found in Chapter 10. A listing of uncompressed audio bit rate and file sizes can be found in Table 10.2.

DAW Software

Probably one of the strongest playing cards in the modern digital audio deck is the digital audio workstation. By their very nature, DAWs (Figures 6.23 through 6.26) are software programs that integrate with computer hardware and functional applications to create a powerful and flexible audio production environment. These programs commonly offer extensive record, edit, and mixdown facilities through the use of such production tools as:

- Extensive sound file recording, edit, and region definition and placement
- MIDI sequencing and scoring
- Real-time, on-screen mixing
- Real-time effects
- Mixdown and effects automation

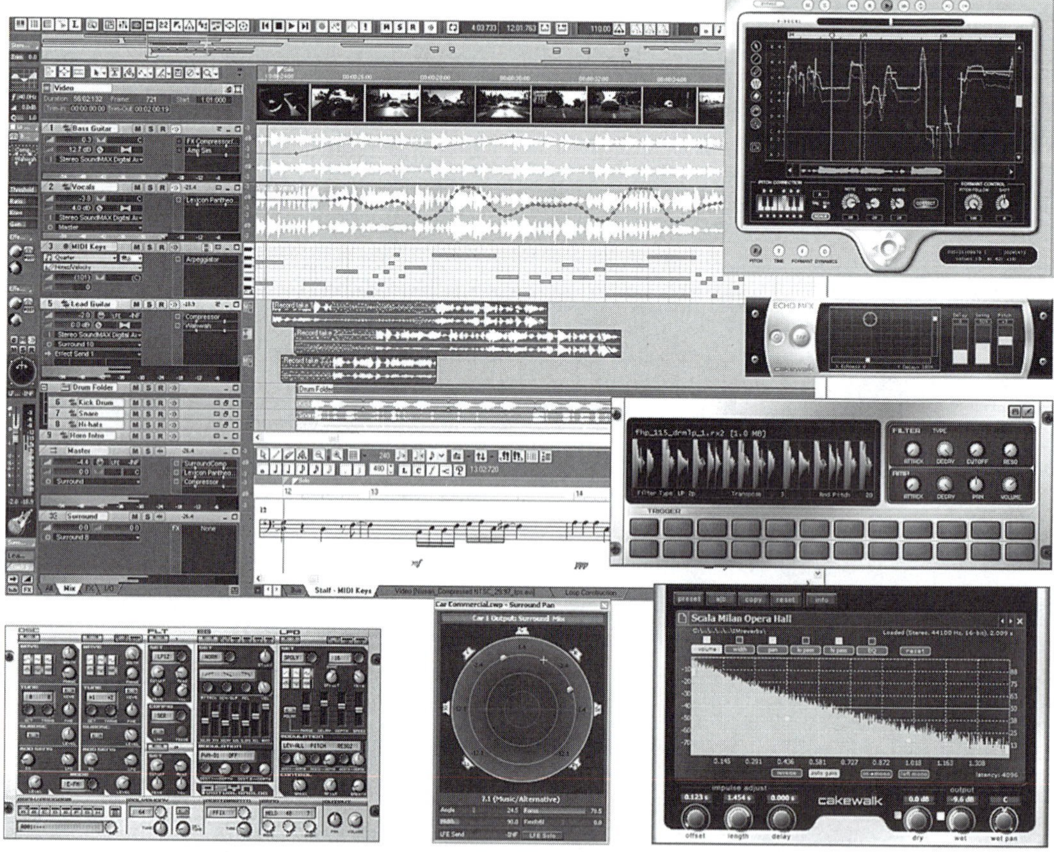

__Figure 6.23.__ Sonar 5 studio edition. (Courtesy of Twelve Tone Systems, www.cakewalk.com.)

- Sound file import/export and mixdown export
- Support for video/picture synchronization
- Systems synchronization
- Audio, MIDI, and sync communications with other audio programs (*e.g.*, ReWire)
- Audio, MIDI, and sync communications with other software instruments (*e.g.*, VST technology)

This list is but a smattering of the functional capabilities that can be offered by an audio production DAW.

Suffice it to say that these powerful software production tools are extremely varied in their form and function. Even with their inherent strengths, quirks, and complexities … their basic look, feel, and operational capabilities have, to some degree, become unified among the major DAW competitors. Having said this, there are enough variations in features, layout, and basic operation that individuals—from aspiring beginner to seasoned professional—will have their favorite

Figure 6.24. *Mark of the Unicorn Digital Performer. (Courtesy of MOTU, Inc.; www.motu.com.)*

Figure 6.25. *Nuendo 3.0 media production DAW. (Courtesy of Steinberg Media Technologies GMBH, www.steinberg.net.)*

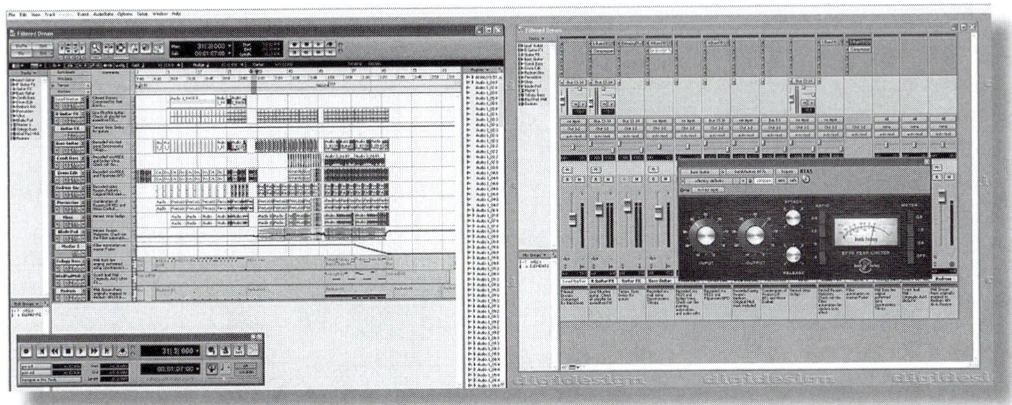

Figure 6.26. *ProTools M-Powered 7 digital audio software. (Courtesy of Digidesign, A Division of Avid Technology, Inc.; www.digidesign.com.)*

DAW make and model. With the growth of the DAW and computer industries, people have begun to customize their computers with features, added power, and peripherals that rival their love for souped-up cars and motorcycles. In the end, though, as with many things in life, it's doesn't matter which type of DAW you use—it's how you use it that counts!

Sound File Recording, Editing, Region Definition, and Placement

Most digital audio workstations are capable of recording sound files in mono, stereo, surround, or multichannel formats (either as individual files or as a single interleaved file). These production environments graphically display sound file information within a main graphic window (Figure 6.27), which contain drawn waveforms that graphically represent the amplitude of a sound file over time in a WYSIWYG (what you see is what you get) fashion. Depending on the system type, sound file length, and the degree of zoom, the entire waveform can be shown on the screen. Or, only a portion will show but will continue to scroll off one or both sides of the screen.

Graphic editing differs greatly from the "razor blade" approach that's used to cut analog tape, in that the waveform gives us both visual and audible cues as to where a precise edit point should

Figure 6.27. *Main edit window within the Cubase SX audio production software. (Courtesy of Steinberg Media Technologies GmbH, A Division of Yamaha Corporation; www.steinberg.net.)*

be. Using this common display technique, any position, cut/copy/paste, gain, and time changes to the waveform will be instantly reflected on the screen. Usually, these edits are nondestructive (a process whereby the original file isn't altered—only the way that the region in/out points are accessed or the file is processed as to gain, spectrum, etc.).

Only when a waveform is zoomed-in fully is it possible to see the individual waveshapes of a sound file (Figure 6.28). At this zoom level, it becomes simple to locate zero-crossing points (points where the level is at the 0, center-level line). In addition, when a sound file is zoomed-in to a level that shows individual sample points, the program might allow the sample points to be redrawn in order to remove potential offenders (such as clicks and pops) or to smooth out amplitude transitions between loops or adjacent regions.

The nondestructive edit capabilities of a DAW refer to a disk-based system's ability to edit a sound file without altering the data that was originally recorded to disk. This important capability means that any number of edits, alterations, or program versions can be performed and saved to disk without altering the original sound file data.

The process of nondestructive editing is accomplished by accessing defined segments of a recorded digital audio file (often called regions) and allowing them to be reproduced in a user-defined order or segment length in a manner other than was originally recorded. In effect, when a specific region is defined, we're telling the program to access the sound file at a point that begins at a specific memory address on the hard disk and continues until the ending address has been reached (Figure 6.29). Once defined, these regions can be inserted into the list (often called a playlist or editlist) in such a way that they can be accessed and reproduced in any order. For example, Figure 6.30 shows a snippet from *Gone With the Wind* that contains the immortal words "Frankly, my dear, I don't give a damn." By segmenting it into three regions we could request that the DAW editor output the words in several ways.

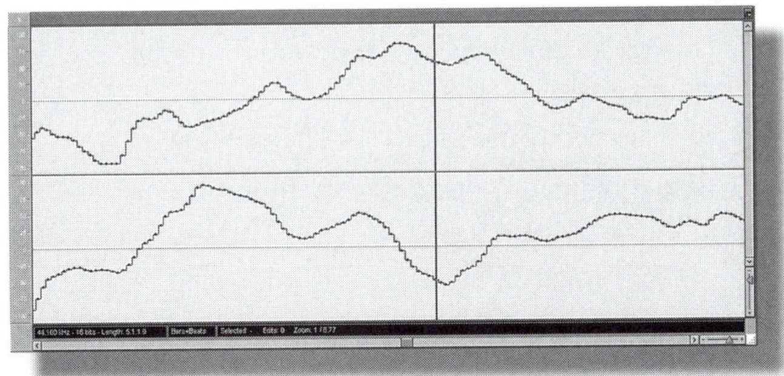

Figure 6.28. *Zoomed-in edit window showing individual samples. (Courtesy of Steinberg Media Technologies GmbH, A Division of Yamaha Corporation; www.steinberg.net.)*

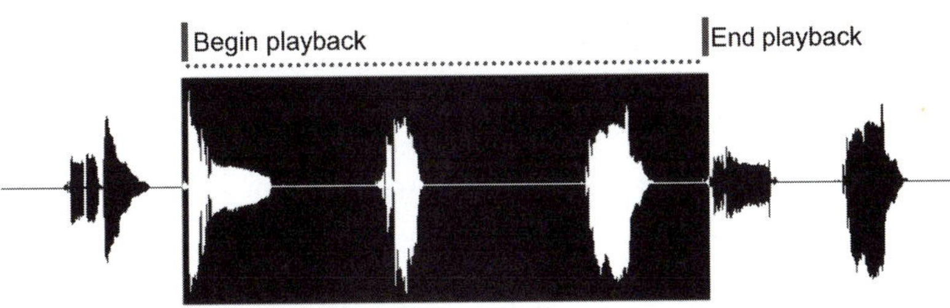

Figure 6.29. *Nondestructive editing allows a region within a larger sound file to begin at a specific point and play until the endpoint is reached.*

Figure 6.30. *Example of how snippets from Rhett's famous Gone with the Wind dialogue can be easily rearranged.*

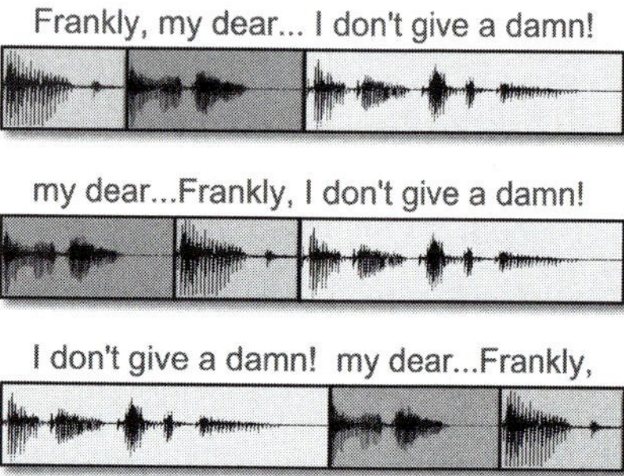
Frankly, my dear... I don't give a damn!
my dear...Frankly, I don't give a damn!
I don't give a damn! my dear...Frankly,

DIY
do it yourself

Recording a Sound File to Disk

- Download a demo copy of your favorite DAW (these are generally available off of the company's website for free).
- Download the workstation's manual and familiarize yourself with its functional operating basics.
- Consult the manual regarding the recording of a sound file.
- Assign a track to an interface input sound source.
- Name the track! It's almost always best to name the track (or tracks) before going into record. In this way, the file will be saved to disk within the session folder under a descriptive name instead of an automatically generated filename (*e.g.*, killerkick.wav instead of track16-01.wav).
- Save the session and assign the input to another track and overdub a track along with the previously recorded track.
- Repeat as necessary until you're having fun!
- Save your final results for the next tutorial.

When working in a graphic editing environment, regions can usually be defined by positioning the cursor over the waveform, pressing and holding the mouse or trackball button, and then dragging the cursor to the left or right, which highlights the selected region for easy identification. After the region has been defined, it can be edited, marked, named, maimed, or otherwise processed.

As one might expect, the basic cut and paste techniques used in hard-disk recording are entirely analogous to those used in a word processor or other graphics-based programs:

- Cut—Places the highlighted region into clipboard memory and deletes the selected data (Figure 6.31)

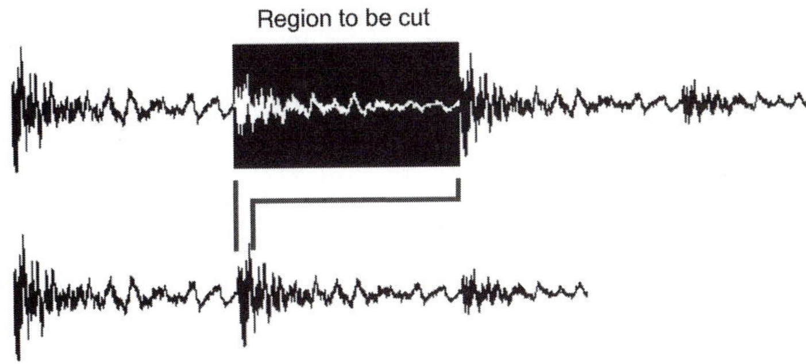

Figure 6.31. *Cutting inserts the highlighted region into memory and deletes the selected data.*

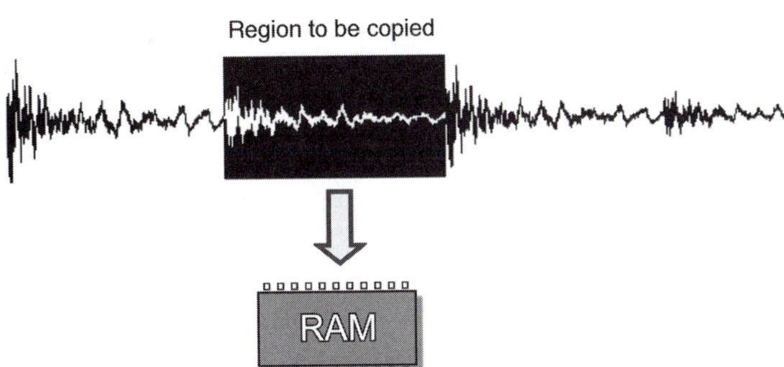

Figure 6.32. *Copying simply places the highlighted region into memory without changing the selected waveform in any way.*

- Copy—Places the highlighted region into memory and doesn't alter the selected waveform in any way (Figure 6.32)
- Paste—Copies the waveform data that's within the system's clipboard memory into the sound file beginning at the current cursor position (Figure 6.33)

Copy and Paste

- Open the session from the above tutorial
- Consult your editor's manual regarding basic cut and paste commands (which are almost always the standard PC and Mac® commands).
- Open a sound file and define a region that includes a musical phrase, lyric, or sentence.
- Cut the region and try to paste it into another point in the sound file in a way that makes sense (musical or otherwise).
- Feel free to cut, copy, and paste to your heart's desire to create an interesting or totally wacky sound file.

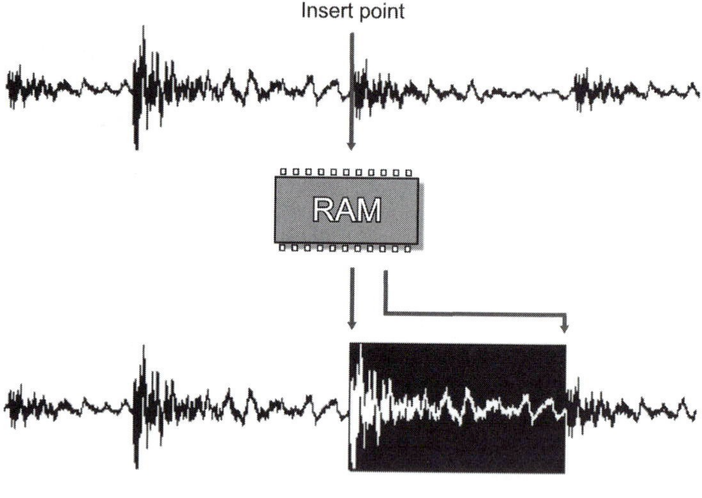

Figure 6.33. Pasting copies the data within the system's clipboard memory into the sound file at the current cursor position.

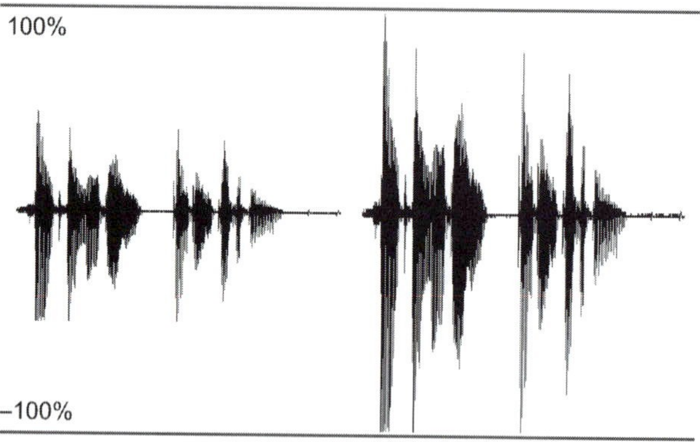

Figure 6.34. Original signal and normalized signal level.

Besides basic nondestructive cut and paste editing techniques, the amplitude processing of a signal is one of the most common types of changes that are likely to be encountered. These include such processes as gain changing, normalization, and fading.

Gain changing relates to the altering of a region or track's overall amplitude level, such that a signal can be proportionally increased or reduced to a specified level (often in dB or percentage value). In order to increase a sound file or region's overall level, a function known as *normalization* can be used. Normalization (Figure 6.34) refers to an overall change in a sound file or defined region's signal level, whereby the file's greatest amplitude will be set to 100% = full scale (or a set percentage level of full scale), with all other levels in the sound file or region being proportionately changed in gain level.

The fading of a region (either in or out) is accomplished by increasing or reducing a signal's relative amplitude over the course of a defined duration; for example, fading in a file (Figure 6.35a) proportionately increases a region's gain from infinity (zero) to full gain. Likewise, a fade-out (Figure 6.35b) has the opposite effect of creating a transition from full gain to infinity. These

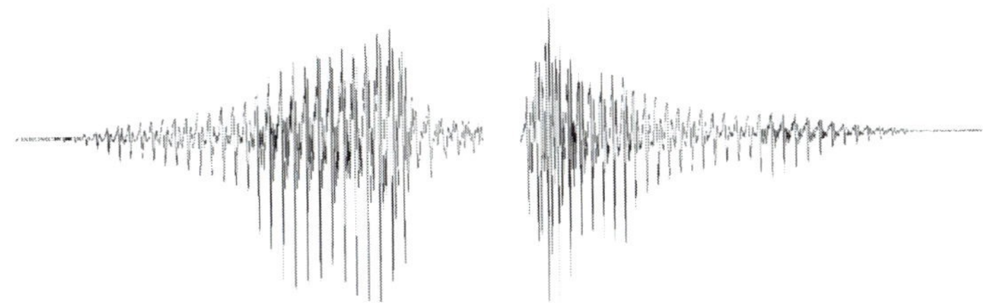

Figure 6.35. Examples of fade-in and fade-out curves.

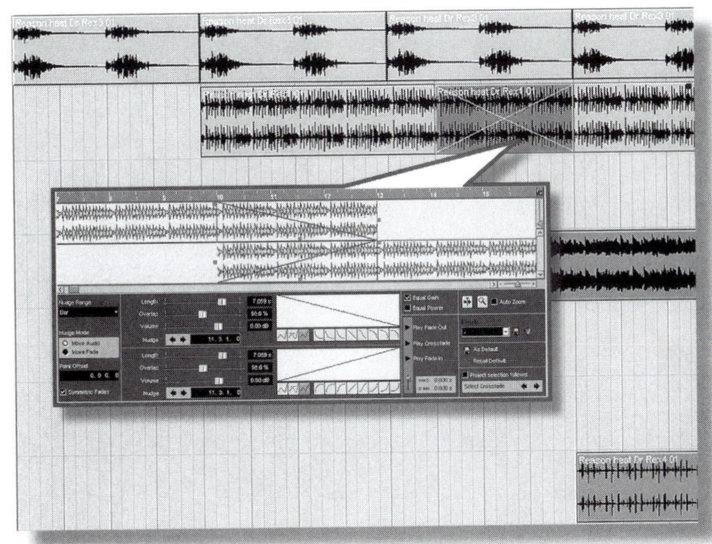

Figure 6.36. Example of a cross-faded sound file. (Courtesy of Steinberg Media Technologies GmbH, A Division of Yamaha Corporation; www.steinberg.net.)

DSP functions have the advantage of creating a much smoother transition than would otherwise be humanly possible when performing a manual fade.

A cross-fade (or X-fade) is often used to smooth the transition between two audio segments that either are sonically dissimilar or don't match in amplitude at a particular edit point (a condition that would otherwise lead to an audible "click" or "pop"). This functional tool basically overlaps a fade-in and fade-out between the two waveforms to create a smooth transition from one segment to the next (Figure 6.36). Technically, this process averages the amplitude of the signals over a user-definable length of time in order to mask the offending edit point.

Real-Time, On-Screen Mixing

In addition to their ability to offer extensive region edit and definition, one of the most powerful cost- and time-effective features of a digital audio workstation is its ability to offer on-screen mixing (Figures 6.37 and 6.38). Essentially, most DAWs include a digital mixer interface that offers most, if not all, of the capabilities that are offered by larger analog and/or digital consoles that are far more cost and space prohibitive. In addition to the basic input strip fader, pan, solo/mute, and select controls, most DAW software mixers offer extensive support for EQ, effects plug-ins (offering a

Figure 6.37. *Nuendo 3 on-screen mixer. (Courtesy of Steinberg Media Technologies GmbH, A Division of Yamaha Corporation; www.steinberg.net.)*

Figure 6.38. *ProTools M-Powered on-screen mixer. (Courtesy of Digidesign, A Division of Avid Technology, Inc.; www.digidesign.com.)*

staggering amount of DSP flexibility), spatial positioning (pan and possibly surround-sound positioning), total automation (both mixer and plug-in automation), external mix, function and transport control from a supported external hardware controller, support for exporting a mixdown to a file ... the list goes on and on and on. Further reading on the process of mixing audio can be found in Chapter 12.

DSP Effects

In addition to being able to cut, copy, and paste regions within a sound file, it's also possible to alter a sound file, track or segment using digital signal processing techniques. In short, DSP works by directly altering the samples of a sound file or defined region according to a program algorithm (a set of programmed instructions) in order to achieve a desired result. These processing functions can be performed either in real time or non-real time (offline):

- Real-time DSP—Commonly used in most modern-day DAW systems, this process makes use of the computer's CPU or additional acceleration hardware to perform complex DSP calculations during actual playback. Because no calculations are written to disk in an offline fashion, significant savings in time and disk space can be realized when working with productions that involve complex or long processing events. In addition, the automation instructions for real-time processing are imbedded within the saved session file, allowing any effect or set of parameters to be changed, undone, and redone without affecting the original sound file.
- Non-real-time DSP—Using this method, signal processing (such as changes in level, EQ, dynamics, or reverb) that is too calculation intensive to be carried out during playback will be calculated (in an offline fashion). In this way, the newly calculated file (containing the effect, sub-mix, etc.) will be played back, without having to use up the extra resources that are now available to the CPU for other functions. DAWs will often have a specific term for tracks or processing functions that have been written to disk, such as "locking" or "freezing" a file. When DSP is performed in non-real time, its almost always wise to save both the original and the effected sound files, just in case you need to make changes at a later time.

Most DAWs offer an extensive array of DSP options, ranging from options that are built into the basic I/O path of the input strip (*e.g.*, basic EQ and gain-related functions) to DSP effects and plug-ins that come bundled with the DAW package to third-party effects plug-ins that can be either inserted directly into the signal path (direct insertion) or offered as a master effect path that numerous tracks can be assigned to and/or mixed into (side chain).

Although the way that effects are implemented in a DAW will vary from one make and model to the next, the basic fundamentals will be much the same. The following notes describe but a few of the possible effects that can be plugged into the signal path of DAW; however, further reading on effects processing can be found in Chapter 12 ("Mixing and Automation").

- *Equalization*—EQ is, of course, a feature that's often implemented at the basic level of a virtual input strip (Figures 6.39 and 6.40). Most systems give full parametric control over

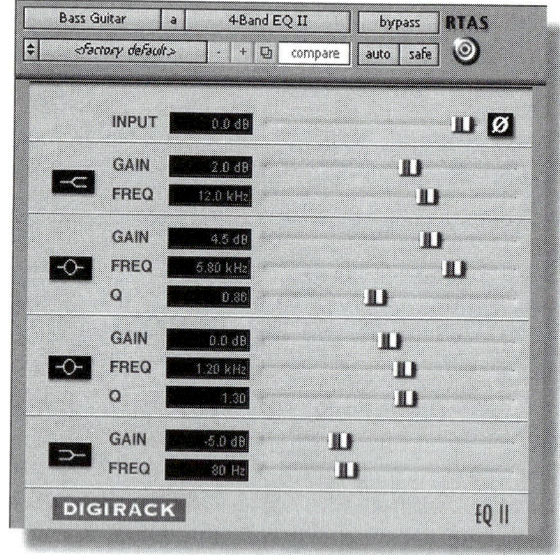

Figure 6.39. *4-Band EQII plug-in for Pro Tools. (Courtesy of Digidesign, A Division of Avid Technology, Inc.; www.digidesign.com.)*

Figure 6.40. *EQ plug-in for Cubase SX/Nuendo. (Courtesy of Steinberg Media Technologies GmbH, A Division of Yamaha Corporation; www.steinberg.net.)*

the entire audible range, offering overlapping control over several bands with a variable degree of bandwidth control (Q). Beyond the basic EQ options, many third-party EQ plug-ins are available on the market that vary in complexity, musicality, and market appeal (Figures 6.41 and 6.42).

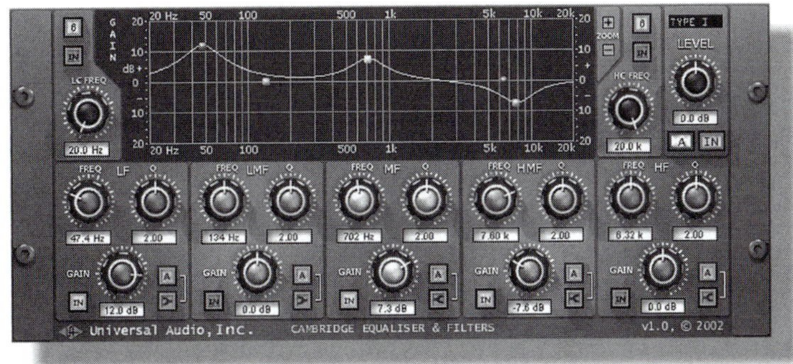

Figure 6.41. Cambridge equalizer and filters for the UAD-1 effects processing card. (Courtesy of Universal Audio, www.uaudio.com.)

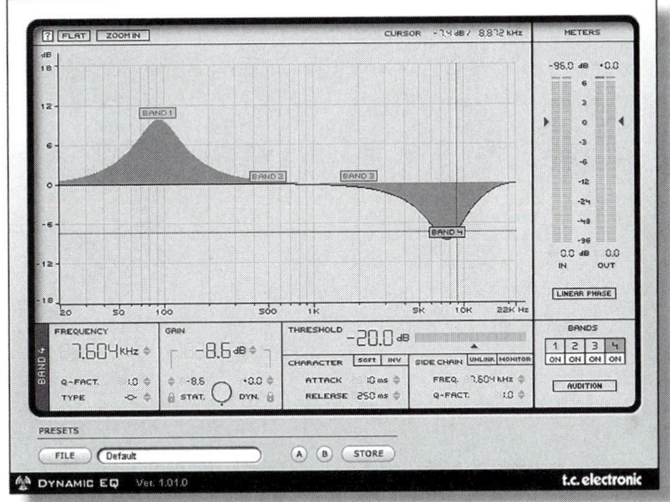

Figure 6.42. Dynamic EQ for the Powercore DSP processor and plug-in system. (Courtesy of TC Electronic A/S, www.tcelectronic.com.)

Figure 6.43. Precision Limiter for the UAD-1 effects processing card. (Courtesy of Universal Audio, www.uaudio.com.)

- *Dynamic range*—Dynamic range processors (Figures 6.43 and 6.44) can be used to change the signal level of a program. Processing algorithms are available that emulate a compressor (a device that reduces gain by a ratio that's proportionate to the input signal), limiter (reduces gain at a fixed ratio above a certain input threshold), or expander (increase the overall dynamic range of a program). These gain changers can be inserted directly into a track, used as a grouped master effect, or inserted into the final output path for use as a master gain processing block.

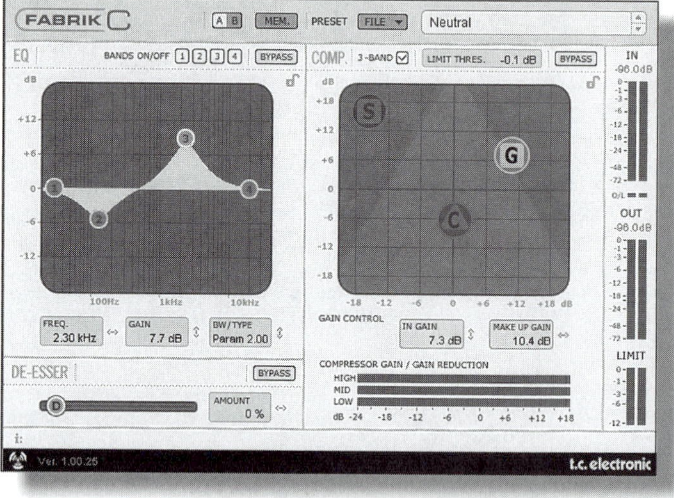

Figure 6.44. *Fabric C for the Powercore DSP processor and plug-in system. (Courtesy of TC Electronic A/S, www.tcelectronic.com.)*

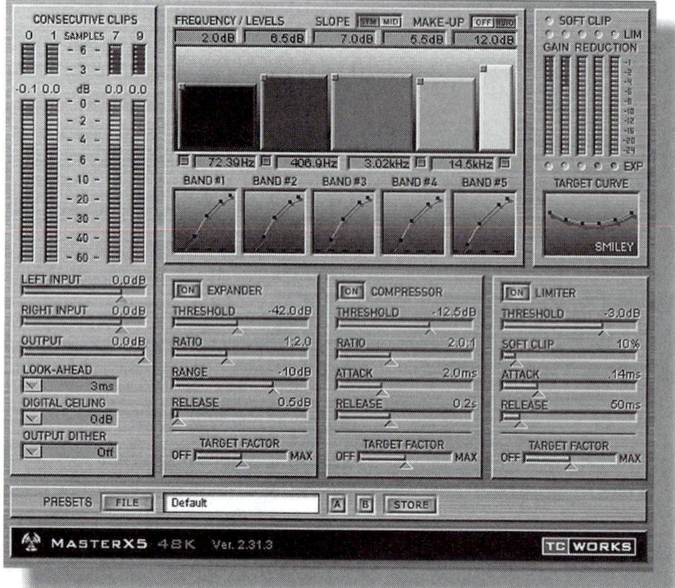

Figure 6.45. *MasterX5 48 k multiband dynamics for the Powercore DSP processor and plug-in system. (Courtesy of TC Electronic A/S, www.tcelectronic.com.)*

In addition to the basic complement of dynamic range processors, wide assortments of multiband dynamic plug-in processors (Figure 6.45) are available for general and mastering DSP applications. These processors allow the overall frequency range to be broken down into various frequency bands. For example, a plug-in such as this could be inserted into a DAW's main output path which allows the lows to be compressed while the mids are lightly limited and the highs are de-essed to reduce sibilance.

- *Delay*—Another important effects category that can be used to alter and/or augment a signal revolves around delays and regeneration of sound over time. These time-based effects use

Figure 6.46. DoubleDelay VST plug-in. (Courtesy of Steinberg Media Technologies GmbH, A Division of Yamaha Corporation; www.steinberg.net.)

Figure 6.47. DM-1 Delay Modulator for the UAD-1 effects processing card. (Courtesy of Universal Audio, www.uaudio.com.)

delay (Figures 6.46 and 6.47) to add a perceived depth to a signal or change the way that we perceive the dimensional space of a recorded sound. A wide range of time-based plug-in effects exist that are all based on the use of delay (and/or regenerated delay) to achieve such results as:

- Delay
- Chorus
- Flanging
- Reverberation

Further reading on the subject of delay (and the subject of signal processing in general) can be found in Chapter 12 ("Mixing and Automation").

- *Pitch and time change*—Pitch change functions make it possible to shift the relative pitch of a defined region or track either up or down by a specific percentage ratio or musical interval. Most systems can shift the pitch of a sound file or defined region by determining a ratio between the current and the desired pitch and then adding (lower pitch) or dropping (raise pitch) samples from the existing region or sound file. In addition to raising or lowering a sound file's relative pitch, most systems can combine variable sample rate and pitch shift techniques to alter the duration of a region or track. These pitch- and time-shift combinations make it possible for such changes as:

  - *Pitch shift only*—A program's pitch can be changed while recalculating the file so that its length remains the same.

- *Change duration only*—A program's length can be changed while shifting the pitch so that it matches that of the original program.
- *Change in both pitch and duration*—A program's pitch can be changed while also having a corresponding change in length.

When combined with shifts in time (delay), changes in pitch make it possible for a multitude of effects to be created (such as flanging, which results from random fluctuations in delay and time shifts that are mixed with the original signal to create an ethereal "phasey" kind of sound).

DSP Plug-Ins

Workstations often offer a number of DSP effects that come bundled with the program; however, a staggering range of third-party plug-in effects can be inserted into a signal path which perform functions for any number of tasks ranging from the straightforward to the wild 'n' zany. These effects can be programmed to seamlessly integrate into a host DAW application that conforms to such plug-in platforms as:

- DirectX—A DSP platform for the PC that offers plug-in support for sound, music, graphics (gaming), and network applications running under Microsoft® Windows® (in its various OS incarnations)
- AU (Audio Units)—Developed by Apple® for audio and MIDI technologies in OS X; allows for a more advanced GUI and audio interface
- VST (Virtual Studio Technology)—A native plug-in format created by Steinberg for use on either a PC or Mac®; all functions of a VST effect processor or instrument are directly controllable and automatable from the host program
- MAS (MOTU Audio System)—A real-time native plug-in format for the Mac® that was created by Mark of the Unicorn as a proprietary plug-in format for Digital Performer; MAS plug-ins are fully automatable and do not require external DSP in order to work with the host program
- AudioSuite—A file-based plug-in that destructively applies an effect to a defined segment or entire sound file, meaning that a new, effected version of the file is rewritten in order to conserve on the processor's DSP overhead (when applying AudioSuite, it's often wise to apply effects to a copy of the original file so as to allow for future changes)
- RTAS (Real-Time Audio Suite)—A fully automatable plug-in format that was designed for Digidesign's Pro Tools LE; available on Digi ToolBox and Digi 001 (any system with Pro Tools LE) and runs on the power of the host CPU (host-based processing) on either the Mac® or PC
- TDM (Time Domain Multiplex)—A plug-in format that can only be used with Digidesign Pro Tools systems (Mac® or PC) that are fitted with Digidesign Farm cards; this 24-bit, 256-channel path integrates mixing and real-time digital signal processing into the system with zero latency and under full automation

These popular software applications (which are being programmed by major manufacturers and third-party startups alike) have helped to shape the face of hard-disk recording by allowing us to pick and choose the plug-ins that best fit our personal production needs. As a result, new companies, ideas, and task-oriented products are constantly popping up on the market, literally on a monthly basis.

ReWire

ReWire and ReWire2 are special protocols that were developed by Propellerhead Software and Steinberg to allow audio to be streamed between two simultaneously running computer applications. Unlike a plug-in, where a task-specific application is inserted into a compatible host program, ReWire allows the audio and timing elements of an independent client program to be seamlessly integrated into another host program. In essence, ReWire provides virtual patch chords (Figure 6.48) that link the two programs together within the computer. A few of ReWire's supporting features include:

- Real-time streaming of up to 64 separate audio channels (256 with ReWire2) at full bandwidth from one program into its host program application
- Automatic, sample accurate synchronization between the audio in the two programs
- An ability to allow the two programs to share a single sound card or interface
- Linked transport controls that can be controlled from either program (provided it has some kind of transport functionality)
- An ability to allow numerous MIDI outs to be routed from the host program to the linked application (when using ReWire2)
- A reduction of the total number of system requirements that would be required if the programs were run independently

This useful protocol essentially allows a compatible program to be plugged into a host program in a tandem fashion. As an example, ReWire could allow Propellerhead's Reason (client) to be "ReWired" into Steinberg's Cubase DAW (host), allowing all MIDI functions to pass through

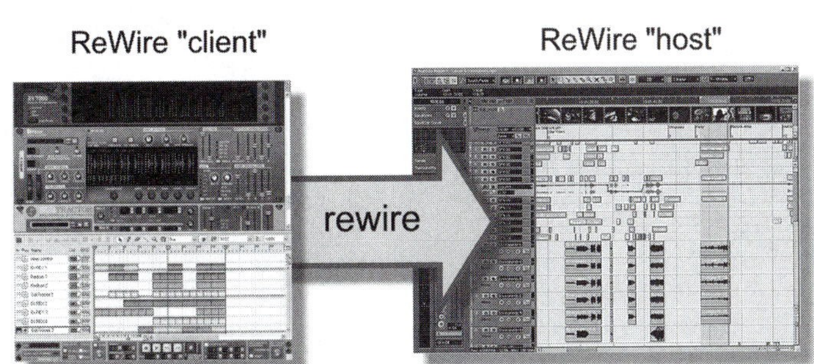

Figure 6.48. *ReWire allows a client program to be inserted into a host program (often a DAW) so they can run simultaneously in tandem.*

Figure 6.49. *The UAD-1 DSP and plug-in card. (Courtesy of Universal Audio, www.uaudio.com.)*

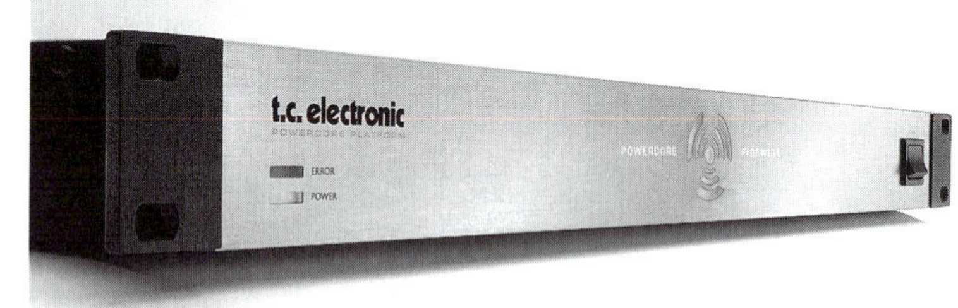

Figure 6.50. *Powercore FireWire® DSP processor and plug-in system. (Courtesy of TC Electronic A/S, www.tcelectronic.com.)*

Cubase into Reason, while patching the audio outs of Reason into Cubase's virtual mixer inputs. For further information on this useful protocol, consult the supporting program manuals.

Accelerator Cards

In most circumstances, the CPU of a host DAW program will have sufficient power and speed to perform all of the DSP effects and processing needs of a project. Under extreme production conditions, however, the CPU might run out of computing steam and choke during real-time playback. Under these conditions, there are a couple of ways to reduce the workload on a CPU: On one hand, the tracks could be "frozen," meaning that the processing functions would be calculated in non-real time and then written to disk as a separate file. On the other hand, an accelerator card (Figures 6.49 and 6.50) that's capable of adding extra CPU power can be added to the system, giving the system extra processing power to perform the necessary effects calculations. It should be noted that in order for the plug-ins to take advantage of the accelerator card they need to be specially coded for that specific DSP card.

Power to the Processor ... Uhhh, People!

Speaking of having enough power and speed to get the job done, there are definitely some tips and tricks that can help you get the most out of your digital audio workstation. Let's take a look at some of the more important items. It's important to keep in mind that keeping up with technology can have its triumphs and its pitfalls. No matter which platform you chose to work with, there's no substitute for reading, research, and talking with your peers about your techno needs. It's generally best to strike a balance between our needs, our desires, the current state of technology, and the relentless push of marketing to grab your money. It's usually best to take a few big breaths before making any important decisions.

1 Get a Computer That's Powerful Enough

With the increased demand for higher bit/sample rate resolution, more tracks, more plug-ins, more of everything, you'll obviously want to make sure that your computer is fast and powerful enough to get the job done in real-time without spitting and sputtering digits. This often means getting the most up-to-date and powerful computer/processor system that your budget can reasonably muster. With the advent of 64-bit OS platforms and dual-core processors (chips that effectively contain two CPUs), you'll want to make sure that your hardware will support these features before making the upgrade plunge. The same goes for your production software and driver availability. If any part of this hardware, software, and driver equation is missing, the system will not be able to make use of these advances. Therefore, one of the smartest things that you can do is to research the type and system requirements that would be needed to operate your production system—and then make sure that your system exceeds these figures by a comfortable margin so as to make allowances for future technological advances and the additional processing requirements that are associated with them. If you have the budget to add some of the extra bells and whistles that go with living on the cutting edge, you should take the time to research whether or not your system will actually be able to deliver the extra goods when these advances actually hit the streets.

2 Make Sure You Have Enough Memory

It almost goes without saying that your system will need to have an adequate amount of random access memory (RAM) and hard disk storage in order for you to take full advantage of your processor's potential and your system's data storage requirements. RAM is used as a temporary storage area for data that is being processed and passed to and from the computer's central processing unit (CPU). Just as there's a "need for speed" within the computer's CPU, the general guidelines for RAM ask that we install memory with the fastest transfer speed that can be supported by the computer. Obviously, it's important that you install as much memory as your computer and budget will allow. Installing too little RAM will force the OS to write this temporary data to and from the hard disk—a process that's *much* slower than transfer to RAM and causes the system's overall performance to slow to a crawl. For those who are making extensive use of virtual sampling technology (whereby samples are transferred to RAM), it's usually a wise idea to throw as much RAM as is possible into the system. It should be noted that although PC computers and Windows® XP allow for up to 4Gb RAM installs, making use of more than 2 Gb can actually be tricky or not may not be accessible at all. Reading and research should be undertaken when attempting to add large amounts of RAM to a system.

Hard-disk requirements for a system are certainly an important consideration. The general considerations include:

- *Need for size*—Obviously, you'll want to have drives that are large enough to meet your production storage needs. With the use of numerous tracks within a session, often at sample rates of 24 bit/96 K, data storage can quickly become an important consideration.
- *Need for speed*—With the current track count and sample rate requirements that can commonly be encountered in a DAW session, it's easy to understand how lower disk access times (the time that's required for the drive heads to move from one place to another on a disk and then output that data) becomes important.

It's interesting to note that, in general, the disk drives with larger storage capacities will often have shorter access times and a higher data transfer rate. These large-capacity drives will often spin at speeds that are at least 7200 rpm, will have access times that are 10 ms or faster, and are able to sustain a transfer rate of 8 to 10 MB/s or faster.

(3) Keep Your Production Media Separate

Whenever possible, it's important to keep your program and operating system data on a separate drive from the one that holds your production data.. This is due to the simple fact that a computer periodically has to check in and interact with both the currently running program and the OS. Should the production media be on the same disk, periodic interruptions in audio data will occur as the disk takes a moment out to go perform a program-related task (and *vice versa*), resulting in a reduction in media and program data access time, as well as throughput. For these reasons, it's also wise to keep the program and media drives off of the same data cable and onto separate data paths.

(4) Update Your Drivers!

In this day and age of software revisions, it's *always* a good idea to go on the web and search for the latest update to a piece of software or a hardware driver. Even if you've just bought a product new out of the box, it might have easily have been sitting on a music store shelf for over a year. By going to the company website and downloading the latest versions, you'll be assured that it has the latest and greatest capabilities. In addition, it's always wise to save these updates to disk in your backup directories. In this way, if you're outstanding in your field and there's a hardware or software problem, you'll be able to reload the software or drivers and be on your way in no time.

(5) Going Dual Monitor

Q: How do you fit the easy visual reference of multiple programs, documents, and a digital audio workstation into a single 17-inch space?

A: You don't. Those of you who rely on your computer for recording and mixing, surfin', writing, etc., should definitely think about doubling your computer's visual real estate by adding an extra monitor to your computer system (Figure 6.51).

Folks who have never seen or thought much about adding a second monitor might be skeptical and ask, "What's the big deal?" But, all you have to do is sit down and start opening programs to see just how fast your screen can get filled up. When using a complicated production program

Figure 6.51. Dual screen … the only way to go!

(such as a DAW or a high-end graphics app), getting the job done with a single monitor can be an exercise in total frustration. There's just too much we need to see and not enough screen real estate to show it on. The ability to quickly make a change on a virtual mixer, edit window, or software plug-in has become more of a necessity than a luxury.

Truth is, in this age of OS X for the Mac® and Windows® XP, adding an extra monitor is a fairly straightforward proposition. You could either spring for a dual-head graphics card that can support two monitors or you can visit your favorite computer store and buy a second video card. Getting hold of a second monitor could be as simple as grabbing an unused one from the attic, buying a standard 17- or 19-inch CRT model, or springing for a new LCD monitor that matches your current screen.

Once you've installed the hardware, the software side of building a dual-monitor system is relatively straightforward. For example, in Windows® XP, right-clicking on the main desktop and selecting Properties/Settings will call up a window that lets you alter the resolution settings for each monitor. Checking the "Extend my Windows desktop into this monitor" will extend your desktop across both monitors and you should be well on your way.

Those of you who use a laptop can also enjoy many of these benefits by plugging the second monitor into the video out and following the setup steps that are recommended by your computer's operating system. You should be aware that many laptops are limited in the way they share video memory and might be restricted in the resolution levels that can be selected.

This might not seem much like a recording tip, but once you get a dual monitor system going your whole approach to producing content (of any type) on a computer will instantly change and you'll quickly wonder how you ever got along without it!

(6) Keeping Your Computer Quiet

Noise! Noise! Noise! It's everywhere! It's in the streets, in the car, and even in our studios. It seems like we spend all those bucks getting the best sound possible, only to gunk it all up by placing this big computer box that's full of noisy fans and whirring hard drives smack in the middle of a critical listening area. Fortunately, a number of companies have begun to find ways to reduce the problem. Here are a few solutions:

- Whenever possible, use larger, low-rpm fans to reduce noise.
- Certain PC motherboards come bundled with a fan speed utility that can monitor the CPU and case heat and adjust the fan speeds accordingly.
- Route your internal case cables carefully. They could block the flow of air, which can add to heat and noise problems.
- A growing number of hard-disk drives are available as *quiet drives*. Check the manufacturer's noise ratings.
- You might consider placing the computer in a well-ventilated area, just outside the production room. Always pay special attention to ventilation (both inside and outside the computer box), as heat is a killer that'll reduce the lifespan of your computer. (*Note:* When building my own studio I designed a special alcove/glass door enclosure that houses my main computer—no muss, no fuss, and very little noise)
- Thanks to gamers and audio-aware buyers, a number of companies exist that specialize in quiet computer cases and components. These are always fun to check out.

Backup and Archive Strategies

It's pretty much always true that it's not a matter of *if* a hard drive will fail but *when*. It's not a matter of *if* something will happen to an irreplaceable hard disk but *when*. When we least expect it, disaster will strike. It's our job to be prepared for the inevitable. This type of headache can of course be partially or completely averted by backing up your active program and media files, as well as by archiving your previously created sessions and then making sure that these files are also backed up.

As was previously stated, it's generally wise to keep your computer's operating system and program data on a separate hard disk (usually the boot drive) and then store your session files on a separate media drive. Let's take this as a practical and important starting point. Beyond this premise, as most of you are quite aware, the basic rules of hard-disk management are extremely personal—and will often differ from one computer user to the next (Figure 6.52). Given these differences, I'd still like to offer up some basic guidelines:

- It's important to keep your data (of all types) well organized, using a system that's both logical and easy to follow. For example, online updates of a program or hardware driver downloads can be placed into their own directories; data relating to your studio can be placed in the "studio" directory and subdirectories; documents, MP3s, and all the trappings of day-to-day studio operations can be also placed on the disk, using a system that's easy to understand.
- Session data should likewise be logical and easy to find. Each project should reside in its own directory on a separate media drive, and each song should likewise reside in its own subdirectory of the session project directory.

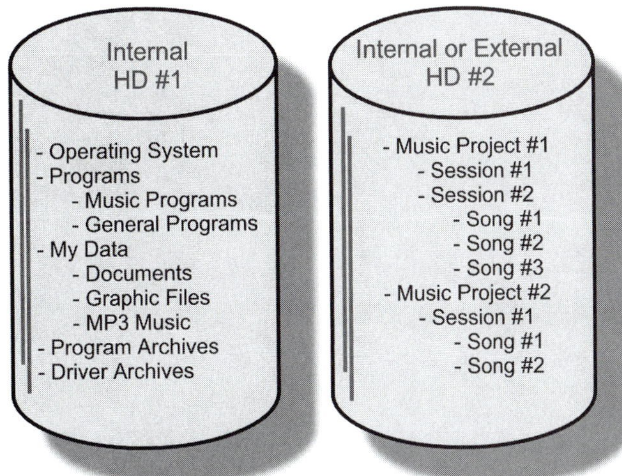

Figure 6.52. *Data and hard-drive management (along with a good backup scheme) are extremely important facets of media production.*

- Remember to save various take versions of a mix. If you just added the vocals to a song, go ahead and save the session under a new version name. This acts as an "undo" function that let's you go back to a specific point in a session. The same goes for mixdown versions. If someone likes a particular mix version or effect, go ahead and save the mix under a new name or version number (my greatest song 1–15.ses) or (my greatest song 1–15—favorite effect.ses). In fact, it's generally a good idea to save versions throughout the mix. The session files are usually small and might save your butt at a later point in time.

With regard to backup schemes, a number of options also exist. Although making backups to optical media (CD and DVD) is a common occurrence, it's been found that the most robust and long-lived backups are those that reside on a hard drive. There's also a saying: "Media isn't truly backed up unless it's copied in three different places." All I can do is lay out a general backup formula that has worked extremely well for me over the years:

- *CD/DVD*—I start off with the general premise that CD and DVD media is temporary media that's to be used for getting media data into the hands of clients, friends, etc.
- *Boot drive*—The boot disk is used to store the operating system data, program data, program- and driver-related downloads, documents, graphics, non-production-related media (such as MP3 music); can also be used as an archive drive for storing all of your past sessions.
- *Second internal or external drive*—Current media drive; can also be used as an archive drive for storing all of your past sessions along with your current sessions.
- *External backup drive 1*—If the disk is large enough, it can be used as a backup drive for both your boot and second media drive. If it isn't large enough, you'll want two backup drives.

Figure 6.53. *External FireWire® or USB2 cases can come in handy, both in the studio and on the go.*

- *External backup drive 2*—If the disk is large enough, it can be used as a second backup drive for both your boot and second media drive. You might strongly consider storing this drive off-site. In the case of a theft, fire, or other unforeseen disaster you'll still be able to recover your precious data.

If all of this sounds like a lot of drives, that's because it can be; however, one method for reducing the number of external drive cases that you'll need is to purchase an external 5-1/2″ FireWire® or USB2 case. If you get two, one case can be fitted with an external CD-RW/DVD-RW drive, while the other can be fitted with a removable IDE hard drive enclosure. If you buy extra drive enclosures (they can be had for a relatively small song), drives can be placed into a case and hot swapped out at will (Figure 6.53).

All of the above are simply suggestions. I rarely give opinions like these in a book; however, they've served me so well and for so long that I had to pass them along.

In Closing

At this time, I'd like to refer you to the section on "Helpful Production Hints" in Chapter 13. I'm doing this in the hope that you'll read this section twice (at least)—particularly the discussion on project preparation, session documentation, and backup and archive strategies. I promise that the time will eventually come when you'll be extremely glad you did.

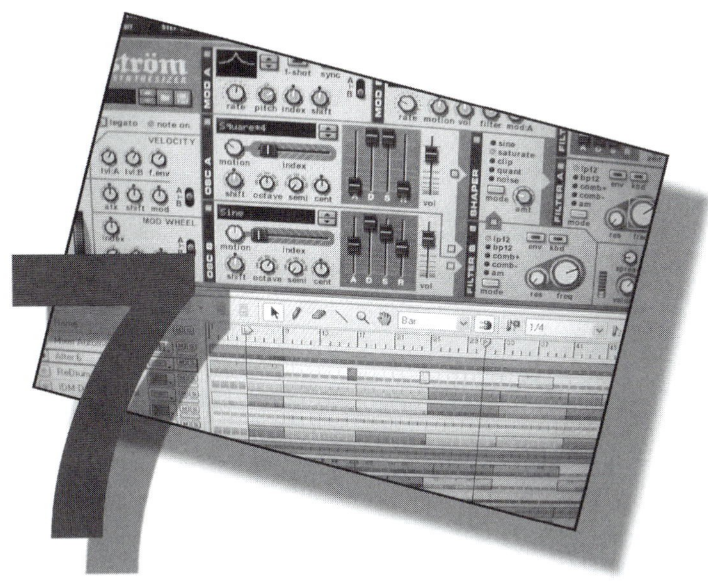

7 Groove Tools and Techniques

The expression "getting into the groove" of a piece of music often refers to a feeling that's derived from the underlying foundation of music—rhythm. With the introduction and maturation of MIDI and digital audio, new and wondrous tools have made their way into the mainstream of music production. These tools that can help us to use technology to forge, fold, mutilate, and create compositions that make direct use of rhythm and other building blocks of music through the use of looping technology. Of course, the cyclic nature of loops can be repeat–repeat–repetitive in nature, but new toys and techniques for looping have injected the notion of flexibility, control, real-time processing, and mixing to new heights that can be used by an artist in wondrously expressive ways.

In this chapter we'll be touching upon many of the approaches and software packages that have evolved (and continue evolve) into what is one of the fastest and most accessible facets of personal music production. It's literally impossible to hit on the finer operational points of these systems; for that, I'll rely upon your motivation and drive to:

- Download many of the software demos that are readily available.
- Delve into their manuals and working tutorials.

- Begin to create your own grooves and songs that can then be integrated with your own music or those of collaborators.

If you do these three things, you'll be shocked and astounded as to how much you've learned. And these experiences can directly translate into skills that'll widen your production horizons and possibly change your music.

The Basics

Because groove-based tools often deal with rhythms and cyclic-based loops, there are a couple of factors that need to be managed:

- Sync
- Tempo and length

The aspect of sync relates to the fact that the various loops in a groove project will need to sync up with each other (or in multiple lengths and timings of each other). It almost goes without saying that multiple loops that are successively or simultaneously triggered must have a synchronous timing relationship with one another—otherwise, it's a jumbled mess of sound.

The next relationship relates to the aspect of tempo. Just as sync is imperative, it's also necessary for the files to be adjusted in pitch (*resampling*) and/or length (*time stretching*), so that they precisely match the currently selected tempo (or are programmed to be a relative multiple of the session's tempo).

Time and Pitch Change Techniques

The process of altering a soundfile to match the current session tempo and to synchronously align them occurs in the software by combining variable sample rates with pitch-shifting techniques. Using these basic digital signal processing (DSP) tools, it's possible to alter a soundfile's duration (varying the length of a program by raising or lowering its playback sample rate) or to alter its relative pitch (either up or down). In this way, three possible combinations of time and pitch change can occur:

- Time change—A program's length can be altered without affecting its pitch.
- Pitch change—A program's length can remain the same while pitch is shifted either up or down.
- Both—Both a program's pitch and length can be altered using resampling techniques.

Such loop-based programs and plug-ins involve the use of recorded sound files that are encoded with headers that include information on their original tempo and length (in both samples and beats). By setting the loop program to a master tempo (or a specific tempo at that point in the song), an audio segment or file can be imported, examined as to sample rate or length, and then recalculated to a new pitch and relative tempo that matches the current session tempo. *Voila*! We now have a defined segment of audio that matches the tempo of all of the other segments, allowing it to play and interact in relative sync with the other defined segments and or loopfiles.

Warping

Newer digital audio workstation (DAW) and loop production tools are capable of altering the playback speed of a soundfile or segment over the course of its duration to match the tempo of a loop to that of the session tempo or to another soundfile. This process, called *warping*, uses various time-shift techniques to match the timing elements of a soundfile by entering hitpoints into the soundfile. These points can be automatically detected or manually placed at percussive transient points or at metric time divisions in the song. Once a hitpoint has been entered, it can be moved in time to a place that matches the timing division marks of another track. By moving this defined point, the software is able to speed or slow the file's playback as it moves from one hitpoint to the next so as to smooth out the relative timing differences, allowing the involved tracks to play back in relative sync with each other.

Certain software DAWs are capable of addressing timing and sync in different ways. Instead of varying the playback speed of various soundfiles so they match the constant tempo of a session … changes in the tempo of a session can be varied to match the timing changes that exist in a recorded track. By placing tempo changes at their appropriate places, the beats per minute (bpm) will be shifted to match timing shifts in the recording. This means that the click will change to reflect the varying recorded tempo—meaning that the MIDI tracks, effects, and instrument timings will automatically be adjusted to the proper session tempo over the course of a song.

Beat Slicing

Most loop-based production tools use a variety of time stretching, pitch-shifting, and format-shifting algorithms to provide for intelligent beat matching in order to alter the length of a loop or defined segment so it matches the timing elements (generally denoted in beats per minute) of the other loops within the session. Another method, called *beat slicing*, makes use of an entirely different process to match the length and timings of a soundfile segment to the session tempo. Rather than changing the speed and pitch of a soundfile, the beat slicing process actually breaks an audio file into a number of small segments (not surprisingly called *slices*) and then changes the length and timing elements of a segment by adding or subtracting time between these slices. This procedure has the advantage of preserving the pitch, timbre, and sound quality of a file but has the potential downside of creating silent breaks in the soundfile.

From this, it's easy to understand why the beat-slicing process often works best on percussive sounds that have silence between the individual hitpoints. This process is usually carried out by detecting the transient events within the loop and then to automatically placing the slices at their appropriate points (often according to user-definable sensitivity and detection controls).

As with so many other things we've discussed, the sky (and your imagination) is the limit when it comes to the tricks and techniques that can be used to match the tempos and various time-/pitch-shift elements between loops, MIDI, and soundfile segments. I strongly urge you to take the time to read the manuals of various loop and DAW software packages to learn more about the actual terms and procedures and then put them into practice. If you take the time, I guarantee that your production skills and your outlook on these tools will be greatly expanded.

For the remainder of this chapter, we'll be looking at several of the more popular groove tools and toys. This is by no means a complete listing, and I recommend that you keep reading the trade magazines, websites, and other resources, as new and exciting technologies come onto the market regularly.

Looping Your DAW

Most digital audio workstations offer various features that can make it possible to incorporate looping into a session, along with other track-based sound, MIDI, and video files. Even when features that could make a specific task much easier aren't available, it's often possible to think through the situation and find work-arounds that can help tackle the problem at hand. For example, a number of workstations allow the beginning point of an edited segment to be manually placed at specific point. And, by simply clicking on and dragging the tail-end of the segment, the soundfile can be time-stretched into the session's proper time relationship (while the pitch remains unchanged). Said another way, if a session has been set to a tempo of 94 bpm, an 88-bpm loop can be imported at a specific measure. Then, by turning on the DAW's snap-to-grid and automatic time-stretch functions, the segment can be time-stretched until it snuggly fits into the session's native tempo (Figure 7.1). Now that the loop fits, it can be manually looped (*i.e.*, copied) to your heart's content.

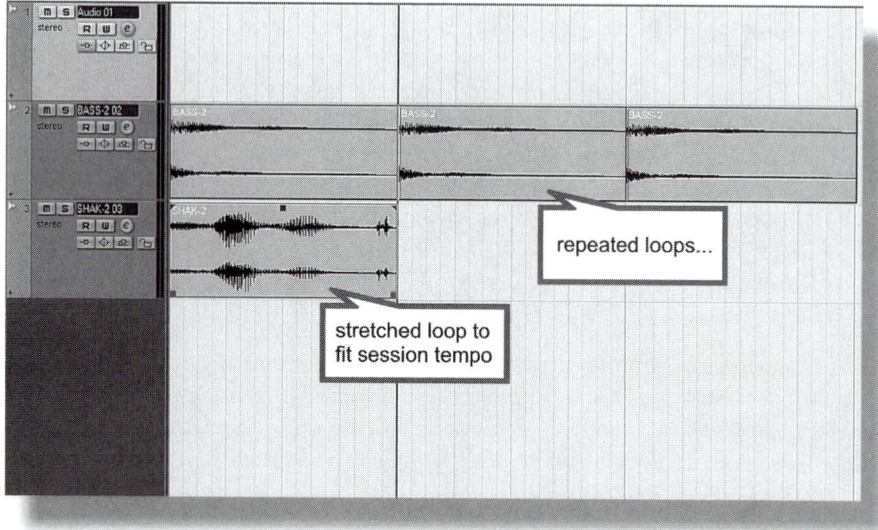

Figure 7.1. *Commonly, a loop or soundfile can be stretched to match the overall session tempo.*

As was said, different DAWs and editing systems will have differing ways of tackling a situation and with varying degrees of ease. One of the best ways to avoid pitfalls is to set your session tempo (or varying tempo map) and click track at the beginning of a session and then manually adjust your loop timings to fit that tempo. Just remember, there are often no hard and fast rules. With planning, ingenuity and your manual's help, you'll be surprised at the number of ways that a looping problem can be turned into an opportunity.

Manual Looping with a DAW

- Go to www.acidplanet.com and download an ACID 8pack set of loops.
- Load the individual loops into your favorite DAW.
- Set the time display to read in bars and beats (tempo).
- Import the loops into the session.
- Try to match the session tempo to a bpm setting that seems to match the 8pack's intended tempo.
- It's possible (but not likely) that the loop lengths won't match up. If this happens, consult your DAW's manual in order to manually time stretch the loops to fit the desired session tempo.
- Copy and duplicate the various loop tracks, until you've made a really great song!
- Save the session to play for your friends!

Loop-Based Audio Software

Loop-based audio editors are groove-driven music programs (Figures 7.2 and 7.3) that are designed to let you drag and drop prerecorded or user-created loops and audio tracks into a graphic multi-track production interface. At their basic level, these programs differ conceptually from their traditional DAW counterpart in that the pitch- and time-shift architecture is so variable and dynamic that, even after the basic rhythmic, percussive, and melodic grooves have been created, their tempo, track patterns, pitch, session key, etc., can be quickly and easily changed at any time. With the help of custom, royalty-free loops (available from many manufacturer and third-party companies), users can quickly and easily experiment with setting up grooves, backing tracks, and creating a sonic ambience by simply dragging the loops into the program's main soundfile view, where they can be arranged, edited, processed, saved and exported.

One of the most interesting aspects of the loop-based editor is its ability to match the tempo of a specially programmed loop to the tempo of the current session. Amazingly enough, this process isn't that difficult to perform, as the program extracts the length, native tempo, and pitch information

Figure 7.2. *Apple GarageBand™. (Courtesy of Apple Computers, Inc.; www.apple.com.)*

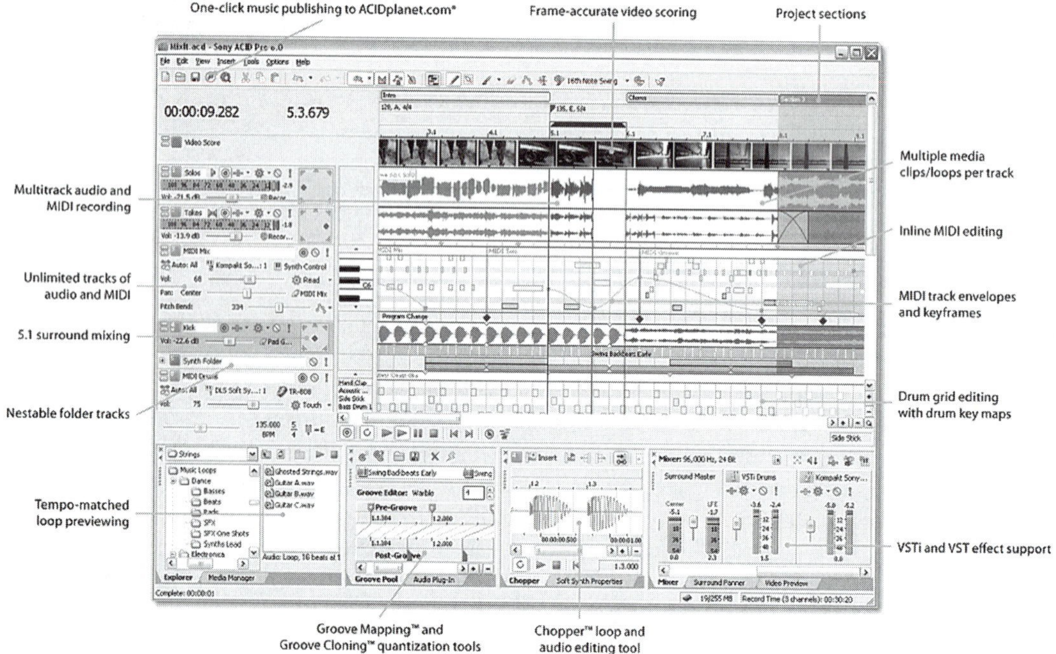

Figure 7.3. *ACID Pro 6 professional music workstation. (Courtesy of Sony Media Software, A Division of Sony Corporation of America; www.sonymediasoftware.com.).*

from the imported file's header and, using various digital time- and pitch-change techniques, adjusts the loop to fit the native time and pitch parameters of the current session. This means that loops of various tempos and musical keys can be automatically adjusted in length and pitch so as to fit in time with previously existing loops … just drag, drop, and go!

Behind the scenes, these shifts in time can be performed in a number of ways. For example, using basic DSP techniques to time stretch and pitch shift, a recorded loop will often work well over a given plus-or-minus percentage range (which is often dependant upon the quality of the program algorithms). Beyond this range, the loop will often begin to distort and become jittery. At such extremes, other playback algorithms and beat slice detection techniques can be used to make the loop sound more natural. For example, drums or percussion can be stretched in time by adding additional silence between the various hitpoints within the loop (slicing), at precisely calculated intervals. In this way, the pitch will remain the same while the length is altered. Of course, such a loop would sound choppy and broken up when played on its own; however, when buried within a mix, it might work just fine. It's all up to you and the current musical context.

Having Fun with a Loop-Based Editor

- Go to www.acidplanet.com and download their free version of ACID Xpress.
- Download an ACID 8pack set of loops.
- Load the individual loops into ACID Xpress (or a loop-based editor that can read ACID files).
- Read the program's manual and begin to experiment with the loops.
- Mess around with the tempo and musical keys.
- Copy and duplicate the various loop tracks to your heart's content until you've made a really great song!
- Save the session to play for your friends!

Of course, the graphic user interfaces (GUIs) between looping software editors and tools can differ greatly. Most layouts use a track-based system that lets you enter or drag a preprogrammed loopfile into a track and then drag it to the right in a way that repeats the loop in a traditional way. Again, it's worth stressing that DAW editors will often include such looping functions that can be basic (requiring manual editing or soundfile processing) or advanced in nature (having any number of automated loop functions).

Other loop programs make use of visual objects that can be triggered and combined into a real-time mix by clicking on an icon or track-based grid. One such program is Live from Ableton Software

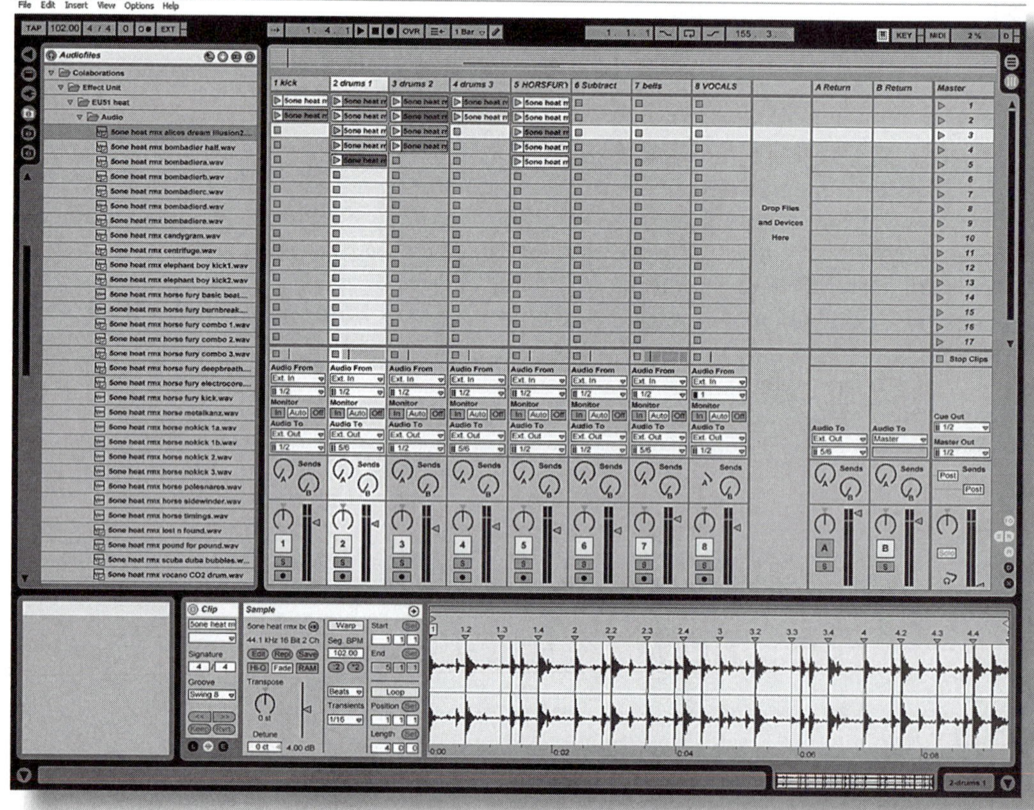

(a)

Figure 7.4. *Ableton Live performance audio workstation: (a) Arrangement View; (b) Session View. (Courtesy of Ableton, www.ableton.com, and M-Audio, A Division of Avid Technology, Inc.; www.m-audio.com.)*

(Figure 7.4). Live is an interactive loop-based program that's capable of recalculating the time, pitch and tempo structure of a soundfile or easily defined segment, and then entering that loop into the session at a defined global tempo. Basically, this means that segment of any length or tempo that's been pulled into the session grid will be recalculated to the master tempo and can be combined, mixed, and processed in perfect sync with all other loops in the project session.

In Live's Arrangement View, as in all traditional sequencing programs, everything happens along a fixed song timeline. The program's Session View breaks this limiting paradigm by allowing mediafiles to be mapped onto a grid as buttons (called *clips*). Any clip can be played at any time and in any order in a random fashion that lends itself to interactive performance both in the studio and on-stage. Functionally, each vertical column, or track, can play only one clip at a time. The horizontal rows are called *scenes*. Any clip can be played in a loop fashion by clicking on its launch button. By clicking on a scene launch at the screen's right, every clip in a row will simultaneously play beyond the wide range of timing, edit, effects, MIDI, and instrument controls. All you need to do to have fun is to start clicking.

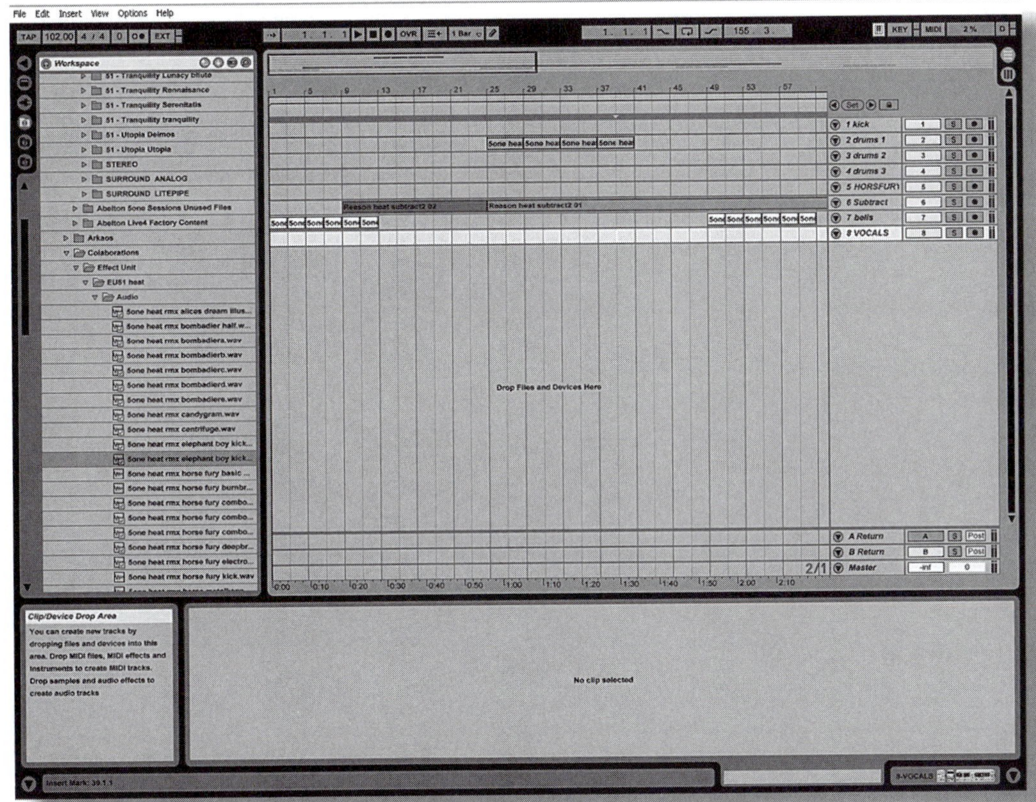

(b)

Figure 7.4. *Continued.*

Of course, most of the looping software packages are capable of incorporating MIDI into a project or live interactive performance. This can be done through the creation of a MIDI sequence or drum pattern track, allowing performance data to be sequenced in a traditional fashion, or by allowing MIDI pattern loops to be mixed in with other media. When a loop DAW or editor is used, external MIDI hardware and various instrument plug-ins can easily play a prominent role in the creative process.

Reason

If there's a software package that has gripped the hearts and minds of electronic musicians in the 21st century, it would have to be Reason from the folks at Propellerheads (Figure 7.5). Reason defies specific classification in that it's an overall music production environment that has many facets. For example, it includes a MIDI sequencer (Figure 7.6, possibly the most widely used sequencer of them all), as well as a wide range of software instrument modules that can be played, mixed, and combined in a comprehensive environment that can be controlled from any external

Figure 7.5. *Reason music production instrument. (Courtesy of Propellerhead Software, www.propellerheads.se.)*

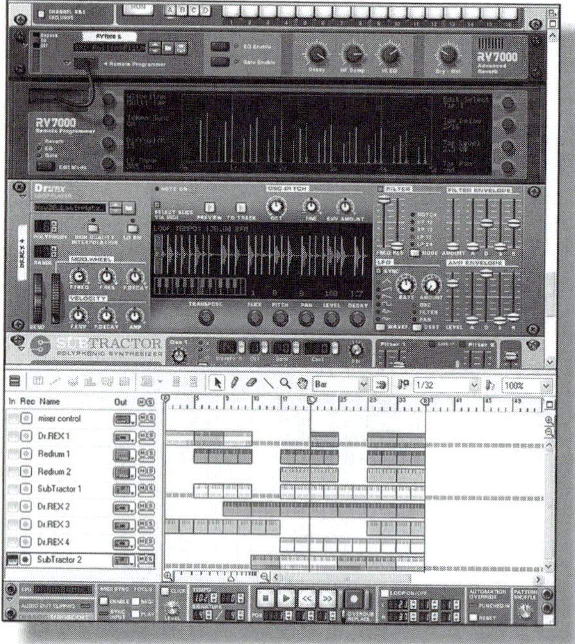

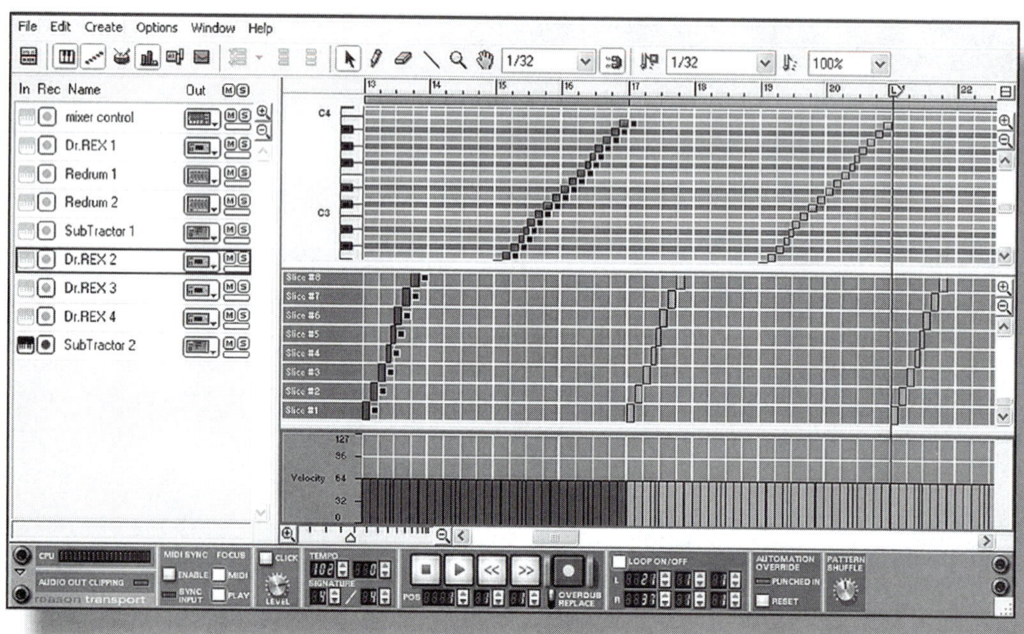

Figure 7.6. *Reason's sequencing window. (Courtesy of Propellerhead Software, www.propellerheads.se.)*

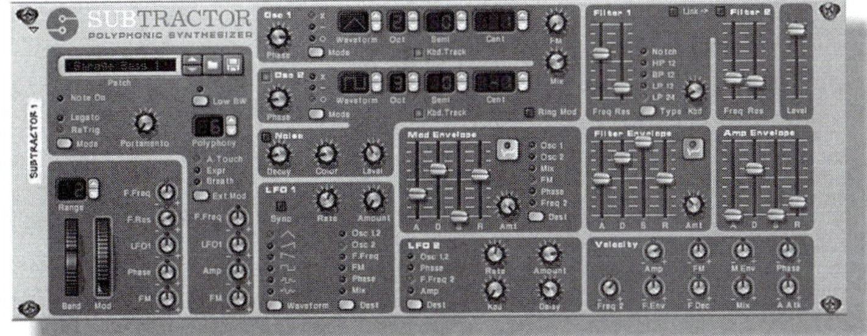

Figure 7.7. *Reason's SubTractor polyphonic synth module. (Courtesy of Propellerhead Software, www.propellerheads.se.)*

Figure 7.8. *Reason's NN-XT sampler module. (Courtesy of Propellerhead Software, www.propellerheads.se.)*

keyboard or MIDI controller. Reason also includes a large number of signal processors that can be applied to any instrument or instrument group, under full and easily controlled automation.

In essence, Reason is a combination of modeled representations of vintage analog synthesis gear, mixed with the latest of digital synthesis and sampling technology (Figures 7.7 and 7.8). Combine these with a modular approach to signal and effects processing, add a generous amount of internal and remote mix and controller management (via an external MIDI controller), top this off with a quirky but powerful sequencer, and you have a software package that's powerful enough for top-flight production and convenient enough that you can build tracks using your laptop in your seat on a crowded plane. I know that it sounds like I read this off of a sales brochure, but these are the basic facts of this program. When asked to explain Reason to others

I'm often at a loss, as the basic structure is so open-ended and flexible that the program can be approached in as many ways as there are people who produce on it. That's not to say that Reason doesn't have a signature sound—it often does; however, it's an amazing tool that can be either used on its own or in combination with other production instruments and tools.

In short, Reason functions by:

- Allowing you to choose a virtual instrument (or combination of instruments)
- Calling up a programmed sample loop or an instrument performance patch
- Allowing these sounds and sequence patterns to be looped and edited in new and unique ways
- Letting you create controller tracks that can vary and process the track in new and innovative ways
- Allowing a wide range of synchronized and controllable effects to be added to the track mix

Of course, once you've finished the outline of one track, the obvious idea is to create new instrument tracks that can be combined in a traditional multitrack building-block approach, until a song begins to form. Since the loops and instruments are built from preprogrammed sounds and patches that reside on the hard disk, the actual Reason file (.rns) is often quite small, making this a perfect production media for publishing your songs on the web or for collaborating with others around the world. Additional sounds, loops, and patches are widely available for sale or for free as "refills" that can be added to your collection to greatly expand the software's palette.

I definitely suggest that you download the demo and take it for a spin. You'll also want to check out the "New to Reason" video clips. Due to its open-ended nature; this program can be a bit hard to grasp and might take a while to get used to, but the journey just might open you up to a whole new world of production possibilities.

ReWire

In Chapter 6, we learned that ReWire and ReWire2 are special protocols that were developed by Propellerhead Software and Steinberg to allow audio to be streamed between two, simultaneously running computer applications. Unlike a plug-in, where a task-specific application is inserted into a compatible host program, ReWire allows the audio and timing elements of a supporting program to be seamlessly integrated into another host program that supports ReWire. For example, both Cubase/Nuendo and Pro Tools are examples of DAWs that support ReWire, while Live and Reason are looping production tools that support the application. As Reason doesn't support the recording of live audio tracks, it could be ReWired into either host DAW, allowing audio and timing elements to be controlled by the host DAW. Likewise, Reason could be plugged into Live, allowing for greatly expanded instrument and production options in a studio or on-stage environment.

If you feel up to the task, download a few program demos, consult the supporting program manuals, and try it out for yourself. The most important rule to remember when using ReWire is that the host program should always be opened first, then the client. When shutting down, the client program should always be closed first.

On a final ReWire note—the virtual instruments within Reason offer up a powerful set of instruments that can be ReWired into the MIDI and performance paths of a DAW. In this way, one or more virtual instruments could be played into the DAW from your controller, allowing you to add more instruments and tonalities to your production. Once you've finished tracking and the MIDI data has been saved to their respective tracks, the ReWired instruments could be exported (bounced) to audio tracks within the DAW—Ta-da!

Groove and Loop Hardware

The software world doesn't actually hold the total patent on looping tools and toys; there are a number of groove keyboards and module boxes on the market. These systems, which range widely in sounds, functionality, and price, can offer up a vast range of unique sounds that can be quite useful for laying a foundation under your production.

In the past, getting a hardware grove tool to sync into a session could be time consuming, frustrating, and problematic, requiring time and tons of manual reading; however, with the advent of powerful time- and pitch-shift processing within most DAWs, the sounds from these hardware devices can be pulled into a session without too much trouble. For example, a single groove loop (or multiple loops) could be recorded into a DAW (at a bpm that's near to the session's tempo), edited, and then imported into the session, at which time the loop could be easily stretched into time sync, allowing it to be looped to your heart's content. Just remember, necessity is the mother of invention—patience and creativity are probably your most important tools in the looping process.

Groove and Loop Plug-ins

Of course, it's a sure bet that for every hardware looping tool, there are far more software plug-in groove tools and toys (Figures 7.9 and 7.10) that can be inserted into your DAW. These amazing software wonders often make life easier by:

- Automatically following the session tempo
- Allowing I/O routing to plug into the DAW's mixer
- Making use of the DAW's automation and external controller capabilities
- Allowing individual or combined groove loops to be imported into a session

These software instruments come in a wide range of sounds and applications that can often be edited and effected using an on-screen user interface. This interface can often be remotely controlled from an external MIDI controller, as shown in Figure 7.11.

Pulling Loops into a DAW Session

When dealing with loops in modern-day production, one concept that needs to be discussed is that of importing loops into a DAW session. As we've seen, it's certainly possible to use the

Figure 7.9. *Stylus RMX real-time groove module. (Courtesy of Spectrasonics, www.spectrasonics.net.)*

Figure 7.10. *Native Instruments' KONTAKT 2 rhythmic groove sampler. (Courtesy of Native Instruments, Inc.; www.nativeinstruments.com.)*

ReWire application to run a supporting client program in conjunction with the host program or even to use an instrument or groove-based plug-in within a compatible session. However, it should be noted that often (but not always) some sort of setup may be required in order to properly configure the application. Or at worst, these applications might use up valuable DSP resources that could otherwise be used for effects and signal processing.

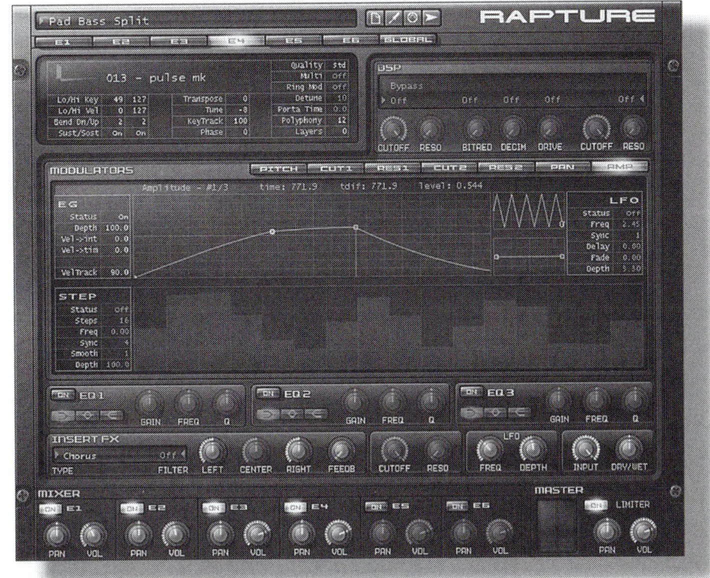

Figure 7.11. Edit screen for Rapture instrument plug-in. (Courtesy of Twelve Tone Systems, Inc.; www.cakewalk.com.)

One of the best options for freeing up these resources is to export the instrument or groove to a new audio track. This can be done in several ways (although with forethought, you might be able to come up with new ways of your own):

- The instrument track can be soloed and exported to an audio track in a contiguous fashion from the beginning "00:00:00:00" of the session to the end of the instrument's performance.
- In the case of a repetitive loop, a defined segment (often of a precise length of 4, 8, or more bars that occurs on the metric boundaries) can be selected for export as a soundfile loop. Once exported to file, it can be imported back into the session and looped into the session.
- In the case of an instrument that has multiple parts or voices, each part can be soloed and exported to it's own track—thus giving a greater degree of control during a mixdown session.

DJ Software

In addition to music production software, there are a growing number of software players, loopers, groovers, effects, and digital devices on the market for the digital DJ of the 21st century. These hardware/software devices make it possible for digital grooves to be created from a laptop, controller, specially fitted turntable, or digital turntable (jog/scratch CD player). Using such hardware-/software-based systems, it's possible to sync scratch and perform with vinyl within a synchronized digital environment with an unprecedented amount of preprogrammable or live performance interactivity that can be used on the floor, on-stage, or in the studio (Figures 7.12 and 7.13).

Figure 7.12. Torq DJ performance/production software with Conectiv 4 × 4 USB DJ interface. (Courtesy of M-Audio, A Division of Avid Technology, Inc.; www.m-audio.com.)

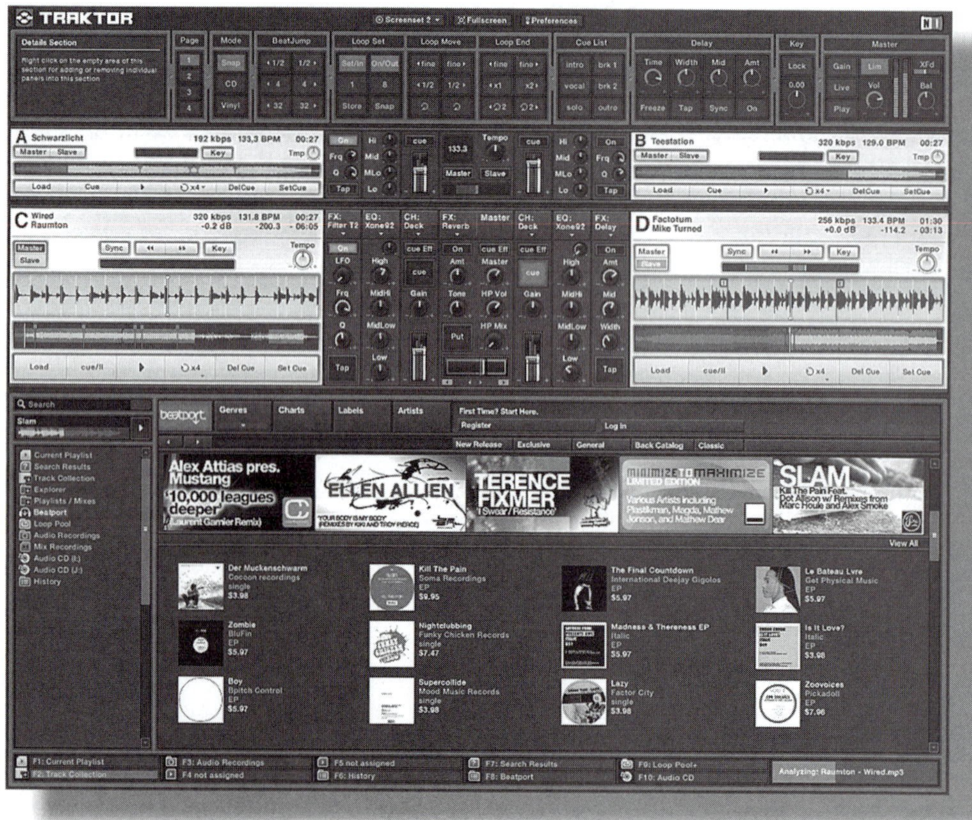

Figure 7.13. Native Instruments' TRAKTOR DJ studio. (Courtesy of Native Instruments, Inc.; www.nativeinstruments.com.)

Obtaining Loopfiles from the Great Digital Wellspring

In this day and age, there's absolutely no shortage of preprogrammed loops that will work with a number of DAWs and groove editors. Some soundfiles will need to be manually edited or programmed to work with your system. Others (such as Sony's ACID format) are already preprogrammed and can be imported into a number of loop-based production editors. Either way, these files can be easily obtained from any number of sources, such as:

- Those that are included for free with newly-purchased software
- The web (both free and for purchase)
- Commercial CDs
- Free files within CDs that are loaded onto demo CDs and magazine CD content
- Rolling your own (creating your own loops can add a satisfying and personal touch)

It's important to note that at any point during the creation of a composition, audio and MIDI tracks (such as vocals or played instruments) can often be easily recorded into a loop session in order to give the performance a fluid and more dynamic feel. It's even possible to record a live instrument into a session with a defined tempo and then edit these tracks into defined loops that can be dropped into the current and future sessions to add a live touch.

As with most music technologies, the field of looping and laying tracks into a groove continues to advance and evolve at an alarming rate. It's almost a sure bet that your current system will support looping in one way or another. Take time to read the manuals, gather up some loops that fit your style and particular interests, and start working them into a session. It might take you some time to master the art of looping—and then again you might be a natural Zen master. Either way, the journey is educational and tons of fun!

Editor/Librarians

As we've seen from Chapter 5, the vast majority of MIDI instruments and devices store their internal *patch preset data* within RAM memory. Synthesizers, samplers, or other devices contain information on how the controlling oscillators, amplifiers, filters, tuning, and other presets are to be configured, in order to create a particular sound timbre or effect. In addition to controlling sound patch parameters, the unit's memory can also store such setup information as effects processor settings, keyboard splits, MIDI channel routing, and controller assignments. These parameters, which can be manually edited from the device's front panel or user interface, can be easily accessed by recalling the patch number or name from the device's bank of preset buttons, alpha dial, or keypad entry (Figure 8.1).

A device's preset patch settings can exist in several forms. The most basic of these is the factory preset. As the name implies, a factory preset is a bank of patches that have been programmed by the manufacturer to give the performer an initial set of sounds that are designed to be useful out of the box and to convince the potential customer to buy the product. In almost every circumstance, hardware devices encode these factory presets into battery-backed random access memory (RAM), where they will be stored until either the battery loses power (generally over a 5-year period) or the patch data is overwritten. In some circumstances, reinitializing the synth can restore these presets, although you should check with the manual to see if this feature is supported.

Figure 8.1. *One of the infinite examples of a bank of preset buttons for storing and recalling patch data.*

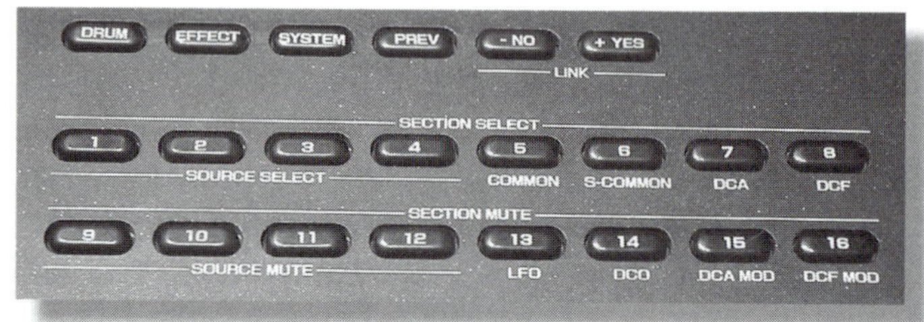

In the final analysis, however, factory patches are simply a collection of programmed patch and parameter settings that reflect the company's own set of tastes and style—and might not reflect your own. Many musicians, in fact, think that you're doing well if you like even half of the initial sounds that are supplied with an instrument. This alone is incentive enough to set even the most technologically timid on the road to editing and creating their own sounds and parameter patches.

Sound- and setup-related patch data can be edited in a number of ways. One of the fastest and simplest ways is to alter the parameters of an already existing factory preset until the desired voicing effect or setup has been achieved. Once done, the new patch setting can be overwritten over the old one with the same name or the edited setting can be saved to a new patch location. This process of editing an existing patch is often easy to do, because half the programming work has already gone into creating the original preset at the factory.

A second, more adventurous, approach is to build a patch entirely from scratch. In this way, a totally new sound can be constructed either by trial and error (by listening to the effect as the parameters are being changed) or by using your own experiences in synthesis and waveform analysis to build a patch from scratch (a process that's not for the faint of heart). For example, a synth patch setting could be programmed in the following way:

- Experiment with playing the various raw (nonprocessed) sounds, until you find one that you like.
- Enter the edit mode, where you can process the way in which the sample, loop or generated sound is played back by adjusting such controls as frequency oscillators, amplifiers, filters, tuning, effects processor settings, keyboard splits, and controller assignments.
- If the unit is polyphonic and can assign and mix more than one voice together to create a patch, you might want to combine new sounds with the original voice and edit them accordingly until you have created a composite sound that you're proud of.

I know that the above process sounds relatively simple, but it often isn't. Editing a patch from scratch can take a lot of time and can be very tedious. However, for those who have become good at it (all things come easier with practice), you'll have the satisfaction of showing your patches off in your work, or, if you're really good at it, you might even consider becoming a professional programmer for a manufacturer.

A Historical Perspective

Earlier analog synthesizers were constructed using a modular building-block approach that made it easy to see and physically control knobs and sliders in a hands-on fashion. For example, a sound could be amplitude modulated by physically inserting an oscillator and voltage-controlled amplifier (VCA) into the circuit using patch chords (thus, the term *patch* was born). Additional modules could be inserted into the path until the desired sound had been achieved. The down side of this approach was that these controls required a large amount of space, were rather expensive, and made it difficult (if not impossible) to store settings for later recall. It's interesting to note, however, that this hasn't stopped the resurgent popularity of devices that have these types of hands-on control surfaces (from both a collector and new product standpoint).

With the advent of digitally controlled analog and fully digital synthesizers, most of the control surfaces have been redesigned to cut cost and to save space in the mass manufacturing process. Newer display types have replaced the vast array of individual controls with a central control panel that's often made up of a liquid crystal display (LCD), buttons, alpha dials, and keypads. The down side of this modern-day approach is that you often have to spend much of your time navigating through the deep layers of programming options from the ever-tedious LCD panel (which can easily limit your view of the techno world to two lines on a 2-inch by 5-inch or smaller screen).

The Patch Editor

Because of this multilayered jungle, electronic musicians now have the option of taking back control of their instrument's parameters by using a computer-based patch editor. A *patch editor* (Figure 8.2) uses software to provide on-screen controls and graphic windows to emulate hands-on controls that can often be varied in real time. A dedicated patch editor is a software device that's been specifically programmed to emulate and vary the editing controls of a single instrument or device type. In certain cases, a patch editor might be designed to control several versions of an instrument or family of instruments (*e.g.*, a Korg WaveStation, WaveStation EX, and WaveStation SR).

In addition to patch editors that are specifically designed to work with a specific instrument, a number of programs are available that can communicate with a wide range of MIDI instruments and effects devices. These programs, known as *universal patch editors* (Figure 8.3 and 8.4), are designed to receive and transmit device-specific data and to provide on-screen control over the programming functions for most (if not every) MIDI device in your system.

Understandably, creating a realistic, graphic representation of the various faders, knobs, and other parameter controls for every possible MIDI instrument is difficult. Therefore, many universal editors are able to conform their graphic user interface (GUI) to the large instruments and devices types that are on the market by using a programmable standard set of sliders, numeric input, and graphic windows. Some programs make the editing process a bit more intuitive by graphically representing (either loosely or with relative accuracy) the basic controls of certain selected devices. More complex parameters might need to be edited using numeric or other graphic display types. A few graphically oriented universal editors let you design your own customized

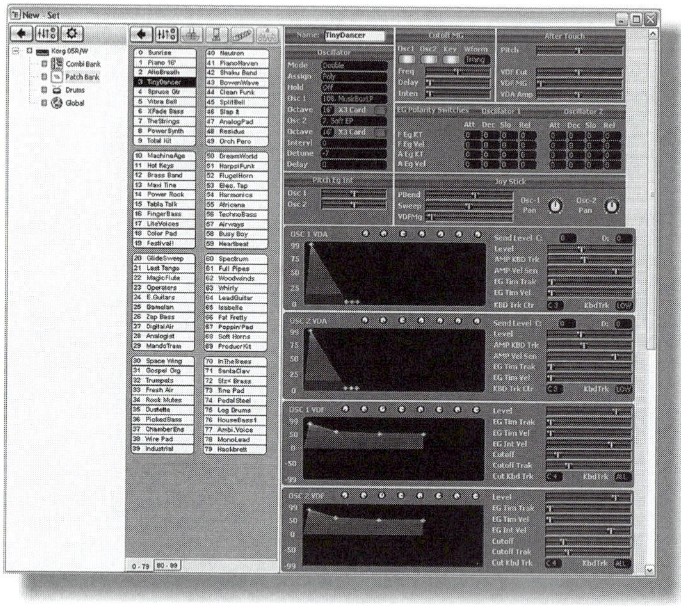

Figure 8.2. Korg 05R/W Editor/Librarian patch editor window. (Courtesy of Sound Quest Inc.; www.squest.com.).

Figure 8.3. MIDI Quest 9 Universal Editor for the Mac and PC. (Courtesy of Sound Quest, Inc.; www.squest.com.).

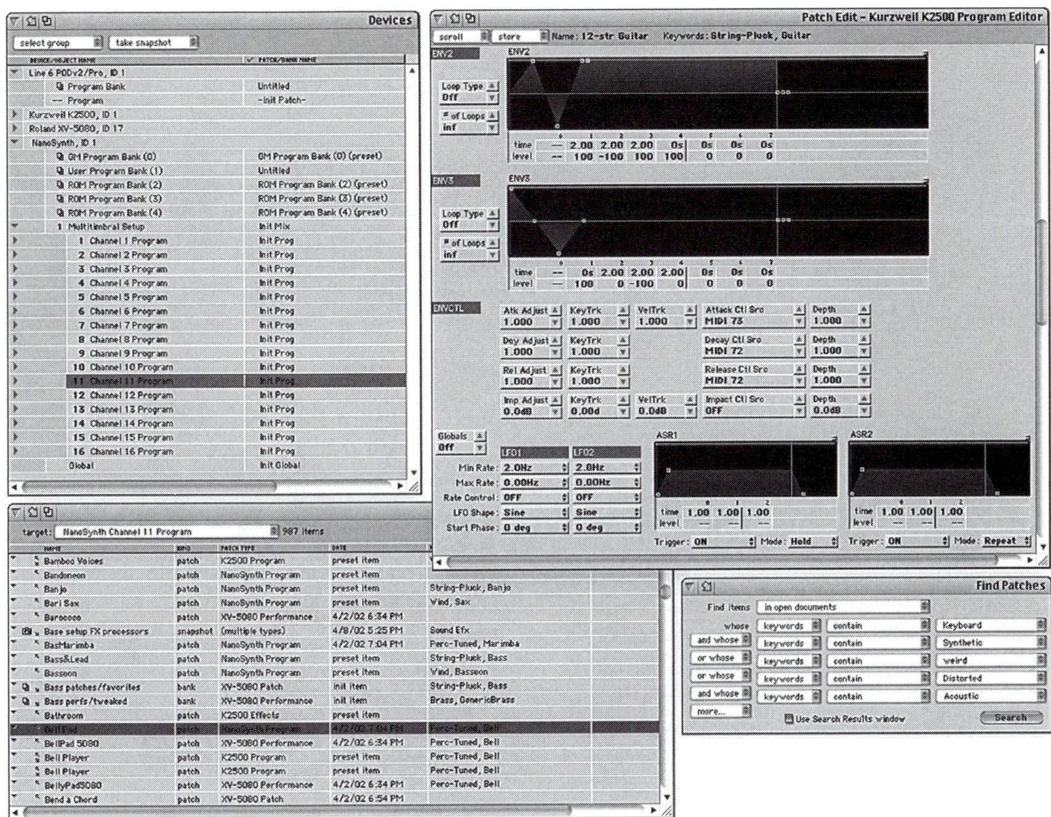

Figure 8.4. *Unisyn Universal Editor for the Mac and PC. (Courtesy of MOTU, Inc.; www.motu.com.).*

controls for newer instruments that have just reached the market or are weird, esoteric devices that no one would otherwise spend the time to create an interface for.

Direct communication between the software/computer system and the device's microprocessor is most commonly accomplished via the real-time transmission and reception of MIDI SysEx messages (Figure 8.6). As noted in Chapter 2, SysEx messages are used to communicate customized MIDI messages (such as patch parameter data) between MIDI devices within a connected system. The format for transmitting these messages includes a SysEx status header, manufacturer's ID number, any number of SysEx data bytes, and an EOX byte. Through the transmission of these device-specific messages, the device's microprocessor can be directly accessed so that a large number of setup parameters can be directly and easily altered in real time. You might note from Figure 8.5 that a MIDI Out line is often required to return data from the device being edited back to the editor/controller. This allows the devices to remain in constant two-way communication (often called a *handshake protocol*). Again, check with your device's manual to see if this two-way handshaking is required.

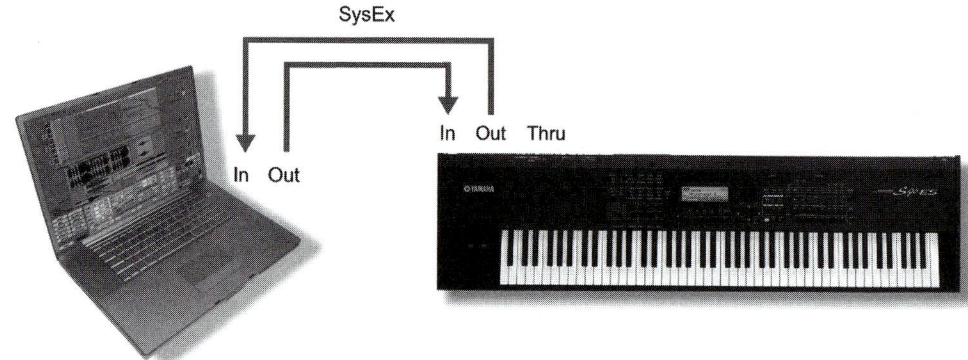

Figure 8.5. *A simple example of SysEx data distribution between a patch editor and a MIDI instrument.*

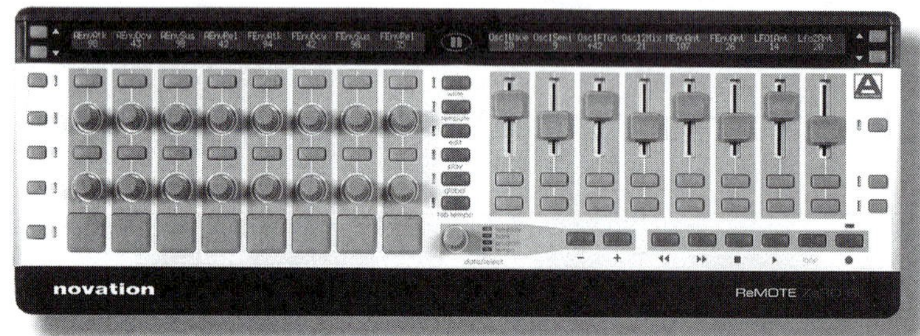

Figure 8.6. *Novation ReMote ZeRO SL controller w/o the keyboard. (Courtesy of Novation Digital Music Systems, Ltd.; www.novationmusic.com.)*

Hardware Patch Editors

In addition to software editing packages, hardware solutions are readily available for gaining quick and easy access to a device's parameters via SysEx in the form of MIDI hardware controllers (Figure 8.6). These devices, which are often equipped with any number of faders, knobs, buttons, or trigger pads can be used to virtually control any or all of a device's mix and sound-shaping parameters. Most modern-day controllers are universal in nature, meaning that they are able to control a wide range of devices. This can be done in either of two ways:

- They can be preprogrammed with a number of SysEx control templates that can be individually called up to allow communication with a specific hardware device.
- The SysEx data stream for each fader, knob, or button can be dynamically assigned to any parameter or control on a hardware instrument, software program or device—leaving each control on the device's surface completely open to being assigned in any way the user sees fit.

The Patch Librarian

After you've edited the original factory patches or created new patches from scratch, sooner or later your instrument or device will run out of preset memory locations. When this happens, these new patches or patch banks should be archived by saving them as SysEx data to either your sequencer or to a *patch librarian* program. The process of saving your instrument or device data via SysEx means that you can begin the process of creating new patches *ad infinitum*. And, when, you want a patch from a previous edit bank, simply reload that SysEx dump back into the device and you're in business.

Before we continue on, it's important to remember a few general guidelines that can help keep your valuable data intact:

- Upon getting a new device, always do a full SysEx dump of the factory's patch data. This can help you reload it to its original sounds, should its internal backup battery go dead or if you just want to reload the patches.
- Always make sure that your present set of edited patches is saved *before* you reload a SysEx patch bank into the device. Permanently losing your current patch set wouldn't be fun!
- Always have a reliable and current set of backups. Remember, it's not "if" a crash will occur—it's "when." In the digital age, the motto is "Be prepared."

Once you've acquired or programmed lots of patches that are spread all over tons of SysEx dumps, it's easy to see how you might end up spending more time loading data dumps than you'd like, just to find the right patch for a song. Simply put, sometimes too much of a good thing is simply too much. In such a case, the library portion of a patch librarian can come in handy. This software tool can be used to take the patches from various banks of data and organize them into a single dump that contains similar patches (Figure 8.7). For example, you could use a patch librarian to organize all of your sustaining pads into a dump called "New Angel" or the hard-hittin' patches into "Heavy Metal." You could also reorganize patches into banks according to sound type (*e.g.*, bass, strings, effects) or in any other way that helps you get the job done easier.

Similar to patch editor software packages, patch librarians come in two flavors: They may be dedicated to organizing and communicating patch data to and from a specific instrument or device, or, more commonly, they can organize and communicate patch data with a wide range of MIDI instruments or devices (this latter type is commonly known as a *universal patch librarian*).

SysEx Dump Utilities

For those who don't need the extensive editing features of a full-featured editor or librarian, a number of commercial, shareware and freeware SysEx patch dump utilities are currently available on the web. Often, these utilities are used as a quick and easy way to transmit preprogrammed patches to your device.

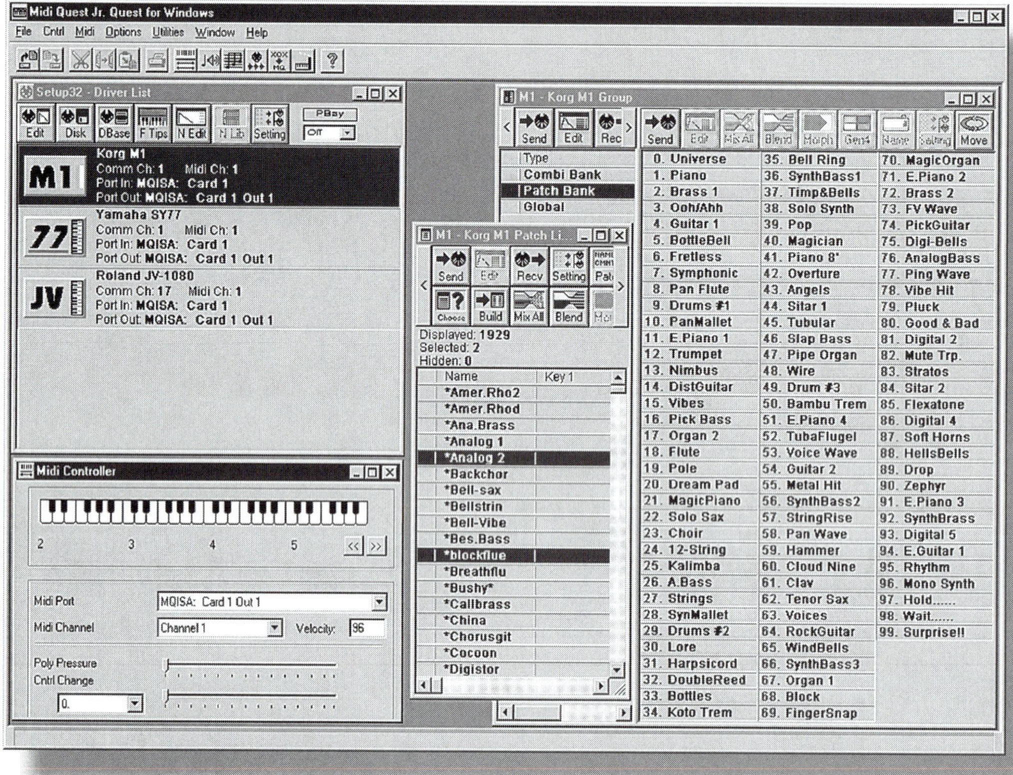

Figure 8.7. *MIDI Quest Jr. universal librarian. (Courtesy of Sound Quest Inc. www.squest.com.).*

SysEx and your DAW

A certain number of DAW/sequencer packages also integrate utilities for importing, managing, editing, and transmitting SysEx data to and from a soft or hardware MIDI device (Figures 8.8).

SysEx patch data can be used to perform such tasks within a production system as:

- Allowing patch data to be efficiently imported and exported, so as to find the right sounds for a project
- Allowing the necessary patch dumps to be automatically transmitted to the appropriate devices at the beginning of a song (a useful feature both in the studio and on-stage)
- Making it possible for backup and archive data dumps to be accessed by a single production program

Even if your DAW or sequencer of choice doesn't include the ability to manage and organize patch data directly, it's often possible to record SysEx data directly into a MIDI track on your

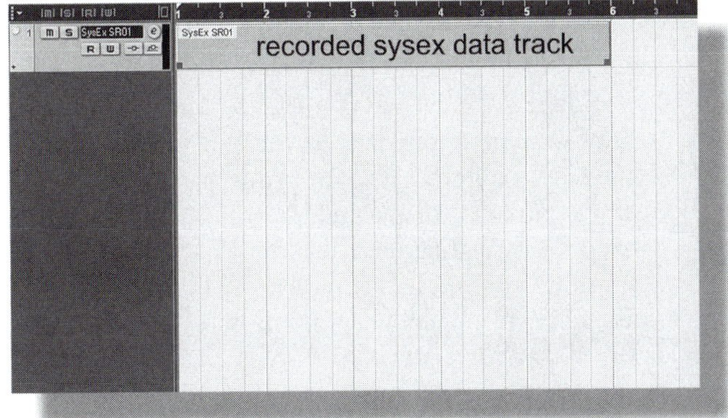

Figure 8.8. *SysEx data can be recorded onto a sequenced track within a DAW, where it can be saved, recalled and imported into a session.*

DAW or sequencer. This simple process has begun to emerge from the fray as a single unified standard that's so simple it's truly amazing that it wasn't universally adopted from the start.

By recording the patch data dumps from one or all instruments and devices in your studio to the program, and then individually saving each dump as a file (often as a standard MIDI or .MID file), you'll be able to archive and organize patch data in ways that can help streamline your work. Using this system, it becomes a simple matter to place the necessary data dumps into a session, at a point before the beginning of a song. Upon pressing the play button, the appropriate data can be automatically dumped to the associated devices in a straightforward no-muss, no-fuss fashion.

Saving a SysEx Dump to a Track

- Read the manual for your DAW or sequencer to verify that it can record SysEx to a MIDI track.
- If so, follow the directions for allowing the program to record a bulk data dump.
- Create a MIDI track, assign it to the proper MIDI channel or ports, and place it into Record Ready.
- Ready the sending device for sending a systemwide data dump (you might want to consult your manual, if this is your first time).
- Begin recording the track, initiate the data transfer, and stop the track at the end of the dump.
- Save the track with an easily identifiable name and in a logical directory path.
- If (and only if) you feel confident about the process, you can now transmit the dump back to the device. Did it receive it?

Alternative Sources for Obtaining Patch Data

In addition to factory- and user-created custom patches, there are literally thousands of opportunities for obtaining patch data, from either commercial or public freeware sources. Preprogrammed patches are readily available for almost every popular MIDI instrument and effects device, spanning almost every imaginable timbre and instrumentation style. Commercially available patch banks are often programmed by working professionals and dedicated gearheads using a wide range of data formats. This data is made available to the public on device-specific ROM/RAM cards, which physically plug into supporting electronic instruments or on CD-ROM. Such professional and homegrown products can be commonly found at a reasonable cost on the web or in the classifieds/back-page sections of most magazines that cater to the electronic musician.

Last, but definitely not least, patch data can be easily found on the Internet. Surfing the web will probably give you instant access to more patch data than you might know what to do with. An easy way to get started is to log onto the web and go to your favorite search engine, then simply type in the name of the instrument for which you want to search (*e.g.*, "Korg X5, patch data"). Most likely, the SysEx data on the Internet will be encoded into a format that's compatible with a generic SysEx Dump Utility or as a standard MIDI file that can be easily loaded into a Mac/PC workstation track for transmission to the instrument or device.

Music Printing Programs

Over the past few decades, the field of transcribing musical scores and arrangements has been strongly affected by both the computer and MIDI technology. This process has been greatly enhanced through the use of newer generations of computer software that makes it possible for music notation data to be entered into a computer either manually (by placing the notes onto the screen via keyboard or mouse movements), by direct MIDI input, or by sheet music scanning technology. Once entered, these notes can be edited in an on-screen environment that lets you change and configure a musical score or lead sheet using standard cut-and-paste editing techniques. In addition, most programs allow the score data to be played directly from the score by electronic instruments via MIDI. A final and important program feature is their ability to quickly print out hard copies of a score or lead sheets in a wide number of print formats and styles.

Entering Music Data

A music printing program (also known as a music notation program) lets you enter musical data into a computerized score in a number of manual and automated ways (often with varying

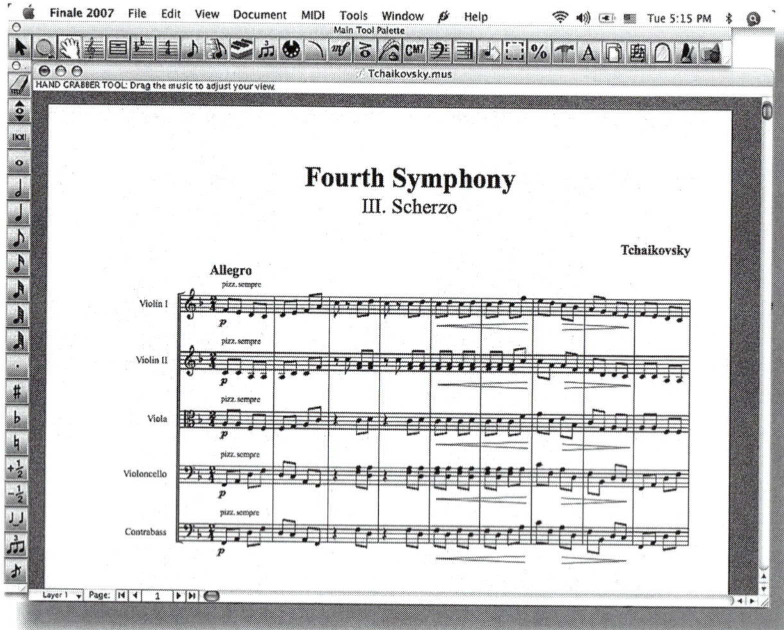

Figure 9.1. Finale 2007 music printing program. (Courtesy of MakeMusic, Inc.; www.finalemusic.com.)

degrees of complexity and ease). Programs of this type (Figure 9.1) offer a wide range of notation symbols and type styles that can be entered either from a computer keyboard or mouse. In addition to entering a score manually, most music transcription programs will generally accept MIDI input, allowing a part to be played directly into a score. This can be done in real time (by playing a MIDI instrument/controller or finished sequence into the program) or in step time (by entering the notes of a score one note at a time from a MIDI controller), or by entering a standard MIDI file into the program (which uses a sequenced file as the notation source).

In addition to dedicated music printing programs, most DAW or sequencer packages will often include a basic music notation application that allows the sequenced data within a track or defined region to be displayed and edited directly within the program (Figure 9.2), from which it can be printed in a limited score-like fashion. However, a number of high-level workstations offer scoring features that allow sequenced track data to be notated and edited in a professional fashion into a fully printable music score.

As you might expect, music printing programs will often vary widely in their capabilities, ease of use and offered features. These differences often center around the graphical user interface (GUI), methods for inputting and editing data, the number of instrumental parts that can be placed into a score, the overall selection of musical symbols, the number of musical staves (the lines that music notes are placed onto) that can be entered into a single page or overall score, the ability to enter text or lyrics into a score, etc. As with most programs that deal with artistic production, the range of choices and general functionality reflect the style and viewpoints of the manufacturer, so care should be taken when choosing the professional music notation program that might be right for you.

Figure 9.2. *Steinberg Cubase/Nuendo score window. (Courtesy of Steinberg Media Technologies GmbH, A Division of Yamaha Corporation; www.steinberg.net.)*

Scanning a Score

Another way to enter music into a score is through the use of an optical recognition program (Figure 9.3). These programs let you use a standard flatbed scanner to download sheet music or a printed score into a music program—and then save the notation, data, and general layout as a standardized Notation Interchange File Format (NIFF) file that can then be entered into a scoring/notation program.

Editing a Score

As was stated earlier, music notation and printing programs allow music-related data to be input in a number of ways (manually, via MIDI, file import, or optical scanning). Once the data is imported into the program, it's generally a simple matter to add, delete, or change individual notes, duration, and markings by using a combination of computer keyboard commands, mouse movements, or MIDI keyboard commands. Of course, larger blocks of music data can also be

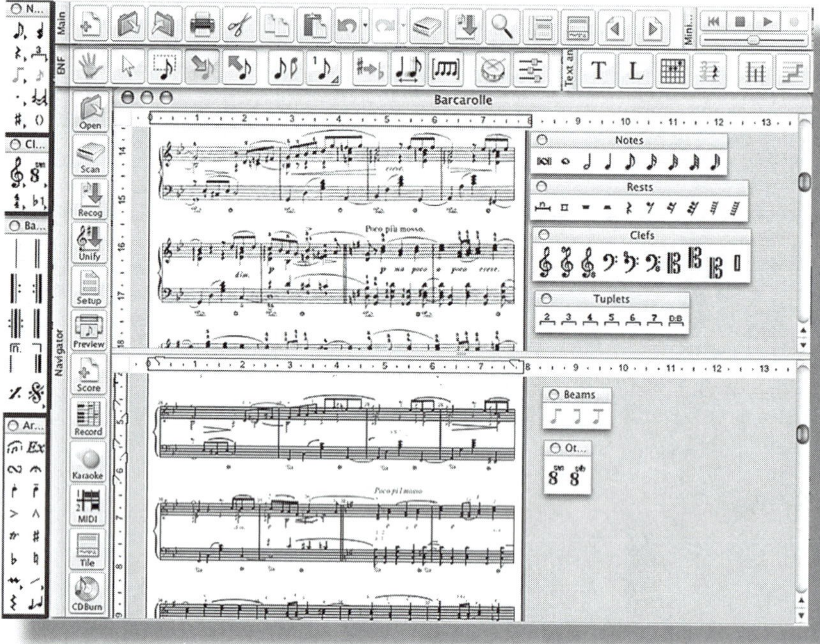

Figure 9.3. *SmartScore Piano Edition 3 music scanning program. (Courtesy of Music Imaging Technologies, www.musitek.com.)*

edited using the standard cut-and-paste approach. As you might expect, a wide selection of musical symbols is commonly available; these can be placed and moved within a score to denote standard note lengths, rest duration markings, accidental markings (flat, sharp, and natural), dynamic markings (*e.g.*, pp, mp, mf, ff), and a host of other important score markings that can range from the commonly used to the obscure (Figure 9.4). In addition to standard notation entry, text can usually be placed into a lead sheet or score for the purpose of adding song lyrics, song titles, special performance markings, and additional header/footer information.

One of the major drawbacks to entering a score into such a program (either as a real-time performance or as an imported file) is the fact that music notation is an interpretive art. As the saying goes, "To err is human," and this human feel is what usually gives music its full range of emotional expression. It's very difficult, however, for a computer program to properly interpret these minute, yet importantly subtle, imperfections and then place them into the score in their exact and proper place. (For example, the program might interpret a section of a passage with a held 1/4 note as one that contains a dotted 1/4 or one that's tied to a 1/32 note.) In short, it will often have no way of actually knowing and will have to make its best guess. As such, even though the algorithms are getting better at interpreting musical data, and quantization can be used to instruct a computer to round a note value to a specified length, a score will still often need to be edited manually in order to correct for misinterpretations.

Before continuing, I'd like to point out the importance of taking time to properly set up the initial default settings of a music notation program. These can be in the form of global settings that can help set such parameters as measure widths, number of stems, instrument layouts, and

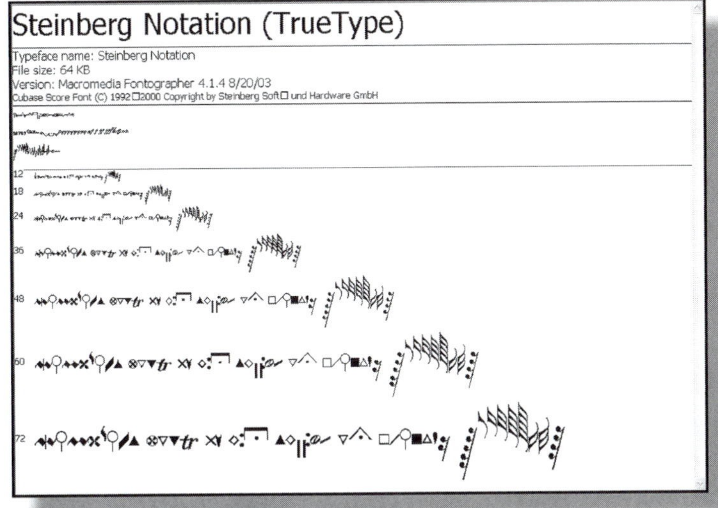

Figure 9.4. *TrueType display of notation font symbols. (Courtesy of Steinberg Media Technologies GmbH, A Division of Yamaha Corporation; www.steinberg.net.)*

default key signature—or they can help take care of the less important decisions, such as title, composer, and instrument name fonts. To save time in the long run, you might consider making a few file setup templates that contain several of the more common time and key signatures, stave and instrument layouts, and other settings that you tend to work with.

Finally, it's often wise to take the time to familiarize yourself with the vast number of notational parameters that are contained in most programs before beginning work on a song or score. This can save you from having real headaches when making changes (even minor ones that might impact a whole piece). Giving a composition some layout thought beforehand can definitely help you avoid unnecessary bandage-type fixes at a later stage in the process.

Playing Back a Score

In the not-so-distant past, classical composers commonly had to wait months or years to hear their finished composition. Orchestras and ensembles were generally far too expensive and often required that the composer have a financial patron or a good knowledge of corporate or state politics in order to get their works heard. One simple solution to this obstacle was the piano reduction, which served to condense a score down into a compromised version that could be played at the piano keyboard.

With the advent of MIDI, however, classical compositions and film scores could be played through various MIDI instruments in a composing/project studio, directly from a DAW's sequencer or notation program (Figure 9.5). In this way, a production system can approximate a working rendition of the final performance with relative ease, allowing the artist to check for errors and final tweaks before being subjected to the expense, time constraints, and the inherent difficulties of working with a live orchestra.

Figure 9.5. *Los Angeles film composer Jeff Rona. (Courtesy of M-Audio, A Division of Avid Technology, Inc.; www.m-audio.com.)*

Figure 9.6. *The Music Pad Pro® Plus, digital sheet music tablet (stand not included). (Courtesy of FreeHand Systems, Inc.; www.freehandsystems.com.)*

Displaying and Printing Out a Score

Once the score has been edited into a final form, the process of creating a hard copy to print is relatively simple. Generally, a notation program can lay out the score in a way that best suits your taste, the producer's taste, or the score's intended purpose. Often, a professional program will let you make final changes to such parameters as margins, measure widths, title, copyright, and other text-based information. Once completed, the final score can then be printed out using a standard, high-quality ink-jet or laser printer.

Figure 9.7. *Download services are springing up that allow electronic downloads of sheet music to be easily accessed for free or as a commercial service. (Courtesy of Yamaha Corporation of America, www.digitalmusicnotebook.com.)*

In this modern age of computer technology, it's actually possible to display sheet music directly onto an LCD screen, bypassing the printing process altogether. Dedicated LCD sheet music displays (Figure 9.6) would allow you to save an entire season's worth of concerts or a semester's worth of class material within its internal memory. In addition, online sites allow for sheet music to be downloaded and taken on the road—all without the need for printing out the music. Of course, printing the music out might save the day in the case of a power blackout or the ever-present technical glitch (you know what they say about being prepared?).

In addition to the growing number of options for storing and displaying sheet music in the digital domain, several online services have begun to show up that offer printed music downloads, using a model that's not unlike iTunes® and other music download services. One such service is www.digitalmusicnotebook.com from the folks at Yamaha (Figure 9.7). This service offers a free player that can be downloaded from the site and allows for purchased downloads of sheet music in a wide range of styles and at a reasonable cost. Some tunes, including classical pieces, public-domain songs, and copyright-free material, are available free of charge.

Multimedia and the Web

It's no secret that modern-day computers have gotten faster, sleeker, and sexier in their overall design. In addition to its ability to act as a multifunctional production workhorse, one of the crowning achievements of the modern computer is the degree of media and networking integration that has worked its way into our collective consciousness in a way that has come to be universally known by what is now known as the household buzzword—multimedia.

The combination of working and playing with multimedia has found its way into modern computer culture through the use of various hardware and software systems that work in a multitasking environment and combine to bring you a unified experience that seamlessly involves such media types as:

- Text
- Graphics
- Audio and music
- Computer animation

- Musical instrument digital interface (MIDI)
- Video

The obvious reason for creating and integrating these media types is the human desire to share and communicate one's experiences with others. This has been done for centuries in the form of books and, in relatively more recent decades, through movies and television. In the here and now, the amazingly powerful and versatile presence of the Web can be added to this communications list. Nothing allows individuals and corporate entities alike to reach millions so easily. Perhaps most importantly, the web is a multimedia experience that each individual can manipulate, learn from, and even respond to in an interactive fashion. The web has indeed unlocked the potential for experiencing multimedia events and information in a way that makes each of us a participant—not just a passive spectator. To me, this is the true revolution occurring at the dawn of the 21st century!

The Multimedia Environment

When you get right down to it, multimedia is nothing more than a unified programming and operating system (OS) environment that allows multiple forms of program data and content media to simultaneously stream and be routed to the appropriate hardware ports for output, playback, and/or processing (Figure 10.1).

> The two most important concepts behind this environment are:
>
> - Multitasking
> - The device driver

Basically, multitasking can be thought of as a modern-day form of illusion. Just as a magic trick can be quickly pulled off with sleight of hand, or a film that switches frames 24 times each second can create the illusion of continuous movement, the multimedia environment deceives us into thinking that all of the separate program and media types are working at the same time. In reality, the computer uses multitasking to quickly switch from one program to the next in a cyclic fashion. Similar to the film example, newer computer systems have gotten so lightning fast at cycling between programs and applications that they give the illusion they're all running at the same time.

Another central concept to multimedia is that of the device driver. Briefly, a driver acts as a device-specific software patch cord that routes media data from the source application to the appropriate hardware output device (also known as a port) and from a port back to the application's input (Figure 10.2). Thus, whenever any form of media playback is requested, it'll be recognized as being a particular data type and will be routed to the appropriate device driver and finally sent out to the selected output port (or ports).

Figure 10.1. Example of a multimedia program environment.

Figure 10.2. Basic interaction between a software application and a hardware device via the device driver.

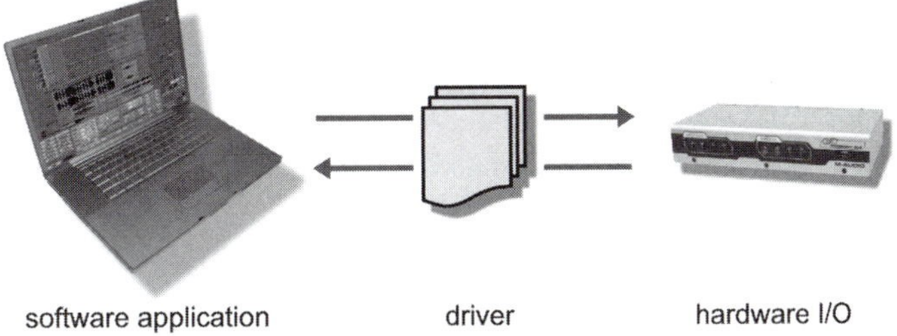

Delivery Media

Although media data can be stored and/or transmitted over a wide range of media storage devices, the most commonly found delivery media at this time of writing are the:

- CD
- DVD
- Web

Table 10.1. CD Format Standards

Format	*Description*
Red Book	Audio-only standard; also called CD-A (Compact Disc Audio)
Yellow Book	Data-only format; used to write/read CD-ROM data
Green Book	CD-I (Compact Disc Interactive) format; never gained mass popularity
Orange Book	CD-R (Compact Disc Recordable) format
White Book	VCD (Video Compact Disc) format for encoding CD-A audio and MPEG-1 video data; used for home video and Karaoke
Blue Book	Enhanced Music CD format (also known as CD Extra or CD+) can contain both CD-A and data
ISO-9660	A data file format that's used for encoding and reading data from CDs of all types across platforms
Joliet	Extension of the ISO-9660 format that allows for up to 64 characters in its file name (as opposed to the 8 file × 3 extension characters allowed by MS-DOS)
Romeo	Extension of the ISO-9660 format that allows for up to 128 characters in the file name
Rock Ridge	Unix style extension of the ISO-9660 format that allows for long file names
CD-ROM/XA	Allows for extended usage for the CD-ROM format—Mode-1 is strictly Yellow Book, while Mode-2 Form-1 includes error correction and Mode-2 Form-2 doesn't allow for error correction; often used for audio and video data
CD-RFS	Incremental packet writing system from Sony that allows data to be written and rewritten to a CD or CD-RW (in a way that appears to the user much like the writing/retrieval of data from a hard drive)
CD-UDF	UDF (Universal Disc Format) is an open incremental packet writing system that allows data to be written and rewritten to a CD or CD-RW (in a way that appears to the user much like the writing/retrieval of data from a hard drive) according to the ISO-13346 standard
HDCD	The High-Definition Compatible Digital system adds 6 dB of gain to a Red Book CD (when played back on an HDCD-compatible player) through the use of a special compansion (compression/expansion) mastering technique
Macintosh HFS	An Apple file system that supports up to 31 characters in a file name; includes a data fork and a resource fork that identify which application should be used to open the file

The CD

One of the most important developments in the mass marketing and distribution of large amounts of media is the compact disc (CD), both in the form of the CD-Audio and the CD-ROM. As most are aware, the CDAudio disc is capable of storing up to 74 minutes of audio at a rate of 44.1 kHz/16 bits. Its close optical cousin, the CD-ROM, is capable of storing up to 700 MB of graphics, video, digital audio, MIDI, text, and raw data. Consequently, these premanufactured and user-encoded media are able to store large amounts of music, text, video, graphics, etc., to such an interactive extent that this medium has become a driving force among all communications media. Table 10.1 details the various CD standards that are currently in use.

The DVD

Similar to their optical cousins, the DVD (which, after a great deal of industry deliberation, simply stands for "DVD") can contain any form of data. Unlike the CD, these discs are capable of storing a whopping 4.7 gigabytes (GB) within a single-sided disc and 8.5 GB on a double-layered disc. This capacity makes the DVD the perfect delivery medium for encoding video in the MPEG-2 encoding format, data-intensive games, DVD-Audio, and numerous DVD-ROM titles. The increased demand for multimedia games, educational products, etc., has spawned the computer-related industry of CD and DVD-ROM authoring. The term *authoring* refers to the creative, design, and programming aspects of putting together a CD/DVD project. At its most basic level, a project can be authored, mastered, and burned to disc from a single commercial authoring program. Whenever the stakes are higher, trained professionals and expensive systems are often called in to assemble, master, and produce the final disc for mass duplication and sales.

The Web

One of the most powerful aspects of multimedia is the ability to communicate experiences either to another individual or to the masses. For this, you need some kind of network connection. The largest and most common network that can be found in the home, studio, office, classroom—you name it—is a connection to the Internet.

Here's the basic gist of how the Internet works:

- The Internet (Figure 10.3) can be thought of as a communications network that allows your computer (or connected network) to be connected to an Internet Service Provider (ISP) server (a specialized computer or cluster of ISP computers designed to handle, pass, and route data between large numbers of user connections).
- These ISPs are connected (through specialized high-speed connections) to an interconnected network of network access points (NAPs). This network essentially forms the connected infrastructure of the World Wide Web (WWW).
- In its most basic form the WWW can be thought of as a unified array of connected of networks.

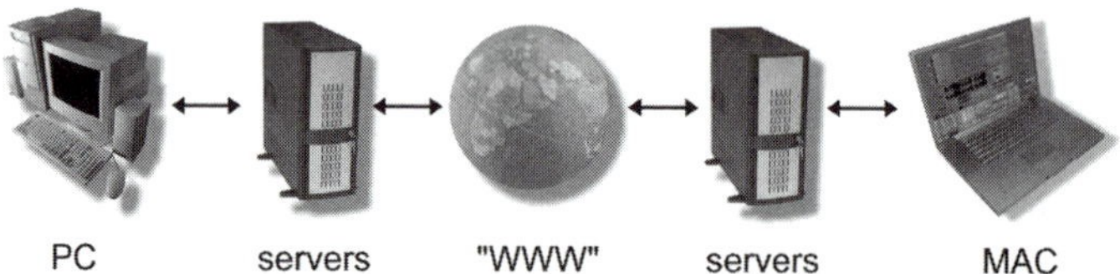

Figure 10.3. *The Internet works by communicating requests and data from a user's computer to a single server that's connected to other servers around the world, which are likewise connected to other users' computers.*

Internet browsers transmit and receive information on the web via a Uniform Resource Locator (URL) address. This address is then broken down into three parts: the protocol (*e.g.*, http), the server name (*e.g.*, www.modrec.com.), and the requested page or file name (*e.g.*, index.htm). The connected server is able to translate the server name into a specific Internet Provider (IP) address. This address is used to connect your computer with the desired server, after which the requests to receive or send data are communicated and the information is passed to your desktop.

E-mail works in a similar data transfer fashion, with the exception that an e-mail isn't sent to or requested from a specific server (which might be simultaneously connected to a large group of users); rather, it's communicated from one specific e-mail address (*e.g.*, myname@myprovider.com.) to a destination address (*e.g.*, yourname@yourprovider.com.).

Delivery Formats

When creating content for the various media systems, it's extremely important that you match the media format and bandwidth requirements to the content delivery system that's being used. In other words, it's always smart to maximize the efficiency of the message (media format and required bandwidth) to match (and not alienate) your intended audience. The following section outlines many standard and/or popular formats for delivering media to a target audience.

Digital Audio

Digital audio is obviously a component that adds greatly to the multimedia experience. It can augment a presentation by adding a dramatic music soundtrack, help us to communicate through speech, or give realism to a soundtrack by adding sound effects. Because of the large amounts of data required to pass video, graphics, and audio from a CD-ROM, the Internet, or other media, the bit- and sample-rate structure of an uncompressed audio file is usually limited, compared to that of a professional-quality sound file. At the "lo-fi" range, the generally accepted sound file standard for multimedia production is either 8-bit or 16-bit audio at a sample rate of 11.025 or 22.050 kHz. This standard had come about mostly because older CD drive and processor systems generally couldn't pass the professional 44.1-kHz rate. With the introduction of faster processing systems and better hardware, these limitations have generally been lifted to include 44.1-kHz/16-bit audio and compressed data formats that offer CD quality and discrete surround-sound playback capabilities. In addition, there are obvious limitations to communicating uncompressed professional-rate sound files over the Internet or from a CD or DVD disc that is also streaming full-motion video. Fortunately, with improvements in codec (encode/decode) techniques, hardware speed, and design, the overall sonic and production quality of compressed audio data has greatly improved in audience acceptance.

Uncompressed Sound File Formats

Although several formats exist for encoding and storing sound file data, only a few have been universally adopted by the industry. These standardized formats make it easier for files to be exchanged between compatible media devices. Probably the most common file type is the Wave (or .wav) format. Developed for the Microsoft® Windows® format, this universal file type supports both mono and stereo files at a variety of uncompressed resolutions and sample rates. Wave files contain pulse code modulation (PCM) coded audio that follows the Resource Information File Format (RIFF) spec, which allows extra user information to be embedded and saved within the file itself. The newly adopted Broadcast Wave format, which has been adopted by the NARAS Producers and Engineers wing (www.grammy.com/Recording_Academy/Producers_and_Engineers) as the preferred soundfile format for DAW production and music archiving, allows for time-code-related positioning information to be directly imbedded within the soundfile's data stream.

Another file format that's most commonly used in production using Mac computers is the Audio Interchange File (AIFF; .aif) format. Like Wave files, AIFF files support mono or stereo, 8-bit or 16-bit audio at a wide range of sample rates—and like Broadcast Wave files, AIFF files can also contain embedded text strings. Table 10.2 details the differences between uncompressed file sizes as they range from 32-bit/192-kHz rates that are used to encode DVD-Audio sound all the way down to voice-quality 8-bit/10-kHz files.

Compressed Codec Sound File Formats

As was mentioned earlier, high-quality uncompressed sound files often present severe challenges to media delivery systems that are restricted in terms of bandwidth, download times, or memory storage. Although the streaming of audio data from various media and high-bandwidth networks (including the web) has improved over the years, memory storage space and other bandwidth limitations have led to the popular acceptance of audio-related data formats known as *codecs.* These datafile formats can encode audio data in a manner that reduces data file size and bandwidth requirements and then decode the information upon playback using a system known as *perceptual coding.*

PERCEPTUAL CODING

The central concept behind perceptual coding is the psycho-acoustic principle that the human ear will not always be able to hear all of the information that's present in a recording. This is largely due to the fact that louder sounds will often mask sounds that are both lower in level and relatively close to another louder signal. These perceptual coding schemes take advantage of this masking effect by filtering out noises and sounds that can't be detected and removing them from the encoded audio stream.

The perceptual encoding process is said to be lossy or destructive, as once the filtered data has been eliminated it can't be replaced or introduced back into the file. For the purposes of audio

Table 10.2. PCM Audio Bit Rate and File Sizes

Sample Rate	*Bit Rate*	*Ch #*	*Data Rate (Kbps)*	*File Size (Kb)*	*MB/min*	*MB/hour*
192	32	2	1536	92,160	92.16	5529.6
192	32	1	768	46,080	46.08	2764.8
192	24	2	1152	69,120	69.12	4147.2
192	24	1	576	34,560	34.56	2073.6
96	32	2	768	46,080	46.08	2764.8
96	32	1	384	23,040	23.04	1382.4
96	24	2	576	34,560	34.56	2073.6
96	24	1	288	17,280	17.28	1036.8
48	32	2	384	23,040	23.04	1382.4
48	32	1	192	11,520	11.52	691.2
48	24	2	288	17,280	17.28	1036.8
48	24	1	144	8,640	8.64	518.4
48	16	2	192	11,520	11.52	691.2
48	16	1	96	5,760	5.76	345.6
44.1	32	2	352.8	21,168	21.16	1270.08
44.1	32	1	176.4	10,584	10.584	635.04
44.1	24	2	264.6	15,876	15.876	952.56
44.1	24	1	132.3	7938	7.938	476.28
44.1	16	2	176.4	10,584	10.584	635.04
44.1	16	1	88.2	5292	5.292	317.52
32	16	2	128	7680	7.68	460.8
32	16	1	64	3840	3.84	230.4
22	16	2	88	5280	5.28	316.8
22	16	1	44	2640	2.64	158.4
22	8	2	44	2640	2.64	158.4
22	8	1	22	1320	1.32	79.2
11	16	2	44	2640	2.64	158.4
11	16	1	22	1320	1.32	79.2
11	8	1	11	660	0.66	39.6

quality, the amount of audio that's to be removed from the data can be selected by the user during the encoding process. Higher bandwidth compression rates will remove less data from a stream (resulting in a reduced amount of filtering and higher audio quality), while low bandwidth rates will greatly reduce the data stream (resulting in smaller file sizes, increased filtering, increased artifacts, and lower audio quality). The amount of filtering that's to be applied to a file will depend upon the intended audio quality and the delivery medium's bandwidth limitations. Due to the lossy character of these encoded files, it's always a good idea to keep a copy of the original, uncompressed sound file as a data archive backup should changes in content or future technologies occur (never underestimate Murphy's law of technology).

The most commonly used perceptual coding schemes that are in use today are:

- MP3
- MWA
- AAC
- Ogg Vorbis
- Real Audio

Many of the listed codecs are capable of encoding and decoding audio using a constant bit rate (CBR) and variable bit rate (VBR) structure:

- CBR encoding is designed to work effectively in a streaming scenario where the end-user's bandwidth is a consideration. With CBR encoding, the chosen bit rate will remain constant over the course of the file or stream.
- VBR encoding is designed for use when you want to create a downloadable file that has a smaller file size and bit rate without sacrificing sound and video quality. This is carried out by detecting which sections will need the highest bandwidth and adjusting the encode process accordingly. When lower rates will suffice, the encoder adjusts the process to match the content. Under optimum conditions, you might end up with a VBR-encoded file that has the same quality as a CBR-encoded file, but with only half the file size.

MP3

MPEG (which is pronounced "M-peg" and stands for the Moving Picture Experts Group [www.mpeg.org]) is a standardized format for encoding digital audio and video into a compressed format for the storage and transmission of various media over the Web. As of this writing, the most popular format is the ISO-MPEG Audio Level-2 Layer-3, commonly referred to as MP3. Developed by the Fraunhofer Institute (www.iis.fhg.de) and Thomson Multimedia in Europe, MP3 has advanced the public awareness and acceptance of compressing and distributing digital audio by creating a codec that can compress audio by a substantial factor while still maintaining quality levels that approach those of a CD (depending on which compression levels are used).

Although faster Web connections are capable of streaming MP3 in real time, this format is most often downloaded to the end consumer for storage to disk, disc, and memory media for the storage and playback of downloaded songs. Once saved, the data can then be transferred to solid-state playback devices (such as portable MP3 players, PDAs, cell phone players—you name it!). In fact, over a billion music tracks are currently being downloaded every month on the Internet using MP3, practically every personal computer contains licensed MP3 software, virtually every song has been MP3 encoded, and some 150 million MP3 players are soon expected to reach the global market, making it the web's most popular audio compression format by far.

In 2001, MP3 Pro was introduced to the public as an encoding system for enhancing sound quality and improving the compression scheme. MP3 Pro works by splitting the encoding process into two parts. The first analyzes the low-frequency band information and encodes it into a normal

Figure 10.4. *Window's Media Player in skin mode; notice that the "Bars and Waves" visualization theme can act as a basic spectral display.*

MP3 stream (which allows for complete compatibility with existing MP3 players). The second analyzes the high-frequency band information and encodes it in a way that helps preserve the high-frequency content. When combined, the MP3 Pro codec creates a file that's more compact than original MP3 files, with equal or better sound quality and complete backward and forward compatibility. Although MP3 Pro professes to offer 128-kbps performance at a 64-kbps encoding rate (effectively doubling the digital music capacity of flash memory and compact discs), the general download community has been slow to adopt this new codec.

WMA

Developed by Microsoft as their corporate response to MP3, Windows Media Audio (WMA) allows for compression rates that can encode high-quality audio at low bit-rate and filesize settings. Designed for ripping (encoding audio from audio CDs) and soundfile encoding/playback from within the popular Window's Media Player (Figure 10.4), this format has grown in general acceptance and popularity. In addition to its high quality at low bit rates, WMA has the advantage of allowing for real-time streaming over the Internet (as witnessed by the large amount of radio and Internet stations that stream to the Windows Media Player at various bit-rate qualities). Finally, content providers often favor MWA over MP3, as it is able to provide for a degree of content copy protection through its incorporation of Digital Rights Management (DRM) coding. The introduction of Windows Media Player Version 9 allowed WMA to encode and deliver audio in discrete surround sound. Using various high-end workstations and encoders it's now possible to deliver surround audio to an ever-increasing number of homes whose computers are equipped with surround-sound playback (Figure 10.5).

AAC

Jointly developed by Dolby Labs, Sony, ATT, and the Fraunhofer Institute, the Advanced Audio Coding (AAC) scheme is touted as a multichannel-friendly format for secure digital music distribution over the Internet. Stated as having the ability to encode CD-quality audio at lower bit rates than other coding formats, AAC not only is capable of encoding 1, 2, and 5.1 surround-sound files but can also encode up to 48 channels within a single bitstream at bit/sample rates of up to 24/96. This format is also SDMI compliant (see Secure Digital Music Initiative section, below), allowing copyrighted material to be protected against unauthorized copy and distribution.

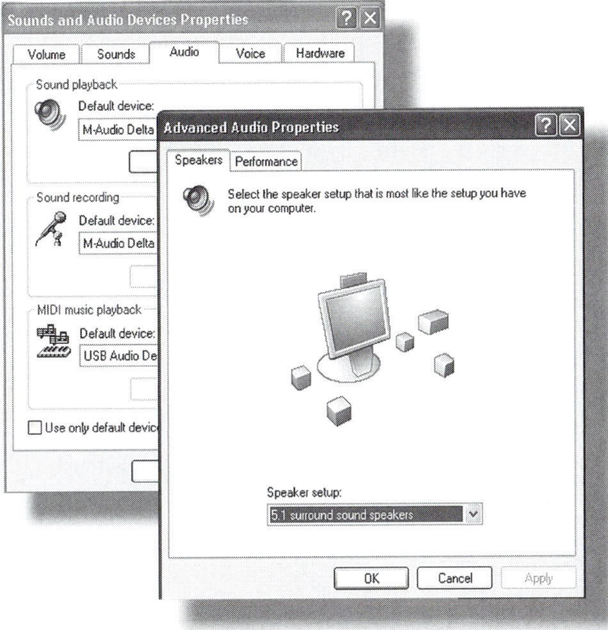

Figure 10.5. *Windows XP speaker setup box for stereo, surround, and other playback schemes.*

REALAUDIO™

With the introduction of their RealPlayer server application, RealNetworks (www.realnetworks.com.) became one of the first companies to provide real-time streaming over the web. RealAudio™ data is transmitted using any of more than 12 proprietary encoding levels that range from transmission rates of 8 kbps (low-fidelity mono voice quality over a 56-k modem) to speeds that exceed the 1.5 Mbps point. Although there are several compression levels to choose from, the most common music mode type compresses data in a way that doesn't introduce extreme artifacts over a wide dynamic range, thereby allowing it to create an algorithm that can faithfully reproduce music with near-FM quality over 56-k or faster lines. At the originating Internet site, the RealAudio server can automatically recognize which modem, cable, or network connection speed is currently in use and then transmits the data in the best possible audio format. This reduced data throughput ultimately means that RealPlayer will take up very little of your computer's resources, allowing you to keep on working while audio is being played.

Tagged MetaData

Within certain types of multimedia file formats it's possible to imbed a wide range of data into the file header. This data can identify and provide extensive information that relates to the content of the file. For example, within the Windows OS file structure, right-clicking on a file name within the Windows Explorer menu and selecting Properties/Summary (or Control-clicking/Get Info on a Mac) will bring up a set of extensive metadata file tags that can be used to enter and identify such media information as artist, album title, year, track number, genre, lyrics (yes, the entire set), title, comments, etc. (Figure 10.6). It should be pointed out that this metatag info

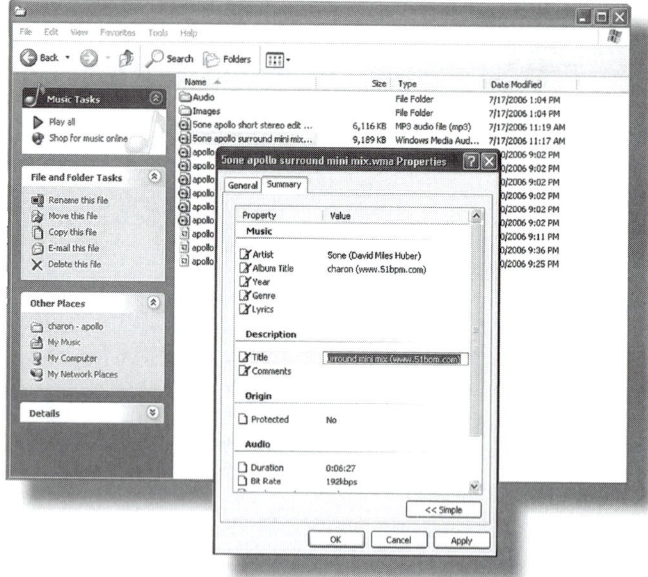

Figure 10.6. *Embedded metatag file tags can be added to a media file within Windows Explorer.*

can also be entered into many DAWs and editors and can often be transferred from one format type to another upon conversion.

MIDI

One of the unique advantages of MIDI as it applies to multimedia is the rich diversity of musical instruments and program styles that can be played back in real time, while requiring almost no overhead processing from the computer's CPU. This makes MIDI a perfect candidate for playing back soundtracks from multimedia games or over the Internet. It's interesting to note that MIDI has taken a back seat to digital audio as a serious music playback format for multimedia. Most likely, this is due to several factors, including:

- A basic misunderstanding of the medium
- The fact that producing MIDI content requires a basic knowledge of music
- The frequent difficulty of synchronizing digital audio to MIDI in a multimedia environment
- The fact that soundcards often include poorly designed FM synthesizers (although most operating systems now include higher quality software synths)

Fortunately, companies such as Microsoft have taken up the banner of embedding MIDI within their media projects and have helped push MIDI a bit more into the web mainstream. As a result, it's becoming more common for your PC to begin playing back a MIDI score on its own or perhaps in conjunction with a game or more data-intensive program.

Standard MIDI Files

The accepted format for transmitting files or real-time MIDI information in multimedia (or between sequencers from different manufacturers) is the *standard MIDI file*. This file type (which is stored with a .mid or .smf extension) is used to distribute MIDI data, song, track, time signature, and tempo information to the general masses. Standard MIDI files can support both single and multichannel sequence data and can be loaded into, edited, and then directly saved from almost any sequencer package. When exporting a standard MIDI file, keep in mind that they come in two basic flavors—type 0 and type 1:

- Type 0 is used whenever all of the tracks in a sequence need to be merged into a single MIDI track. All of notes will have a channel number attached to them (*i.e.*, will play various instruments within a sequence); however, the data will have no definitive track assignments. This type might be the best choice when creating a MIDI sequence for the Internet (where the sequencer or MIDI player application might not know or care about dealing with multiple tracks).
- Type 1, on the other hand, will retain its original track information structure and can be imported into another sequencer type with its basic track information and assignments left intact.

General MIDI

One of the most interesting aspects of MIDI production is the absolute uniqueness of each professional and even semipro project studio. In fact, no two studios will be alike (unless they've been specifically designed to be the same or there's some unlikely coincidence). Each artist will be unique as to his or her own favorite equipment, supporting hardware, way of routing channels and tracks, and assigning patches. The fact that each system setup is unique and personal has placed MIDI at odds with the need for system and setup compatibility in the world of multimedia. For example, after importing a MIDI file over the Net that's been created on another studio, the song will most likely attempt to play with a totally irrelevant set of sound patches (it might sound interesting, but it won't sound anything like what was originally intended). If the MIDI file is loaded into another setup, the sequence will again sound completely different, with patches that are so irrelevant that the guitar track might sound like a bunch of machine-gun shots from the planet Gloop.

In order to eliminate (or at least reduce) the basic differences that exist between systems, a standardized set of patch settings, known as *General MIDI* (GM), was created. In short, General MIDI assigns a specific instrument patch to each of the 128 available program change numbers. Since all electronic instruments that conform to the GM format must use these patch assignments, placing GM program change commands at the header of each track will automatically instruct the sequence to play with its originally intended sounds and general song settings. In this way, no matter what sequencer and system setup is used to play the file back, as long as the receiving instrument conforms to the GM spec, the sequence will be heard using its intended instrumentation.

Tables 10.3 and 10.4 detail the program numbers and patch names that conform to the GM format (Table 10.3 for nonpercussion and Table 10.4 for percussion instruments). These patches include sounds that imitate synthesizers, ethnic instruments, or sound effects that have been derived from early Roland synth patch maps. Although the GM spec states that a synth must

Table 10.3. GM Non-Percussion Instrument Patch Map with Program Change Numbers

1. Acoustic Grand Piano	33. Acoustic Bass	65. Soprano Sax	97. FX 1 (rain)
2. Bright Acoustic Piano	34. Electric Bass (finger)	66. Alto Sax	98. FX 2 (soundtrack)
3. Electric Grand Piano	35. Electric Bass (pick)	67. Tenor Sax	99. FX 3 (crystal)
4. Honky-tonk Piano	36. Fretless Bass	68. Baritone Sax	100. FX 4 (atmosphere)
5. Electric Piano 1	37. Slap Bass 1	69. Oboe	101. FX 5 (brightness)
6. Electric Piano 2	38. Slap Bass 2	70. English Horn	102. FX 6 (goblins)
7. Harpsichord	39. Synth Bass 1	71. Bassoon	103. FX 7 (echoes)
8. Clavi	40. Synth Bass 2	72. Clarinet	104. FX 8 (sci-fi)
9. Celesta	41. Violin	73. Piccolo	105. Sitar
10. Glockenspiel	42. Viola	74. Flute	106. Banjo
11. Music Box	43. Cello	75. Recorder	107. Shamisen
12. Vibraphone	44. Contrabass	76. Pan Flute	108. Koto
13. Marimba	45. Tremolo Strings	77. Blown Bottle	109. Kalimba
14. Xylophone	46. Pizzicato Strings	78. Shakuhachi	110. Bag pipe
15. Tubular Bells	47. Orchestral Harp	79. Whistle	110. Fiddle
16. Dulcimer	48. Timpani	80. Ocarina	112. Shanai
17. Drawbar Organ	49. String Ensemble 1	81. Lead 1 (square)	113. Tinkle Bell
18. Percussive Organ	50. String Ensemble 2	82. Lead 2 (sawtooth)	114. Agogo
19. Rock Organ	51. SynthStrings 1	83. Lead 3 (calliope)	115. Steel Drums
20. Church Organ	52. SynthStrings 2	84. Lead 4 (chiff)	116. Woodblock
21. Reed Organ	53. Choir Aahs	85. Lead 5 (charang)	117. Taiko Drum
22. Accordion	54. Voice Oohs	86. Lead 6 (voice)	118. Melodic Tom
23. Harmonica	55. Synth Voice	87. Lead 7 (fifths)	119. Synth Drum
24. Tango Accordion	56. Orchestra Hit	88. Lead 8 (bass 1 lead)	120. Reverse Cymbal
25. Acoustic Guitar (nylon)	57. Trumpet	89. Pad 1 (new age)	121. Guitar Fret Noise
26. Acoustic Guitar (steel)	58. Trombone	90. Pad 2 (warm)	122. Breath Noise
27. Electric Guitar (jazz)	59. Tuba	91. Pad 3 (polysynth)	123. Seashore
28. Electric Guitar (clean)	60. Muted Trumpet	92. Pad 4 (choir)	124. Bird Tweet
29. Electric Guitar	61. French (muted) Horn	93. Pad 5 (bowed)	125. Telephone Ring
30. Overdriven Guitar	62. Brass Section	94. Pad 6 (metallic)	126. Helicopter
31. Distortion Guitar	63. SynthBrass 1	95. Pad 7 (halo)	127. Applause
32. Guitar harmonics	64. SynthBrass 2	96. Pad 8 (sweep)	128. Gunshot

respond to all 16 MIDI channels, the first 9 channels are reserved for instruments, while GM restricts the percussion track to MIDI channel 10.

Graphics

Graphic imaging occurs on the computer screen in the form of pixels. These are basically tiny dots that blend together to create color images in much the same way that dots are combined to

Table 10.4. GM Percussion Instrument Patch Map (Channel 10)

35. Acoustic Bass Drum	51. Ride Cymbal 1	67. High Agogo
36. Bass Drum 1	52. Chinese Cymbal	68. Low Agogo
37. Side Stick	53. Ride Bell	69. Cabasa
38. Acoustic Snare	54. Tambourine	70. Maracas
39. Hand Clap	55. Splash Cymbal	71. Short Whistle
40. Electric Snare	56. Cowbell	72. Long Whistle
41. Low Floor Tom	57. Crash Cymbal 2	73. Short Guiro
42. Closed Hi-Hat	58. Vibraslap	74. Long Guiro
43. High Floor Tom	59. Ride Cymbal 2	75. Claves
44. Pedal Hi-Hat	60. Hi Bongo	76. Hi Wood Block
45. Low Tom	61. Low Bongo	77. Low Wood Block
46. Open Hi-Hat	62. Mute Hi Conga	78. Mute Cuica
47. Low-Mid Tom	63. Open Hi Conga	79. Open Cuica
48. Hi Mid Tom	64. Low Conga	80. Mute Triangle
49. Crash Cymbal 1	65. High Timbale	81. Open Triangle
50. High Tom	66. Low Timbale	

Note: In contrast to Table 10.3, the numbers in Table 10.4 represent the percussion keynote numbers on a MIDI keyboard, not program change numbers.

give color and form to your favorite comic strip. Just as word length affects the overall amplitude range of a digital audio signal, the number of bits in a pixel's word will affect the range of colors that can be displayed in a graphic image. For example, a 4-bit word only has 16 possible combinations. Thus, a 4-bit word will allow your screen to have a total of 16 possible colors; an 8-bit word will yield 256 colors; a 16-bit word will give you 65,536 colors; and a 24-bit word will yield a whopping total of 16.7 million colors! These methods of displaying graphics onto a screen can be broken down into several categories:

- *Raster graphics*—In raster graphics, each image is displayed as a series of pixels. This image type is what is used utilized when a single graphic image is used (*i.e.*, bitmap, JPEG, GIF, or TIFF format). The sense of motion can come from raster images only by successively stepping through a number of changing images every second (in the same way that standard video images create the sense of motion).
- *Vector graphics*—This process often creates a sense of motion by projecting a background raster image and then overlaying one or more objects that can be animated according to a series of programmable vectors. By instructing each object to move from point A to point B to point C according to a defined script, a sense of animated motion can be created without the need to project separate images for each frame. This script form reduces a file's data size dramatically and is used with several image animation programs (including Macromedia's Flash™, Shockwave™, and Director®).

- *Wireframe animation*—This form of animation uses a computer to create a complex series of wireframe image vectors of a real or imaginary object. Once programmed, these stick-like objects can be filled in with any type of skin, color, shading, etc., and then programmed to move with a staggering degree of realism. Obviously, with the increased power of modern computers and supercomputers, this graphic art form has attained higher degrees of artistry or realism within modern-day film, video, and desktop visual production.

Desktop Video

With the proliferation of digital VCRs, video interface hardware, and video editing software systems, desktop and laptop video has begun to play an increasingly important role in multimedia production and content. Video is encoded into a datastream as a continuous series of successive frames, which are refreshed at rates that vary from 12 or fewer frames/second (fr/sec) to the standard broadcast rates of 29.97 and 30 fr/sec. As with graphic files, a single full-sized video frame can be made up of a gazillion pixels, which are themselves encoded as a digital word of *n* bits. Multiply these figures by nearly 30 frames and you'll come up with a rather impressive data file size.

Obviously, it's more common to find such file sizes and data throughput rates on higher-end desktop systems and professional video editing workstations; however, several options are available to help bring video down to data rates that are suitable for multimedia and even the Internet:

- *Window size*—The basics of making the viewable picture smaller is simple enough: Reducing the frame size will reduce the number of pixels in a video frame, thereby reducing the overall data requirements during playback.
- Frame rate—Although standard video frame rates run at around 30 fr/sec (United States and Japan) and 25 fr/sec (Europe), these rates can be lowered to 12 fr/sec in order to reduce the encode file size or throughput.
- *Compression*—In a manner similar to that which is used for audio, codecs can be applied to a video frame to reduce the amount of data that's necessary to encode the file by filtering out and smoothing over pixel areas that consume data or by encoding data that doesn't change from frame to frame into a shorthand to reduce data. In situations where high levels of compression are needed, it's common to accept degradations in the video's resolution in order to reduce the file size and/or data throughput to levels that are acceptable to a restrictive medium (*e.g.*, the web).

From all of this, it's clear that there are many options for encoding a desktop video file. When dealing with video clips, tutorials, and the like it's common for the viewing window to be medium in size and encoded at a medium to lower frame rate. This middle ground is often chosen in order to accommodate the standard data throughput that can be streamed off of most CD-ROMs and the web. These files are commonly encoded using Microsoft's Audio-Video Interleave (AVI) format, QuickTime (a common codec that was developed by Apple and can be played by either a Mac or PC), or MPEG 1, 2, or 4 (codecs that vary from lower multimedia resolutions to higher ones that are used to encode DVD movies). Both the Microsoft Windows and Apple OS platforms include built-in or easily obtained applications that allow all or most of these file types to be played without additional hardware or software.

Multimedia and the Web in the "Need for Speed'' Era

The household phrase "surfin' the web" has become synonymous with jumping onto the Net, browsing the sites, and grabbin' onto all of those hot songs, videos, and graphics that might wash your way. Dude, with improved audio and video codecs and faster data connections (Table 10.5), the ability to search on any subject, download files, and stream audio or radio stations (Table 10.6) from any point in the world has definitely changed our perception of modern-day communications.

Table 10.5. Internet connection speeds

Connection	*Speed (kbps)*	*Description*
56 k dial-up	56 kbps (usually less)	Common modem connection
ISDN	128 kbps; older technology	—
DSL	384 kbps or higher high bandwidth	Phone line technology
Cable	384 kbps and higher high bandwidth	Cable technology
T1	1.5 Mbps	—
T3	45 Mbps	—
OC-1	52 Mbps	Optical fiber
OC-3	155 Mbps	Optical fiber
OC-12	622 Mbps	Optical fiber
OC-48	2.5 Gbps	Optical fiber
Ethernet	10 Mbps	Local area network (LAN); not an Internet connection
Fast Ethernet	00 Mbps	Local area network (LAN); not an Internet connection

Table 10.6. Streaming Data File Sizes

Data Rate (kbps)	*File Size (MB/min)*	*File Size (MB/hr)*	*Minutes on a 650-MB CD*	*Hours on a 650-MB CD*
64	480	28.8	1354	23
96	720	43.2	903	15
128	960	57.6	677	11
160	1200	72	542	9
256	1920	115.2	339	6
384	2880	172.8	226	4

Thoughts on Being (and Getting Heard) in Cyberspace

Most of us have grown up in this age of the supermarket, where everything is wholesaled, processed, packaged, and distributed to a single clearinghouse that's gotten so big that older folks can only shop there with the aid of a motorized shopping cart. For more than six decades, the music industry has largely worked on a similar principle: Find artists who'll fit into an existing marketing formula (or, more rarely, create a new marketing image), produce and package them according to that formula, and put tons of bucks behind them to get them heard and distributed. Not a bad thing in and of itself; however, for independent artists the struggle has been, and continues to be, one of getting themselves heard, seen, and noticed—without the aid of the well-oiled megamachine. With the creation of cyberspace, not only are established record industry forces able to work their way onto your desktop screen (and into your multimedia speakers), but independent artists also have a new medium for getting heard. Through the creation of a dedicated website, search engines, links from other sites, and independent music dot-coms, as well as through creative gigging and marketing, new avenues have begun to open up for the web-savvy independent artist.

Uploading to Stardom!

If you build it, they will come! This overly simplistic concept definitely doesn't apply to the web. With an ever-increasing number of dot-whatevers going online every month, expecting people to come to your site just because it's there simply isn't realistic. Like anything that's worthwhile, it takes connections, persistence, a good product, and good ol'-fashioned dumb luck to be seen as well as heard! If you're selling your music, T-shirts, or whatever at gigs, on the streets, and to family and friends, cyberspace can help increase sales by making it possible (and even easy) to get your band, music, or clients onto several independent music websites that offer up descriptions, downloadable samples, direct sales, and a link that goes directly to your main website. Such a site could definitely help to get the word out to a potentially new public—and help clue your audience in to what you and your music are all about.

Cyberproducts can be sold and shipped via the traditional mail or phone-in order channels; however, it's long been considered hip in the web world to flash the silver, gold, or platinum credit card to make your purchase. Because attaining your own credit card processing and authorization system can be costly, a number of cyber companies have sprung up that offer secure credit card authorization, billing, and artist payment plans for an overall sales percentage fee (Figure 10.7).

The preview and/or distribution format choice for releasing all or part of your music to the listening audience will ultimately depend on you and the format/layout style that's been adopted by the hosting site. For example, you could do any or all of the following:

- Place short, low-fidelity segments onto a site that can be streamed or downloaded to entice the listener to buy.

Figure 10.7. *iTunes® music website. (Courtesy of Apple Computers, Inc.; www.apple.com.)*

- Provide free access to the entire project (at low or medium fidelity) while encouraging the listener to buy the CD.
- Place several high-fidelity cuts on your site for free as a teaser or as a gift to your fan base.
- Place the music on a secure site that's SDMI compliant, using the pay-per-download system.
- Sell the completed CD package on the site.
- Create a fanzine to keep your fans up to date on goings-on, upcoming releases, diaries, etc.

No matter what or how many cyber distribution methods you choose for getting your music out, always take the time to read through the contractual fine print. Although most are above board and offer to get your music out on a nonexclusive basis ... *caveat emptor* (let the buyer—and content provider—beware)! In your excitement to get your stuff out there, you might not realize that you are signing away the rights for free distribution of that particular project (or worse). This hints at the fact that you're dealing with the music business, and, as with any business, you should always tread carefully in the cyber jungle.

Copyright Protection: Wanna Get Paid?

It seems that, at least in recent years, many (but by no means all) of the problems of piracy have been addressed. With the shutdown of illegal music download sites and peer-to-peer networks and the risen-from-ashes version of pay-per-download sites (such as www.iTunes.com and the reborn www.napster.com.), many of the major labels and larger independent artists are on the verge of seeing a light at the end of the online tunnel. Even so, the technology that allows music to be shared online is still subject to abuse that can lead to lost revenues.

Secure Digital Music Initiative

With the vast number of software (and hardware) systems that are able to rip CDs to MP3s (or any other format) and MP3s back to audio and CD-R, the powers that be in the recording industry have grown increasingly fearful of the rising prevalence of copyright infringement. Although many online music sites legally use these formats to allow potential buyers to preview music before buying or to simply put unreleased cuts onto the Web as a freebie, a number of sites still exist that connect online users to databases of music that has been illegally ripped and posted. Obviously, neither the artists nor the record companies are being compensated for this distribution of their music.

As a result, the Recording Industry Association of America (RIAA), major record labels, and industry organizations have helped to form the Secure Digital Music Initiative (SDMI; www.sdmi.org). SDMI is an independent forum that brings together the worldwide recording, consumer electronics, and information technology industries to develop open specifications for protecting digital music distribution. As a result of these efforts, the *Digital Rights Management* (DRM) system was developed as a secure and encrypted means of solving the issue of unauthorized copying. DRM functions by digitally locking the content and limiting its distribution to only those who pay for the content. In short, it acts as a digital watermark that identifies the copyright owner and provides an electronic key that allows access to the music or information once the original copy has been legally purchased.

One such DRM-compliant online and digital distribution system is Weed (www.weedshare.com.). This system, which was developed primarily as a distribution tool for the independent artist, allows the end-user to listen to a song three times for free. After three listens, the built-in DRM will alert the user that it's time to pay up (an amount that's set by the artist or copyright owner). Once the payment is made into a PayPal account, the artist receives 50% of the sale. If that end-user shares the Weed file with a friend who also buys the song, the artist gets 50% of that sale and the original end-user gets a percentage of the sale … and so on. In the end, the artist, Weed, and those down the musical food chain get paid (Figure 10.8).

Internet Radio

Due to the increased bandwidth of many Internet connections and improvements in audio streaming technology, many of the world's radio stations have begun to broadcast on the web. In addition to offering a worldwide platform for traditional radio listening audiences, a large number of corporate and independent web radio stations have begun to spring up that can help to

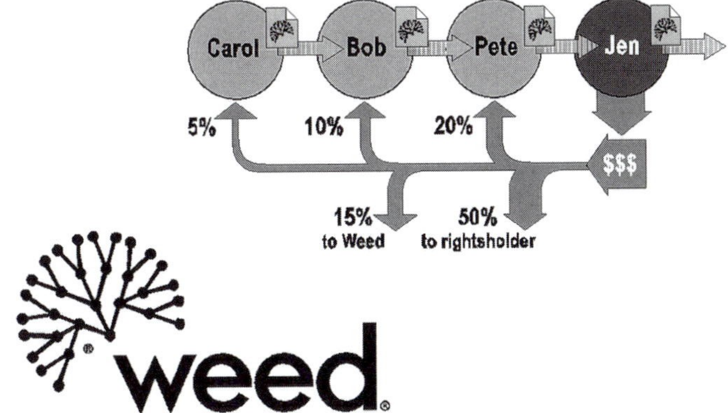

Figure 10.8. The Weed distribution/payment model. (Courtesy of Shared Media Licensing, Inc.; www.weedshare.com.)

increase the fan and listener base of musicians and record labels. Go ahead, get on the web and listen to your favorite Mexicano station *en vivo* ... catch the latest dance craze from London ... or chill to reggae rhythms streaming in from the islands, man.

The Virtual E-Dub

In addition to providing another vehicle for getting an artist's music out to the public at large, the Internet is making it easier for artists to e-collaborate over cyberspace. One approach is to E-dub across the web. Using this approach, you can create a rough session mix of all or a portion of a song and export it to a medium-resolution MP3 or preferred codec. This file (along with sheet music jpgs, descriptive docs, etc.) could be sent to a collaborative buddy—across the street or across the world—who will then load the file into his or her DAW. A track or set of tracks can then be over-dubbed to the original mix in the traditional fashion. The resulting file (or files) should then be encoded without compression or at a high resolution (at least 192 kbps) and then e-mailed back to the artist or producer for re-import back into the original session at the proper time and position. (If your DAW doesn't automatically convert the file back into the session's native file format, you'll need to manually convert the file yourself.) Using this system, the world could be your cost-effective E-dub oyster.

On a Final Note

One of the most amazing things about multimedia, cyberspace, and their related technologies is the fact that they're ever changing. By the time you read this book, many new developments will have occurred. Old concepts fade away; new and possibly better ones will take over and then begin to take on a new life of their own. Although I've always had a fascination with crystal balls and have often had a decent sense about new trends in technology, there's simply no way to foretell the many amazing things that lie ahead in the fields of music, music technology, multimedia—and especially cyberspace. As with everything techno, I encourage you to read the trades and surf the Web to keep abreast of the latest and greatest tools that have recently arrived or are about to rise on the techno horizon.

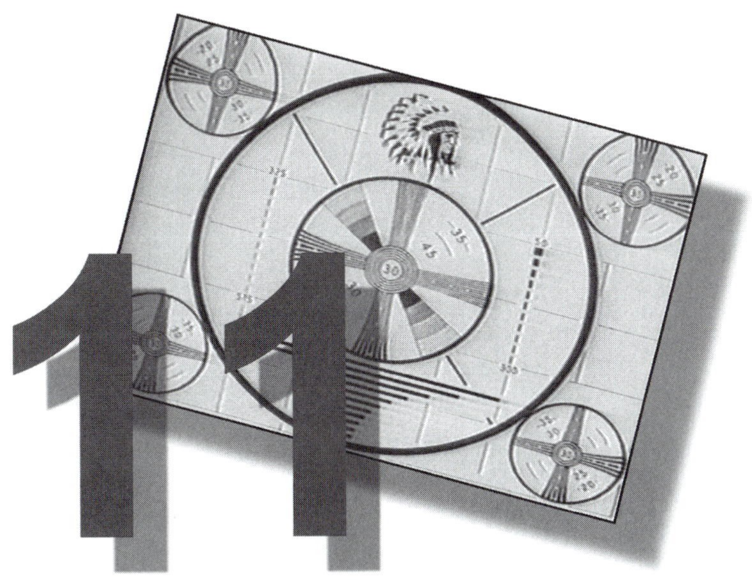

11 Synchronization

Over the years, electronic music has evolved into an indispensable production tool within almost all forms of media production. In video postproduction, for example, audio and video transports, digital audio workstations (DAWs), automated console systems, and electronic musical instruments routinely work together to help create a soundtrack and refine it into its finished form (Figure 11.1). The technology that allows multiple audio and visual media to operate in tandem so as to maintain a direct time relationship is known as *synchronization*, or *sync*.

Strictly speaking, synchronization occurs when two or more related events happen at precisely the same time. With respect to analog audio and video systems, sync is achieved by interlocking the transport speeds of two or more machines. For computer-related systems (such as digital audio, MIDI, and digital video), synchronization between devices is often achieved through the use of a timing clock that can be fed through a separate line or can be directly embedded within the digital data line itself. Frequently, it's necessary for analog and digital devices to be synchronized together; as a result, a number of ingenious forms of systems communication and data translation have been developed. In this chapter, the various forms of synchronization used for both analog and digital devices are discussed, as well as current methods for maintaining sync between media types.

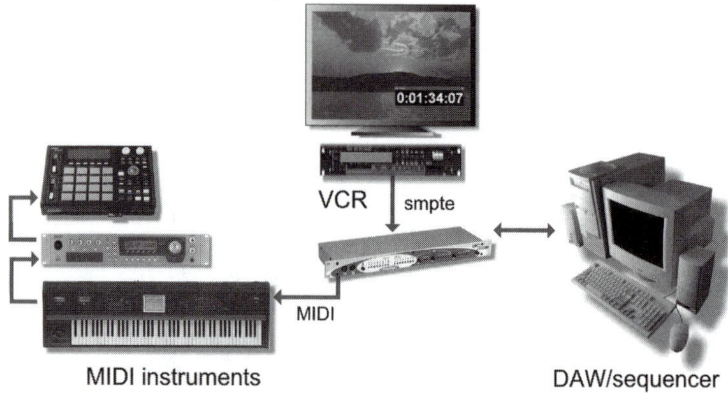

Figure 11.1. *Example of an integrated audio production system.*

Synchronization Between Media Transports

Maintaining relative sync between media devices doesn't require that all transport speeds involved in the process be constant; however, it's important that they maintain the same relative speed and position over the course of a program. Physical analog devices, for example, have a particularly difficult time achieving this. Due to differences in mechanical design, voltage fluctuations, and tape slippage, it's a simple fact of life that analog tape devices aren't able to maintain a constant playback speed, even over relatively short durations. For this reason, accurate sync between analog and digital machines would be nearly impossible to achieve over any reasonable program length without some form of timing lock. It therefore quickly becomes clear that if production is to utilize multiple forms of media and record/playback systems, a method for maintaining sync is essential.

Time Code

The standard method of interlocking audio, video, and film transports makes use of a code that was developed by the Society of Motion Picture and Television Engineers (SMPTE; www.smpte.org). This time code (or SMPTE time code) identifies an exact position on a tape or media form by assigning a digital address that increments over the course of a program's duration. This address code can't slip and always retains its original location, allowing for the continuous monitoring of tape position to an accuracy of between 1/24th and 1/30th of a second (depending on the media type and frame rates being used). These divisional segments are called *frames*, a term taken from film production. Each audio or video frame is tagged with a unique identifying number, known as a *time code address*. This eight-digit address is displayed in the form 00:00:00:00, whereby the successive pairs of digits represent hours:minutes:seconds:frames—HH:MM:SS:FF (Figure 11.2).

The recorded time code address is then used to locate a position on magnetic tape (or any other recorded media) in much the same way that a letter carrier uses a written address to match up,

Figure 11.2. Readout of a SMPTE time code address in HH:MM:SS:FF (www.smpte.org).

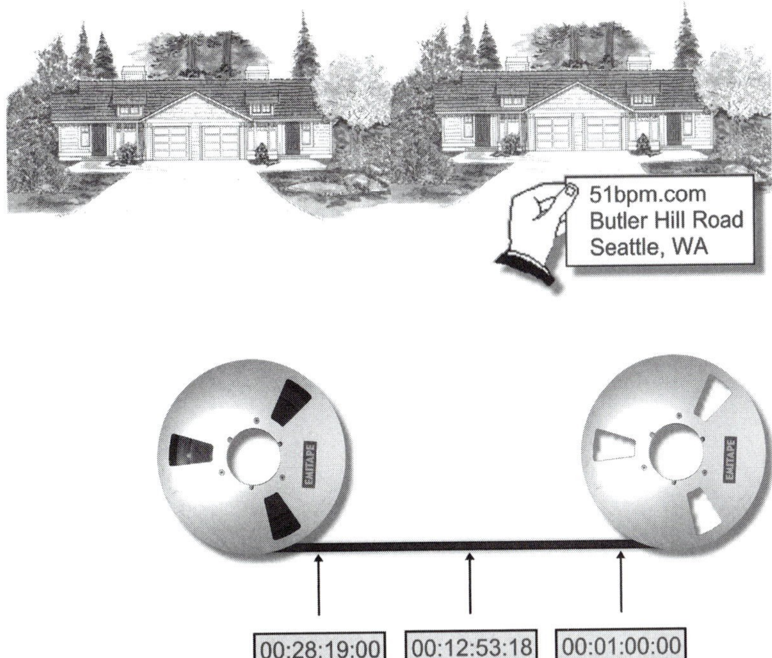

Figure 11.3. Location of relative addresses: (a) postal address; (b) time code addresses and a cue point on longitudinal tape.

locate, and deliver a letter to a specific, physical residence (*i.e.*, by matching the address, you can then find the desired physical location point, as shown in Figure 11.3a). Let's suppose that a time-encoded multitrack tape begins at time 00:01:00:00, ends at 00:28:19:00, and contains a specific cue point (such as a glass shattering) at 00:12:53:18 (Figure 11.3b). By monitoring the time code readout, it's a simple matter to locate the precise position that corresponds to the cue point on the tape and then perform whatever function is necessary, such as inserting an effect into the sound track at that specific point … CRASH!

It should be noted that the standard method for encoding time code in analog audio production is to record (stripe) SMPTE time code onto the highest available track (*e.g.*, track 24). This track can then be read directly from the track in either direction and at a wide range of transport tape

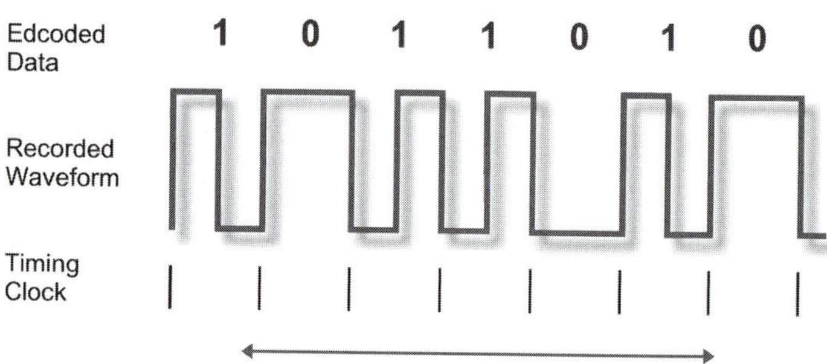

Figure 11.4. Biphase modulation encoding.

speeds. As we'll see later, digital audio devices often deal with time code and its distribution in various ways.

Time Code Word

The total of all time-encoded information that's encoded into each audio or video sync frame is known as a *time code word*. Each word is divided into 80 equal segments, which are numbered consecutively from 0 to 79. One word covers an entire audio or video frame, such that for every frame there is a unique and corresponding time code address. Address information is contained in the digital word as a series of bits that are made up of binary 1's and 0's, which (in the case of an analog SMPTE signal) are electronically encoded in the form of a modulated square wave. This method of encoding information is known as *biphase modulation*. Using this code type, a voltage transition in the middle of a half-cycle of a square wave represents a bit value of 1, while no transition within this same period signifies a bit value of 0 (Figure 11.4). The most important feature about this system is that detection relies on shifts within the pulse and not on the pulse's polarity. Consequently, time code can be read in either the forward or reverse direction, as well as at fast or slow shuttle speeds.

Do-It-Yourself Tutorial: SMPTE Time Code

1. Go to the "Tutorial section" of www.modrec.com, click on "SMPTE Audio Example," and play the time code sound file. Not my favorite tune, but it's a useful one!
2. The 80-bit time code word is subdivided into groups of 4 bits (Figure 11.5), whereby each grouping represents a specific coded piece of information. Each 4-bit segment represents a binary-coded decimal (BCD) number that ranges from 0 to 9. When the full frame is scanned, all eight of these 4-bit groupings are read out as a single SMPTE frame number (in hours, minutes, seconds, and frames).

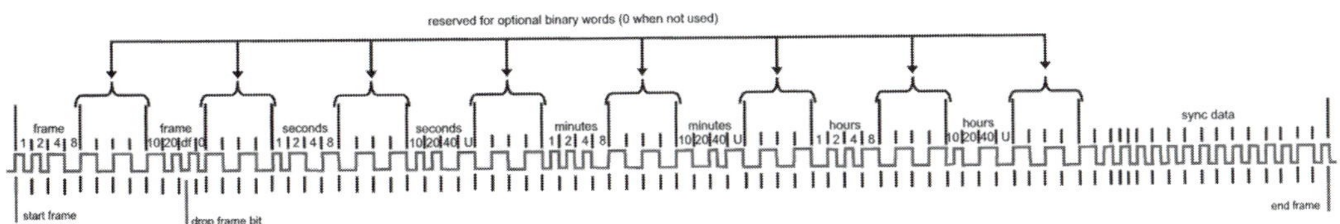

Figure 11.5. *Biphase representation of the SMPTE time code word.*

Sync Information Data

Another form of information that's encoded into the time code word is sync data. This information exists as 16 bits at the end of the time code word, which are used to define the end of each frame. Because time code can be read in either direction, sync data is also used to tell the device which direction the tape or digital device is moving.

Time Code Frame Standards

In productions using time code, it's important that the readout display be directly related to the actual elapsed time of a program, particularly when dealing with the exacting time requirements of broadcasting. For this reason, time code frame rates may vary from one medium, production house, or country of origin to another.

- 30 fr/sec (monochrome U.S. video)—In the case of a black and white (monochrome) video signal, a rate of exactly 30 frames per second (fr/sec) is used. If this rate (often referred to as *non-drop code*) is used on a black-and-white program, the time code display, program length, and actual clock-on-the-wall time would all be in agreement.
- 29.97 fr/sec (drop-frame time code for color NTSC video)—This simplicity was eliminated, however, when the National Television Standards Committee (NTSC) set the frame rate for the color video signal in the United States and Japan at 29.97 fr/sec. Thus, if a time code reader that's set up to read the monochrome rate of 30 fr/sec were used to read a color program, the time code readout would pick up an extra 0.03 frame for every second that passes. Over the duration of an hour, the time code readout would differ from the actual elapsed time by a total of 108 frames (or 3.6 seconds). In order to correct this difference and bring the time code readout and the actual elapsed time back into agreement, a series of frame adjustments were introduced into the code. Because the goal is to drop 108 frames over the course of an hour, the code used for color has come to be known as *drop-frame code*. In this system, two frame counts for every minute of operation are omitted from the code, with the exception of minutes 00, 10, 20, 30, 40, and 50. This has the effect of adjusting the frame count so that it agrees with the actual elapsed duration of a program.
- 29.97 fr/sec (non-drop frame code)—In addition to the color 29.97-drop-frame code, 29.97 non-drop-frame color standard can also be found in video production. When using

non-drop time code, the frame count will always advance one count per frame, without any drops. As you might expect, this mode will result in a disagreement between the frame count and the actual clock-on-the-wall time over the course of the program. Non-drop, however, has the distinct advantage of easing the time calculations that are often required in the video editing process (because no frame compensations need to be taken into account).

- 25-fr/sec EBU (standard rate for PAL video)—Another frame rate format that's used throughout Europe is the European Broadcast Union (EBU) time code. EBU utilizes SMPTE's 80-bit code word but differs in that it uses a 25-fr/sec frame rate. Because both monochrome and color video EBU signals run at exactly 25 fr/sec, an EBU drop-frame code isn't necessary.
- 24-fr/sec (standard rate for film work)—The medium of film differs from all of these as it makes use of an SMPTE time code format that runs at 24 fr/sec.

From the above, it's easy to understand why confusion often exists as to which frame rate should be used on a project. Basically, if you are working on an in-house project that doesn't incorporate time-encoded material that comes from the outside world, you should choose a rate that both makes sense for you and is likely to be compatible with an outside facility (should the need arise). For example, American electronic musicians who are working in-house will often choose to work at 30 fr/sec. Those in Europe have it easy, because in that continent 25 fr/sec is the logical choice for all music and video productions. On the other hand, those who work with projects that come through the door from other production houses will need to take special care to reference their time code rates to those used by the originating media house. This can't be stressed enough: If care isn't taken to keep your time code references at the proper rate, while keeping degradation to a minimum from one generation to the next, the various media might have trouble syncing up when it comes time to put the final master together—and that can spell BIG trouble.

LTC and VITC Time Code

Currently, two major systems exist for encoding time code onto magnetic tape:

- Longitudinal time code (LTC)
- Vertical interval time code (VITC)

Time code recorded onto an analog audio or video cue track is known as longitudinal time code (LTC). LTC encodes a biphase time code signal onto the analog audio or cue track in the form of a modulated square wave at a bit rate of 2400 bits/second. The recording of a perfect square wave onto a magnetic audio track is difficult, even under the best of conditions. For this reason, the SMPTE standard has set forth an allowable rise time of 25±5 microseconds for the recording and reproduction of valid code. This tolerance requires a signal bandwidth of 15 kHz, which

Figure 11.6. Video image showing a burned-in time code window.

is well within the range of most professional audio recording devices. Variable-speed time code readers are often able to decode time code information at shuttle rates ranging from 1/10th to 100 times normal playing speed. This is effective for most audio applications; however, in video postproduction it's often necessary to monitor videotape at slow or still speeds.

As LTC can't be read at speeds slower than 1/10th to 1/20th normal playspeed, two methods can be used for reading time code. The first of these uses a character generator to burn time code addresses directly into the video image of a worktape copy. This superimposed readout allows the time code to be easily identified, even at very slow or still picture shuttle speeds (Figure 11.6). In most situations, LTC code is preferred for audio, electronic music, and mid-level video production, as it's a more accessible and cost-effective protocol.

A second method—one that is used by major video production houses—is the vertical interval time code (VITC). VITC makes use of the same SMPTE address and user code structure as LTC but is encoded onto videotape in an entirely different manner. VITC actually encodes the time code information into the video signal itself, inside a field known as the vertical blanking interval. This field is located outside the visible picture scan area. Because the time code data is encoded into the video signal itself, professional helical scan video recorders are able to read time code at very slow and even still-frame speeds. Since time code is encoded in the video signal, an additional track can be opened up on a video recorder for audio or cue information, while also eliminating the need for a burned window dub.

Time Code Refreshment/Jam Sync

Longitudinal time code operates by recording a series of square-wave pulses onto magnetic tape. As you now know, it's somewhat difficult to record a square waveform onto analog magnetic tape without having the signal suffer moderate to severe waveform distortion (Figure 11.7). Although time code readers are designed to be relatively tolerant of waveform amplitude fluctuations, such distortions are severely compounded when code is copied from tape to tape by one or more

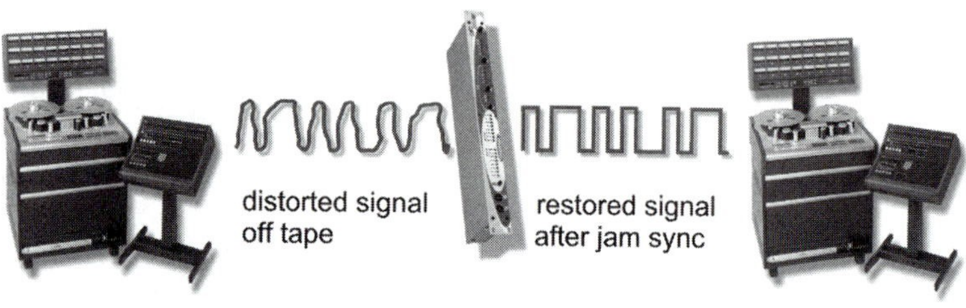

***Figure 11.7.** Jam sync is used to restore distorted SMPTE when copying code from one machine to another.*

generations. For this reason, a time code refresher has been incorporated into most time code synchronizers and MIDI interface devices that have sync capabilities. Basically, this process reads the degraded time code information from a previously recorded track and then amplifies and regenerates the square wave back into its original shape so it can be freshly recorded to a new track or read by another device.

Should the quality of a SMPTE signal degrade to the point where the synchronizer can't differentiate between the pulses, the code will disappear and the slaves will stop unless the system includes a feature known as *jam sync*. Jam sync refers to the synchronizer's ability to output the next time code value, even though one has not appeared at its input. The generator is said to be working in a freewheeling fashion, since the generated code may not agree with the actual recorded address values; however, if the dropout occurs for only a short period, jam syncing works well to detect or refresh the signal. (This process is often useful when dealing with dropouts or undependable code from VHS audio tracks.) Two forms of jam sync options are available:

- Freewheeling
- Continuous

In the freewheeling mode, the receipt of time code causes the generator's output to initialize when a valid address number is detected. The generator then begins to count in an ascending order on its own, ignoring any deterioration or discontinuity in code and producing fresh, uninterrupted SMPTE address numbers. Continuous jam sync is used in cases where the original address numbers must remain intact and shouldn't be regenerated as a continuously ascending count. After the reader has been activated, the generator updates the address count for each frame in accordance with incoming address numbers and outputs an identical, regenerated copy.

Synchronization Using SMPTE Time Code

In order to achieve a frame-by-frame time code lock between multiple audio, video, and film analog transports, it's necessary to use a device or integrated system that's known as a *synchronizer* (Figure 11.8). The basic function of a synchronizer is to control one or more tape, computer-based, or film transports (designated as slave machines) so their speeds and relative positions are made to accurately follow one specific transport (designated as the master). Although the lines of distinction often break down, synchronization as a whole can be divided into two basic system types: those that are used in project or electronic music production facilities

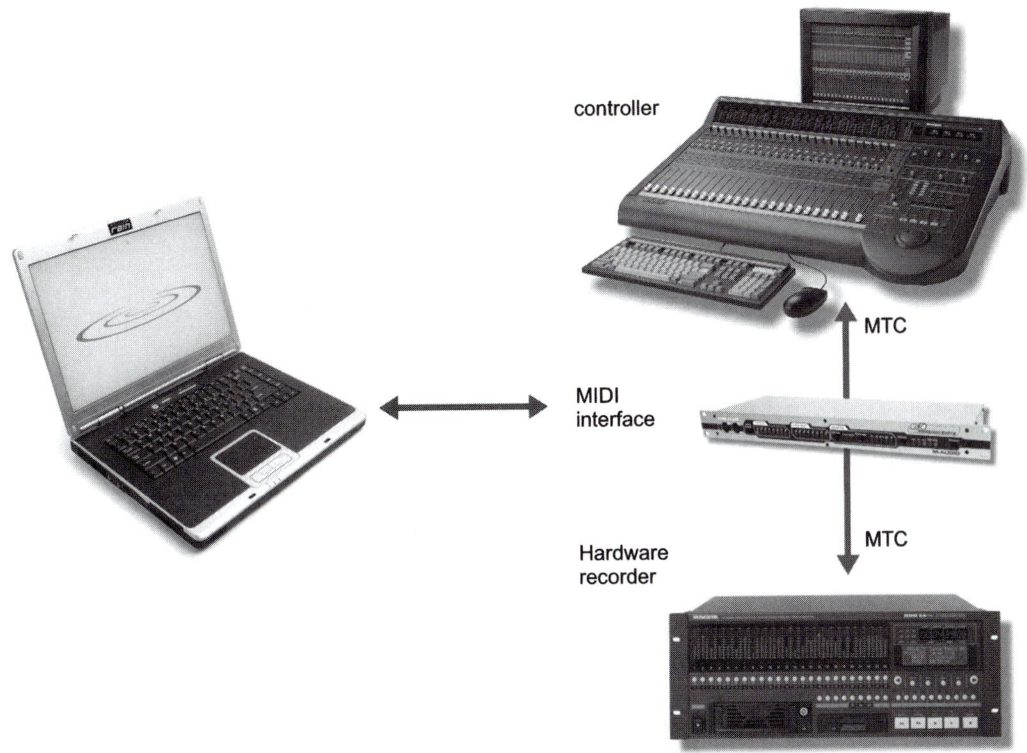

Figure 11.8. Example of time code sync production using a simple MIDI interface synchronizer within a project studio.

and those that can be found in larger audio and video production and postproduction facilities. The greatest reason for this division is not so much a system's performance as its price and the types of devices that are used in the process.

The use of a synchronizer within a project studio environment often involves a multiport MIDI interface that includes provisions for locking an analog audio or video transport to a digital audio, MIDI, or electronic music system by translating LTC SMPTE code into MIDI time code (more on this later in the chapter). In this way, one simple device can cost-effectively serve multiple purposes to achieve lock with a high degree of accuracy. Systems that are used in video production and in higher levels of production will often require a greater degree of control and remote-control functions throughout the studio or production facility. Such a setup will often require a more sophisticated device, such as a control synchronizer or an edit decision list (EDL) controller.

SMPTE Offset Times

In the real world of audio production, programs or songs don't always begin at 00:00:00:00. Let's say that you were handed a recording that needed a synth track laid down onto track 7 of a song that goes from 00:11:24:03 to 00:16:09:21. Instead of inserting more than 11 minutes of empty bars into a MIDI track on your synched DAW, you could simply insert an offset start time of 00:11:24:03. This means that the sequenced track will begin to increment from measure 1 at 00:11:24:03 and will maintain relative offset sync throughout the program.

Offset start times are also useful when synchronizing devices to an analog or videotape source that doesn't begin at 00:00:00:00. As you're probably aware, it takes a bit of time for an analog audio transport to settle down and begin playing (this wait time often quadruples whenever videotape is involved). If a program's time code were to begin at the head of the tape, it's extremely unlikely that you would want to start a program at 00:00:00:00, since playback would be delayed and extremely unstable at this time. Instead, most programming involving an analog audio or video media is striped with an appropriate pre-roll of anywhere from 10 seconds to 2 minutes. Such a pre-roll gives all of the transports ample time to begin playback and sync up to the master TC source.

In addition, it's often wise to start the actual production or first song at an offset time of 01:00:00:00 (some facilities begin at 00:01:00:00). This minimizes the possibility that the synchronizer will become confused by rolling over at midnight; that is, if the content starts at 00:00:00:00, the pre-roll would be in the 23:59:00:00 range and the synchronizer would try to rewind to zero (rolling the tape off the reel) instead of rolling forward. Not always fun in the heat of a production!

Loops

SMPTE in/out times can be programmed into the DAW or sequencer to play, record, or punch-in at a specific address and then loop back to the beginning and start over. This process is common in music studios and audio for visual production, as well as in theme parks where a short demonstration or show is repeatedly played under time code lock in a continuous loop.

Distribution of SMPTE Signals

In a basic audio production system, the only connection that's usually required between the master machine and a synchronizer is the LTC time code track. Generally, when connecting analog slave devices, two connections will need to be made between each transport and the synchronizer. These include lines for the time code reproduce track and the control interface (which often uses the Sony 9-pin remote protocol for giving the synchronizer full logic transport and speed-related feedback information). LTC signal lines can be distributed throughout the production system in the same way that any other audio signal is distributed. They can be routed directly from machine to machine or patched through audio switching systems via balanced, shielded cables or unbalanced cables, or a combination of both. Because the time code signal is biphase or symmetrical, it's immune to cable polarity problems.

Time Code Levels

One problem that can plague systems using time code is crosstalk. Such problems arise from having a high-level time code signal leak into adjacent signal paths or analog tape tracks. Currently, no industry standard levels exist for the recording of time code onto magnetic tape or digital tape

Table 11.1. Optimum Time Code Recording Levels.

Tape Format	*Track Format*	*Optimum Recording Level*
ATR	Edge track (highest number)	−5 to −10 VU
3/4-inch VTR	Audio 1 (L) track or time	−5 to 0 VU code
1-inch VTR	Cue track or audio 3	−5 to −10 VU
MDM	Highest number track	−20 Db

Note: If the VTR is equipped with automatic gain compensation (AGC), override the AGC and adjust the signal gain controls manually.

track; however, the levels shown in Table 11.1 can help you get a good signal level while keeping distortion and analog crosstalk to a minimum.

Synchronization in the Pre-MIDI Era

Before the MIDI specification was implemented, electronic instruments and devices used other types of synchronization methods. Although sync between these non-MIDI and MIDI instruments was a source of mild to major aggravation, a number of these older devices can still be found in MIDI setups because of their distinctive and wonderful sounds.

Click Sync

Click sync or click track refers to the metronomic audio clicks that are generated to communicate tempo. These are produced once per beat or once per several beats (as occurs in cut time or compound meters). Often, a click or metronome is designed into a MIDI interface or sequencing software to produce an audible tone or to trigger MIDI instrument notes that can be used as a tempo guide. These guide clicks help a musician keep in tempo with a sequenced composition. Certain sync boxes and older drum machines can sync a sequence to a live or recorded click track. They can do this by determining the beat based on the tempo of the clicks and then output a MIDI start message (once a sufficient number of click pulses has been received for tempo calculation). A MIDI stop message might be transmitted by such a device whenever more than two clicks have been missed or whenever the tempo falls below 30 beats/minute. Note that this sync method doesn't work well with rapid tempo changes, because chase resolutions are limited to one click per beat (1/24th the resolution of MIDI clock). Thus, it's best to use a click source that is relatively constant in tempo.

TTL and DIN Sync

One of the most common ways to lock sequencers, drum machines, and instruments together, before the adoption of MIDI, was TTL 5-volt sync. In this system, a musical beat is divided into a specific number of clock pulses per quarter note (PPQN), which varies from device to device;

for example, DIN sync (a form of TTL sync, which is named after the now famous 5-pin DIN connector) is transmitted at a rate of 24 PPQN. TTL can be transmitted in either one of two ways. The first and simplest way uses a single conductor that passes a 5-volt clock signal. Quite simply, once the clock pulses are received by a slave device, it will start playing and synchronize to the incoming clock rate. Should these pulses stop, the devices will also stop and wait for the clock to again resume. The second method uses two conductors, both of which transmit 5-volt transitions; however, with this system, one line is used to constantly transmit timing information, while the other is used for start/stop information.

FSK

As technology developed, musicians discovered that it was possible to lock instruments that used TTL 5-volt sync to an analog tape recorder. This was done by recording a master sync square-wave pulse onto tape (Figure 11.9a). Since the most common pulse in use was 24 and 48 PPQN, the recorded square wave consisted of an alternating 24- or 48-Hz signal. Although this system worked, it wasn't without its difficulties, because the synchronized devices relied on the integrity of the square wave's sharp transition edges to provide the clock. Because tape is notoriously bad at reproducing a square wave (Figure 11.9b), the poor frequency response and reduced reliability at low frequencies mandated that a better system for synchronizing MIDI to tape be found. The initial answer was in *frequency shift keying*, better known as FSK.

FSK works in much the same way as the TTL sync track. However, instead of recording a low-frequency square wave onto tape, FSK uses two, high-frequency square-wave signals for marking clock transitions (Figure 11.9c). In the case of the MPU-401/compatible interface, these two frequencies are 1.25 and 2.5 kHz, the rate at which these pitches alternate determines the master timing clock to which all slaved devices are synched. These devices are able to detect a change in modulation, convert these into a clock pulse, and advance their own clocks accordingly.

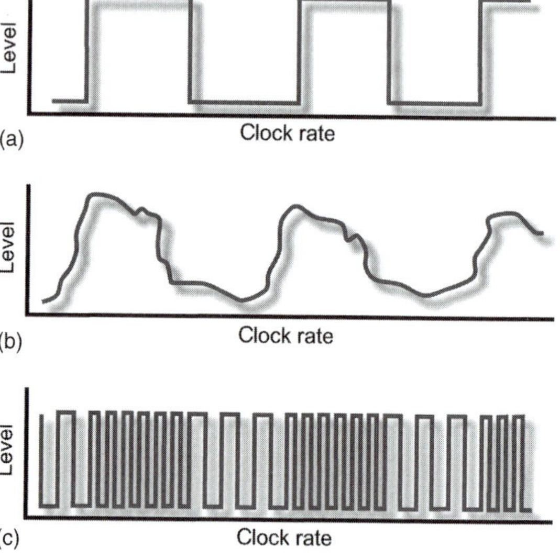

Figure 11.9. *TTL and FSK sync waveforms: (a) original TTL square-wave pulse; (b) playback of TTL sync pulse from tape; (c) modulated FSK sync pulse.*

Unlike most other forms of sync, FSK triggers and plays the sequence relative to the initial clock that's recorded onto tape. As such, the sequence must be positioned at its beginning point and the tape must be cued to a point before the beginning of the song. Once the initial sync pulse is received, the sequencer will begin playback. Should a mistake happen, you'll need to recue the song back to its beginning point ... Now can you see why this form of sync died out?

MIDI-Based Synchronization

With the acceptance of MIDI and digital audio within media production there came an immediate need for a new and cost-effective synchronization protocol that could be easily used in project, mid- and large-scaled production environments. Over the years, devices such as DAWs, MIDI sequencers, digital mixing consoles, and effects devices have become increasingly integrated and networked into the studio environment. These advances saw the rise of an easy-to-use and inexpensive standard that uses MIDI to transmit sync and time code data throughout a connected production system (Figure 11.10). The following sections outline the various forms of synchronization that are often encountered in a MIDI-based production environment. Simply stated, most current forms of synchronization use the MIDI protocol itself for the transmission of sync messages. These messages are transmitted along with other MIDI data over standard MIDI cables, with no need for additional or special connections.

MIDI Real-Time Messages

While MIDI isn't related to SMPTE time code or any external reference, it's important to note that MIDI has a built-in (and often transparent) protocol for synchronizing all of the tempo and timing elements of each MIDI device in a system to a master clock. This protocol operates by

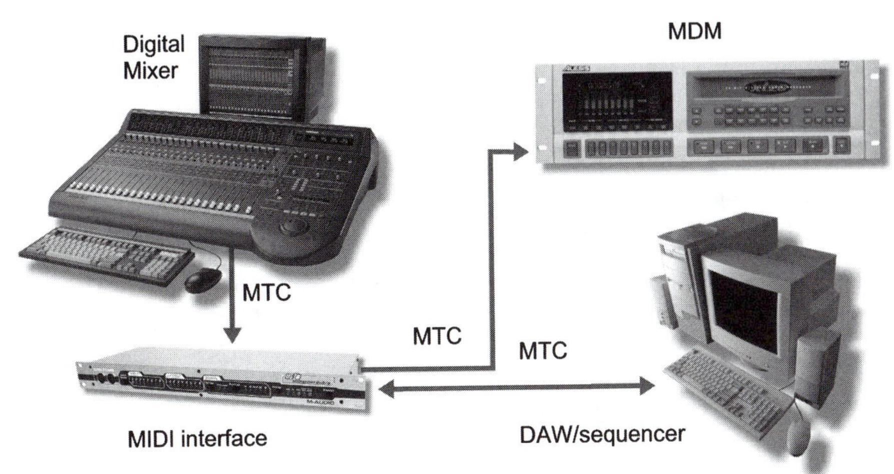

Figure 11.10. *Many time-based media devices in the studio can be cost effectively connected via MIDI time code (MTC).*

transmitting real-time messages to the various devices through standard MIDI cables, USB, and internal CPU paths. Although these relationships are often automatically defined within a system setup, one MIDI device must be designated as the master device in order to provide the timing information to which all other slaved devices are locked. MIDI real-time messages consist of four basic types that are each 1 byte in length:

- *Timing clock*—A clock timing that's transmitted to all devices in the MIDI system at a rate of 24 pulses per quarter note (ppq). This method is used to improve the system's timing resolution and simplify timing when working in nonstandard meters (*e.g.*, 3/8, 5/16, 5/32).
- *Start*—Upon receipt of a timing clock message, the start command instructs all connected devices to begin playing from the beginning of their internal sequences. Should a program be in mid-sequence, the start command repositions the sequence back to its beginning, at which point it begins to play.
- *Stop*—Upon the transmission of a MIDI stop command, all devices in the system stop at their current positions and wait for a message to follow.
- *Continue*—Following the receipt of a MIDI stop command, a MIDI continue message instructs all instruments and devices to resume playing from the precise point at which the sequence was stopped. Certain older MIDI devices (most notably drum machines) aren't capable of sending or responding to continue commands. In such a case, the user must either restart the sequence from its beginning or manually position the device to the correct measure.

Song Position Pointer

In addition to MIDI real-time messages, the *song position pointer* (SPP) is a MIDI system common message that isn't commonly used in current-day production. Essentially, SPP keeps track of the current position in the song by noting how many measures have passed since the beginning of a sequence. Each pointer is expressed as multiples of 6 timing-clock messages and is equal to the value of a 1/16 note. The song position pointer can synchronize a compatible sequencer or drum machine to an external source from any position within a song containing 1024 or fewer measures; thus, when using SPP, it is possible for a sequencer to chase and lock to a multitrack tape from any measure point in a song.

Using such a MIDI/tape setup, a specialized sync tone is transmitted that encodes the sequencer's SPP messages and timing data directly onto tape as a modulated signal. Unlike SMPTE time code, the encoding method wasn't standardized between manufacturers. This lack of standardization prevents SPP data written by one device from being decoded by another device that uses an incompatible proprietary sync format.

Unlike SMPTE, where tempos can be easily varied by inserting a tempo change at a specific SMPTE time, once the SPP control track is committed to tape, the tape and sequence are locked into this predetermined tempo or tempo change map. SPP messages are usually transmitted only while the MIDI system is in the stop mode, in advance of other timing and MIDI continue messages. This is due to the relatively short time period that's needed to locate the slaved device to the correct measure position.

MIDI Time Code

MIDI time code (MTC) was developed to allow electronic musicians, project studios, video facilities, and virtually all other production environments to cost-effectively and easily translate time code into time-stamped messages that can be transmitted via MIDI. Created by Chris Meyer and Evan Brooks, MIDI time code enables SMPTE-based time code to be distributed throughout the MIDI chain to devices or instruments that are capable of synchronizing to and executing MTC commands. MIDI time code is an extension of MIDI 1.0, which makes use of existing message types that were either previously undefined or were being used for other, nonconflicting purposes.

Since most modern recording devices include MIDI in their design, there's often no need for external hardware when making direct connections. Simply chain the MIDI cables from the master to the appropriate slaves within the system (via physical cables, USB, or virtual internal routing). Although MTC uses a reasonably small percentage of MIDI's available bandwidth (about 7.68% at 30 fr/sec), it's customary (but not necessary) to separate these lines from those that are communicating performance data when using MIDI cables. As with conventional SMPTE, only one master can exist within an MTC system, while any number of slaves can be assigned to follow, locate, and chase to the master's speed and position. Because MTC is easy to use and is often included free in many system and program designs, this technology has grown to become the most straightforward and commonly used way to lock together such devices as DAWs, modular digital multitracks, and MIDI sequencers, as well as analog and videotape machines (by using a MIDI interface that includes a SMPTE-to-MTC converter).

MIDI Time Code Control Structure

The MIDI time code format can be divided into two parts:

- Time code
- MIDI cueing

The time code capabilities of MTC are relatively straightforward and allow devices to be synchronously locked or triggered to SMPTE time code. MIDI cueing is a format that informs a MIDI device of an upcoming event that's to be performed at a specific time (such as load, play, stop, punch in/out, reset). This protocol envisions the use of intelligent MIDI devices that can prepare for a specific event in advance and then execute the command on cue.

MIDI time code is made up of three message types: quarter-frame messages, full messages, and MIDI cueing messages:

- *Quarter-frame messages*—These are transmitted only while the system is running in real or variable speed time, in either forward or reverse direction. True to its name, four quarter-frame messages are generated for each time code frame. Since 8 quarter-frame messages are required to encode a full SMPTE address (in hours, minutes, seconds, and frames—00:00:00:00), the complete SMPTE address time is updated once every two frames. In other words, at 30 fr/sec, 120 quarter-frame messages would be transmitted per second, while the full time code address would be updated 15 times in the same period. Each quarter frame message contains 2 bytes. The first byte is F1, the quarter-frame common header; the second byte contains a nibble (four hits) that represents the message number (0 through 7) and a nibble for encoding the time field digit.

- *Full messages*—Quarter-frame messages are not sent in the fast-forward, rewind, or locate modes, as this would unnecessarily clog a MIDI data line. When the system is in any of these shuttle modes, a full message is used to encode a complete time code address within a single message. After a fast shuttle mode is entered, the system generates a full message and then places itself in a pause mode until the time-encoded slaves have located to the correct position. Once playback has resumed, MTC will again begin sending quarter frame messages.
- *MIDI cueing messages*—MIDI cueing messages are designed to address individual devices or programs within a system. These 13-bit messages can be used to compile a cue or edit decision list, which in turn instructs one or more devices to play, punch in, load, stop, and so on, at a specific time. Each instruction within a cueing message contains a unique number, time, name, type, and space for additional information. At the present time, only a small percentage of the possible 128 cueing event types has been defined.

SMPTE/MTC Conversion

A SMPTE-to-MIDI converter is used to read incoming SMPTE time code and convert it into MIDI time code (and *vice versa*). These conversion systems are available as a stand-alone device or as an integrated part of a multiport MIDI interface/patch bay/synchronizer system (Figure 11.11). Certain analog and digital multitrack systems include a built-in MTC port within their design, meaning that the machine can be synchronized to a DAW/sequencing system (with a MIDI interface) without the need for any additional hardware.

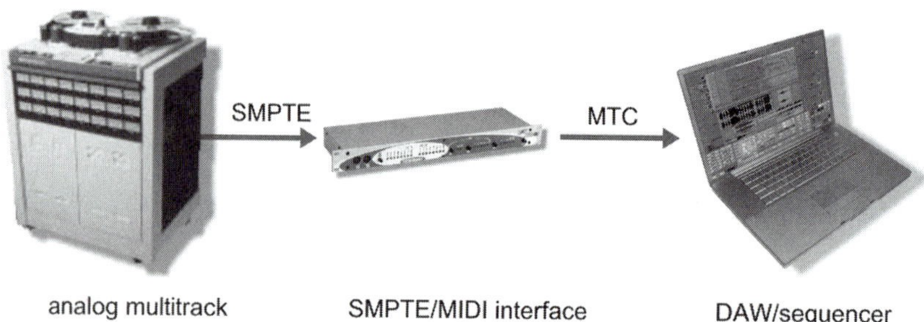

Figure 11.11. *SMPTE time code can be easily converted to MTC for distribution throughout a production system.*

Proprietary Synchronization Systems for Modular Digital Multitrack Recorders

Modular digital multitrack (MDM) recorders, such as the Tascam DA-98HR and Alesis ADAT, encode a proprietary form of time-encoded sync data onto tape, along with the audio information. This sync coding can be used to lock several MDMs together in order to give us more tracks. It can also be used to lock these devices to an external SMPTE source via an interface that can translate MTC or SMPTE into the MDM's native sync code (and *vice versa*). In this way, one or more digital multitracks can be easily rigged to act as a master or slave within a production system.

Video's Need for a Stable Timing Reference

Whenever a video signal is copied from one machine to another, it's essential that the scanned data (containing timing, video, and user information) be copied in perfect sync from one frame to the next. Failure to do so will result in severe picture breakup or, at best, the vertical rolling of a black line over the visible picture area. Copying video from one machine to another generally isn't a problem, as the VCR or VTR that's doing the copying obtains its sync source from the playback machine. Video postproduction houses, however, often simultaneously use any number of video decks, switchers, and edit controllers during the production of a single program. Mixing and switching between these sources will usually result in nonsynchronous chaos—with the end result being a very unhappy client.

Fortunately, referencing all of the video, audio, and timing elements to an extremely stable timing source (called a black burst or house sync generator) will generally resolve this sync nightmare. This reference clock serves to synchronize the video frames and time code addresses that are received or transmitted by every video-related device in a production facility, so the leading frame edge of every video signal occurs at exactly the same instant in time (Figure 11.12). By resolving all video and audio devices to a single black burst reference, you're assured that relative frame transitions and speeds throughout the system will be consistent and stable. This even holds true for slaved analog machines, as their transport's wow and flutter can be smoothed out when locked to such a stable timing reference.

Digital Audio's Need for a Stable Timing Reference

The process of maintaining a synchronous lock between digital audio devices or between digital and analog systems differs fundamentally from the process of maintaining relative speed between analog transports. This is due to the fact that a digital system generally achieves synchronous lock

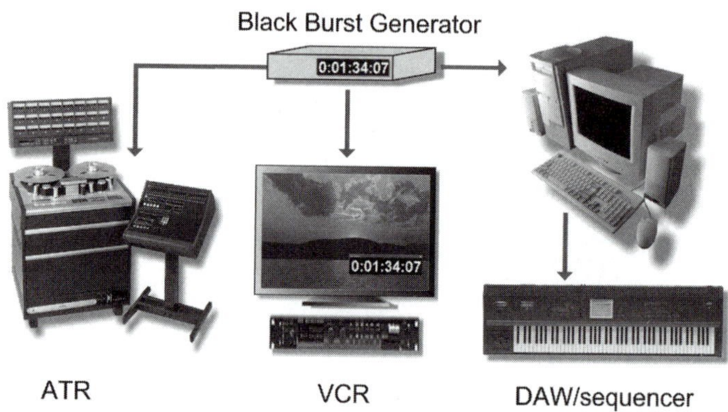

Figure 11.12. Example of a system whose overall timing elements are locked to a black burst reference signal.

by adjusting its playback sample rate (and thus its speed and pitch ratio) so as to precisely match the relative playback speed of the master transport. Whenever a digital system is synchronized to a time-encoded master, a stable timing source is extremely important in order to keep jitter (in this case, an increased distortion due to rapid pitch shifts) to a minimum. In other words, the source's program speed should vary as little as possible to prevent any degradation in the digital signal's quality. For example, all analog tape machines exhibit speed variations, which are caused by tape slippage and transport irregularities (a basic fact of analog life known as wow and flutter). If we were to synchronize a digital device to an analog master source that contains excessive wow and flutter, the digital system would be required to constantly speed up and slow down to precisely match the transport's speed fluctuations. One way to avoid such a problem would be to use a source that's more stable, such as a video deck, DAW, or MDM.

Digital Audio Synchronization

Coverage of synchronization would be incomplete without discussing it in the context of digital audio and hard-disk-based systems. Because digital audio is an important part of modern-day audio and audio-for-visual production, an understanding of digital sync becomes important when working in an environment where digital audio devices are to be synchronized together or to video and analog media.

Real-World Sync Applications for Using Time Code and MIDI Time Code

Before we delve into the many possible ways that a system can be set up to work in a time code environment, it needs to be understood that each system will often have its own particular personality and that the connections, software, and operation of one system might differ from those

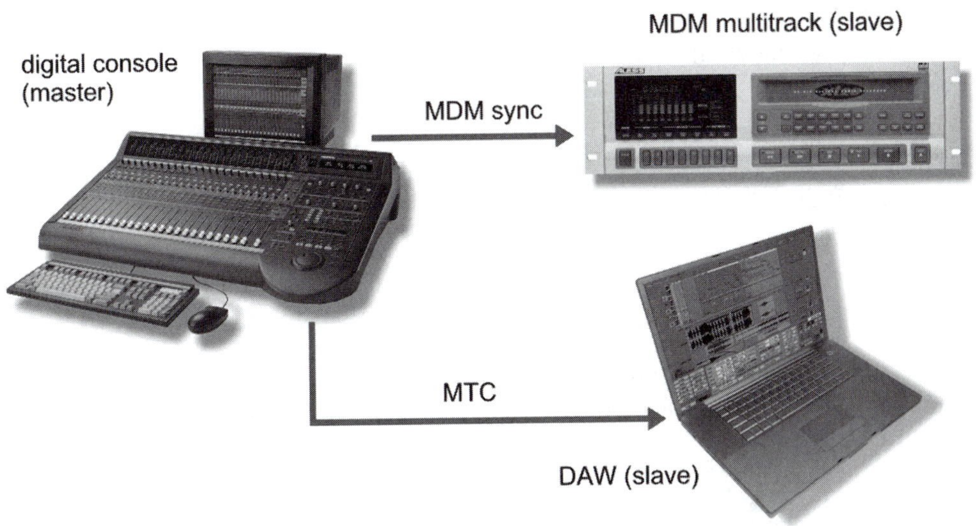

Figure 11.13. *There can be only one master in a synchronized system; however, there can be any number of slaves.*

of another. This is often due to factors such as system complexity and the basic hardware types that are involved, as well as the type of hardware and software systems that are installed in a DAW. Larger, more expensive setups that are used to create television and film soundtracks will often involve extensive time code and system interconnections that can easily get complicated.

Fortunately, the use of MIDI time code has greatly reduced the cost and complexity of connecting and controlling a synchronous electronic music and project studio down to levels that can be easily managed by both experienced and novice users. Having said these things, I'd still like to stress that solving synchronization problems will often require as much intuition, perseverance, insight, and art as it will technical skill. For the remainder of this chapter, we'll be looking into some of the basic concepts and connections that can be used to get your system up and running. Beyond this, the next best course of action will be to consult your manuals, seek help from an experienced friend, or call the tech department for the particular hardware or software that's giving both you and your system the willies.

Master/Slave Relationship

Since synchronization is based upon the timing relationship between two or more devices, it follows that the logical way to achieve sync is to have one or more devices (known as slaves) follow the relative movements of a single transport or device (known as the master). The basic rule to keep in mind is that there can be only one master in a connected system; however, any number of slaves can be set to follow the relative movements of a master transport or device (Figure 11.13).

Generally, the rule for deciding which device will be the master in a production system can best be determined by asking a few questions:

- What type of media is the master time code media recorded on?
- Which device will easily serve as the master?
- Which device will provide the most stable timing reference?

If the master comes to you from an outside source, the answer to the first question will most likely solve your dilemma. If the project is in-house and you have total say in the matter, you might want to research your options more fully. The following sections can help give you insights into which devices will want to serve as the master within a particular system.

Audio Recorders

In many audio production situations, whenever an analog tape recorder is connected in a time code environment, this machine will want to act as the master as costly hardware is often required to lock an analog machine to an external time source. This is due to the fact that the machine's speed regulator (generally a DC capstan servo) must be connected into a special feedback control loop in a way that allows it to continuously adjust its speed by comparing its present location with that of the master SMPTE time code. As a result, it is often far easier and less expensive to set the analog device as the master, especially if the slave device is a DAW or other digital device. When starting a new session, the course of action is to stripe the highest track on a clean roll of tape (with ascending code that continues from the tape's beginning to its end). Once done, the reproduced code can be routed to the SMPTE input on your MIDI interface or synchronizer. If you don't have a multiport interface or if your interface doesn't have a SMPTE input, you'll need to get hold of a box that converts SMPTE to MTC which can then be plugged into a MIDI In port for assignment to a DAW or sequencer device.

VCRs

Since video is often an extremely stable timing source, an analog VCR or VTR should almost invariably act as a system master. In fact, without expensive hardware, a VCR can't easily be set to act as a slave at all, since the various sync references within the machine would be thrown off and the picture would immediately break up or begin to roll. From a practical standpoint, locking other devices to a standard VCR is done in much the same way as with an analog tape machine. Professional video decks generally include a separate track that's dedicated to time code (in addition to other tracks that are dedicated to audio). As with the earlier analog scenario, the master time code track must be striped with SMPTE before beginning the project. This process shouldn't be taken lightly, as the time code must conform to the time code addresses on the original video master or working copy (see the discussion of jam sync).

Basically, the rule of thumb is: If you're working on a project that was created out of house, always use the code that was provided by the original production team. Striping your own code or erasing over the original code with your own would render the timing elements useless, as the new code wouldn't relate to the original addresses or include any variations that might be a part of the original master source. In short, make sure that your working copy includes SMPTE that's a regenerated copy of the original code! Should you overlook this, you can expect to run into timing and sync troubles, either immediately or later in the postproduction phase, while putting the music or dialog back together with the final video master—factors that'll definitely lead to premature hair and client loss.

MDMs

As a digital device, a modular digital multitrack machine is also an extremely stable timing reference and often works well as a master in an audio setting. Because of extensive sync detection and pitch-shifting technology, these devices can also be slaved within a system without too much difficulty. As with an analog machine, it's possible to record SMPTE onto the highest available track and then route this track to a valid SMPTE sync input; however, if you don't feel like losing a physical track to SMPTE, you might want to pick up a sync interface that can translate the MDM's proprietary sync code into SMPTE or MTC.

Software Applications

In general, a MIDI sequencer will be programmed to act as a slave device. This is due to the fact that a digital sequencer can easily chase a master MIDI time code (MTC) source and lock to a location point within a production with extreme ease (*i.e.*, since MIDI is digital in nature and has low data overhead requirements, locating to any point within a sequence is technologically easy to accomplish).

Digital Audio Workstations

A computer-based DAW can often be set to act as either a master or slave. This will ultimately depend on the software, as most professional workstations can be set to chase (or be triggered by) a master time code source or can generate time code (often in the form of MIDI time code).

DAW Support for Video and Picture Sync

Most modern DAWs include support for displaying a video track within a session, both as a video window that can be displayed on the monitor desktop and in the form of a video thumbnail track that will often appear in the track view as a linear guide track. Both of these provide important visual cues for tracking live music, sequencing MIDI tracks, and accurately placing automation moves and effects (sfx) at specific hitpoints within the scene (Figure 11.14). This feature allows audio to be built up within a DAW environment without the need to sync to an external device at all. It's easily conceivable that through the use of recorded tracks, software instruments, and internal mixing capabilities, tracks could be built up, spotted, and mixed—all inside the box.

Figure 11.14. *Most high-end DAW systems are capable of importing a videofile directly into the project session window.*

Routing Time Code to and from Your Computer

From a connections standpoint, most DAW, MIDI, and audio application software packages are flexible enough to let you choose from any number of available sync sources (whether connected to a hardware port, MIDI interface port, or virtual sync driver). All you have to do is assign all of the slaves within the system to the device driver that's generating the system's master code (Figure 11.15). In most cases, where the digital audio and MIDI sequencing applications are operating within the same computer, it's best to have your DAW or editor generate the master code for the system.

From time to time, you might run into an application or editor that's unable to generate time code in any form. When faced with such an all-slave software environment, you'll actually need a physical time code master that can be routed to your editor, MIDI sequencer, etc. In practice,

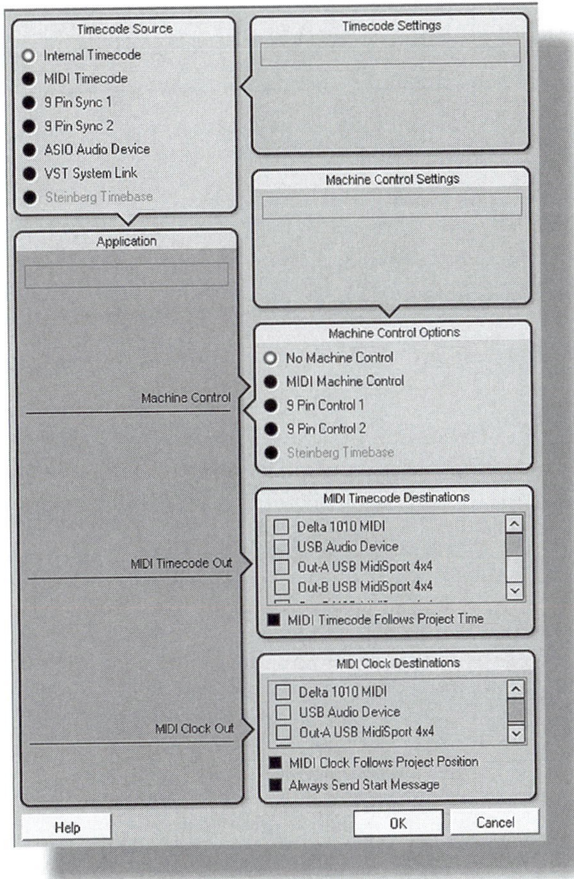

Figure 11.15. *Cubase/Nuendo Sync Preferences dialog box. (Courtesy of Steinberg Media Technologies GMBH, www.steinberg.net.)*

this source could be an MDM or analog recorder, but if you simply want to sync the two pieces of software together without a tape machine, the easiest solution is to use a multiport MIDI interface that includes a software applet for generating time code. In such a situation, all you need to do is to select the interface's sync driver as your sync source for all slave applications. Pressing the Generate SMPTE button in the interface's application window or from its front panel will lock the software to the generated code, beginning at 00:00:00:00 or at any specified offset address.

As more and more DAWs and digital mixing boards become locked to SMPTE/MTC, the issue of locking the word clock of a digital device directly to the SMPTE time code stream will become more and more important. For example, slaving a DAW to external time code under a full time code lock (as opposed to a triggered free-run start) will usually require that specialized sync hardware be used to maintain a frame-by-frame lock.

Keeping Out of Trouble

Here are a few guidelines that can help save your butt when using SMPTE and other time code translations during a project:

- When in doubt about frame rates, special requirements, or anything else, for that matter … ask! You (and your client) will be glad you did.
- Fully document your time code settings, offsets, start times, etc.
- If the project isn't to be used in-house, ask the producer what the proper frame rate should be. Don't assume or guess it.
- When beginning a new session (when using a tape-based device), always stripe the master contiguously from the beginning to end before the session begins. It never hurts to stripe an extra tape, just in case. This goes for both analog and digital MDM devices.
- Start generating new code at a point 1 to 2 minutes before 01:00:00:00 or 00:01:00:00 (to allow for a pre-roll). If the project isn't to be used in-house, ask the producer what the start times should be. Don't assume or guess it.
- Never dub (copy) time code directly. Always make a refreshed (jam synched) copy of the original time code (from an analog master) before the session begins.
- Never use slow videotape speeds. In EP mode, a VHS deck runs too slowly to record or reproduce a reliable code.
- Disable noise reduction on analog audio tracks (on both audio and video decks).
- Work with copies of the original production video, and make a new one when sync troubles appear.
- It's not unusual for the time code to be read incorrectly (when short dropouts occur on the track, usually on videotape). When this happens, you might set the synchronizer to freewheel once the transports have initially locked.

In closing, I'd like to point out that synchronization can be a simple procedure or it can be a fairly complex one, depending on your experience and the type of equipment that's involved. A number of books and articles have been written on this subject. If you're serious about production, I suggest that you do your best to keep up on it. Although the fundamentals stay the same, new technologies and techniques are constantly emerging. As always, the best way to learn is simply by reading and then jumping in and doing it.

12 Mixing and Automation

In the past, almost all commercial music was mixed by a professional recording engineer under the supervision of a producer and/or artist. With the emergence of electronic music and the project studio revolution, the vast majority of production facilities have become much more personal and cost effective in nature. And with the maturation of the digital revolution, MIDI and digital audio workspaces are now being owned by individuals, small businesses, and artists who are taking the time to become experienced in the commonsense rules of creative and commercial mixing.

Within the music production industry, it's a well-known fact that most professional mixers have to earn their "ears" by logging countless hours behind the console. Although there's no substitute for this experience, the mixing abilities and ears of electronic musicians are improving more quickly as equipment quality gets better and as musicians become more knowledgeable about proper mixing environments and techniques—often by mixing their own compositions.

In this chapter, we'll be taking a basic look at the technology and art of mixing, as well as gaining insights into how the digital audio workstation, MIDI sequencing, and modern production equipment can work together to improve both your techniques and your sound.

Console and Mixer Basics

The basic role of a mixer or audio production console is to give us full control over volume, tone, blending, and spatial positioning of any or all signals that are applied to its inputs from microphones, electronic instruments, effects devices, recording systems, and other audio devices. By analogy, a console can be likened to an artist's palette in that it provides a creative control surface that allows an engineer to experiment and blend all the possible variables onto a sonic canvas.

When you come right down to it, most audio production consoles and mixers are designed with similar controls and functional capabilities. They differ mostly in appearance, control location, effects processing, signal routing, and grouping capacities, and how they incorporate automation (if at all).

In the modern-day production world, it goes without saying that mixing systems can take on many shapes, sizes, and configurations. The greatest of these differences are most apparent between mixing systems that are hardware in nature (Figures 12.1 and 12.2), and those that occur in the virtual world of software (Figure 12.3). In this day and age, the argument for a virtual vs. hardware mixer debate is fairly academic. Anyone working in electronic music will quickly be confronted with software mixers that are extremely versatile in their I/O, routing, effects structure, and capacity for automation. Therefore, in order to grasp the basic principles of both hardware and software mixing systems, it's important to understand that a good system of either type can get the job done. The difference lies solely in how they function under the hood.

Figure 12.1. *Xlogic X-Rack mixer. (Courtesy of Solid State Logic, www.solid-state-logic.com.)*

Figure 12.2. *EuroRack Pro RX1202FX mixer. (Courtesy of Behringer International GMBH, www.behringer.com.)*

Understanding the Concept of "The Mixer"

In order to understand the process of mixing, it's important to understand one of the most important concepts in all of audio technology: the *signal chain* (also known as the signal path). As is true with literally any audio system, the recording console can be broken down into functional components that are chained together into a larger (and hopefully manageable) number of signal paths. By identifying and examining the individual components that work together to form this chain, it becomes easier to understand the basic layout of any mixing system, no matter how large or complex.

To better understand the layout of a mixer, let's start with the concept that it's built of numerous building-block components, each having an output (source) that moves to an input (destination). In such a signal flow chain, the output of each source device must be connected to the input of the device that follows it, until the end of the audio path is reached. Whenever a link in this source-to-destination path is broken, no signal will pass. Although this might seem like a simple concept, keeping it in mind can save your sanity when paths, devices, and cables that look like tangled piles of spaghetti get out of hand.

Now, using some of the most versatile environments in use today as examples, let's take a conceptual look at various mixing systems—from the analog hardware mixer to the general layout within a virtual mixing environment. In a traditional hardware mixer (which also goes by the name of board, desk, or console) design, the signal flow for each input travels vertically down a plug-in strip known as an *I/O module* (Figure 12.4) in a manner that generally flows:

- From the input section
- Through a sends section (which taps off to external processing/monitoring devices)
- Into an equalizer (and other processing functions, such as dynamics)

Figure 12.3. *Reason's mixer 14 × 4 and Line Mixer 6 × 2. (Courtesy of Propellerhead Software, www.propellerheads.se.)*

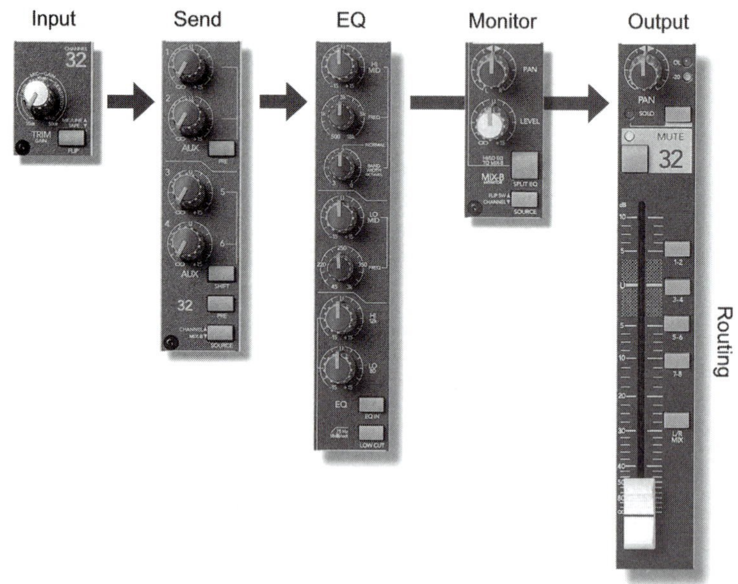

Figure 12.4. *General anatomy of an input strip on the Mackie 8-bus analog console. (Courtesy of Loud Technologies, Inc.; www.mackie.com.)*

- Passing through a monitor mix section (which taps off to external monitoring devices)
- To an output fader that includes pan positioning
- Into a routing section that can send signal to selected mix/signal destinations

Although the layout of a traditional (analog) hardware mixer won't hold true for the graphical user interface (GUI) layout of all virtual mixers, the signal flow will most likely follow along the same or similar paths; therefore, grasping the concept of the signal chain is equally important in the virtual world. The virtual mixer is likewise built from numerous building-block components, each having an output (source) and an input (destination). The output of each source device must be virtually connected to the input of device that follows it—and so on until the end of the audio path is reached. Again, as with all of audio technology, keeping this simple concept in mind is important when paths, plug-ins, and virtual paths seem to meld together into a ball of confusion. When the going gets rough, slow down, take a deep breath, read the manual (if you have the time)—and above all be patient and keep your wits about you.

Figures 12.5 through 12.7 show the general I/O stages of several virtual mixing systems, so as to give you a better understanding of their basic operating structure. It's important that you take the time to familiarize yourself with the inner workings of your own DAW (or those that you might come in contact with) by reading the manual, pushing buttons and by diving in and having fun with your own projects.

Now that we've looked at several hardware and software mixing system layouts, lets discuss the various stages in greater detail as they flow through the process, starting with a channel's input, on through the various processing and send stages, and then out to the final mix destination.

Figure 12.5. Mixer strip layout for Pro Tools. (Courtesy of Digidesign, A Division of Avid Technology, Inc.; www.digidesign.com.)

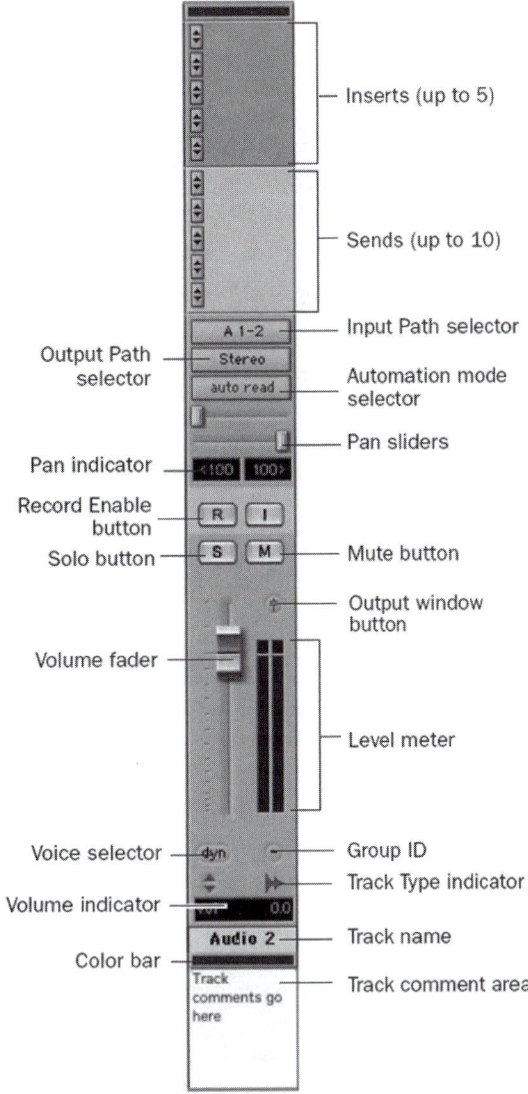

① Channel Input

On a hardware mixer or audio interface that includes mic preamps, either a mic or line input can be selected to be the signal source (Figure 12.8). This *channel input* serves as a preamp section to optimize the signal gain levels at the input of an I/O module before the signal is processed and routed. These levels can be continuously varied in gain by means of shared or independent level controls (often called *gain trims*). Gain trims are a necessary component in the signal path, as the output level of a microphone is typically very low (+45 to +55 dB) and requires that a

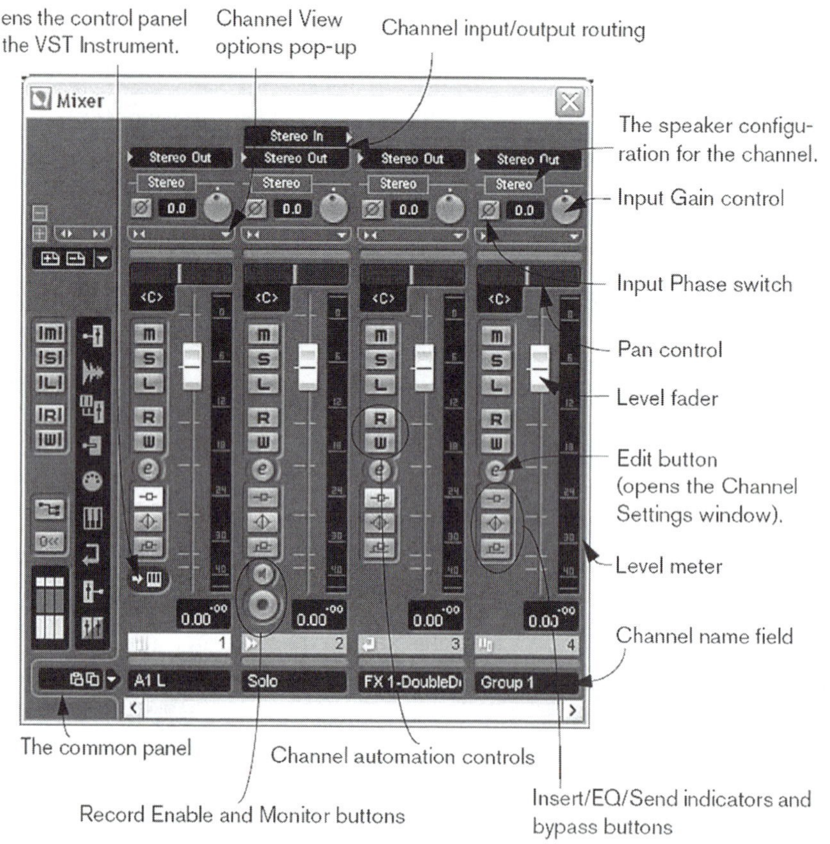

Figure 12.6. *Steinberg's Cubase SX/Nuendo virtual mixer. (Courtesy of Steinberg Media Technologies GmbH, A Division of Yamaha Corporation; www.steinberg.net.)*

Figure 12.7. *Input strip for Reason's Mixer 14:2. (Courtesy of Propellerhead Software, www.propellerheads.se.)*

Aux sends
EQ
Pan
Fader
Mute/solo

Analog Input

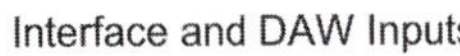

Figure 12.8. Analog and interface input sections. (Courtesy of Loud Technologies, Inc., www.mackie.com, and Steinberg Media Technologies GMBH, www.steinberg.net).

high-quality, low-noise preamp be used to raise or match the levels in order for the signal to be passed through the signal chain at an optimum level (as determined by its design and standard operating levels).

In contrast, an audio interface that doesn't include a mic/line pre-amp section can only accept a line level signal. Because of this, low-level (*e.g.*, microphone or guitar) signals must be amplified through a mixer before being routed to the interface. If the mixer or interface is so equipped, digital input signals can be routed into the signal path. If this is the case, care must be taken to ensure that word sync issues are addressed so as to reduce possible timing conflicts and distortion between the digital devices. (Further info on word clock can be found in Chapter 6.)

Insert Point

Many mixer and audio interface designs provide a break in the signal chain that follows after the channel input (although these insert points can be taken at other places within many hardware console designs). A *direct send/return* or *insert point* (often referred to simply as direct or insert) can then be used to send the strip's line level audio signal out to an external processing or recording device. The external device's output signal can then be inserted back into the signal path, where it can be routed to a destination or mixed back into the audio program. It's important to note that plugging a signal processor into an insert point will only affect the audio on that channel. Should you wish to affect a number of channels, the auxiliary send and group output section can be used to process a combined set of input channels. Physically, the send and return insert jacks of a hardware mixer or audio interface can be accessed as two separate jacks on the top or back of a mixer/console (Figure 12.9a), as a single, stereo jack in the form of a tip–ring–sleeve (TRS) connector that carries the send, return, and common signals (as shown in Figure 12.9b), or as access points on a console's patch bay.

Within a workstation environment, inserts are an extremely important in that they allow audio or MIDI processing/effects plug-ins to be directly inserted into the virtual path of that channel (Figure 12.10). Often, a workstation allows multiple plug-ins to be inserted into a channel in a stacked fashion, allowing complex and unique effects to be built up. Of course, it should be kept in mind that the extensive use of insert plug-ins can eat up processing power. Should the stacking

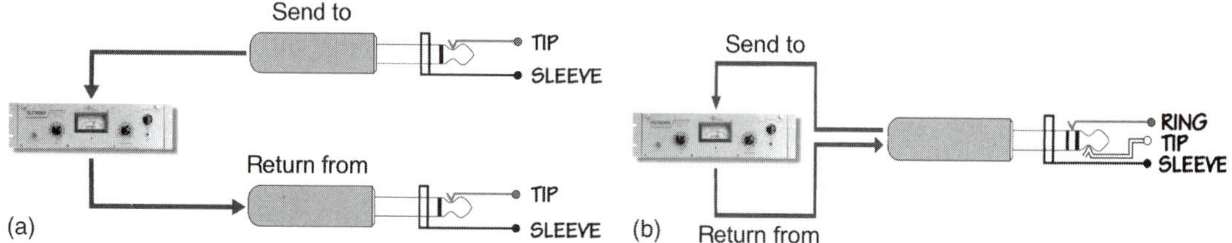

Figure 12.9. *Direct send/return signal paths. (a) Two jacks can be used to send signals to and return signals from an external device. (b) A single TRS (stereo) jack can be used to insert an external device into an input strip's path.*

Figure 12.10. *An effects plug-in can be inserted into the virtual path of a DAW's channel strip. (Courtesy of Steinberg Media Technologies GMBH, www.steinberg.net, and Universal Audio, www.uaudio.com.)*

of multiple plug-ins become a drain on your CPU (something that can be monitored by watching your processor usage meter—aka "busy bar"), many DAWs allow the track to be frozen, meaning that the total sum of the effects is written to an audio file, allowing the effects to be played back without causing undue strain on the CPU.

(2) Auxiliary Send

In a hardware or virtual setting, the *auxiliary (aux) sends* are used to route and mix signals from one or more input strips to the various effects output sends and monitor/headphone cue sends of a console. These sections are used to create a submix of any (or all) of the various console input signals to a mono or stereo send, which can then be routed to any signal processing or signal destination (Figure 12.11).

It's not uncommon for up to eight individual aux sends to be found on an input strip. An auxiliary send can serve many purposes. For example, one send could be used to drive a reverb unit, signal processor, etc., while another could be used to drive a speaker that's placed in that great-sounding bathroom down the hall. A pair of sends (or a stereo send) could be used to provide a headphone mix for several musicians in the studio, while another send could feed a separate mix to the hard-of-hearing drummer. Hopefully, you can see how a send can be used for virtually any

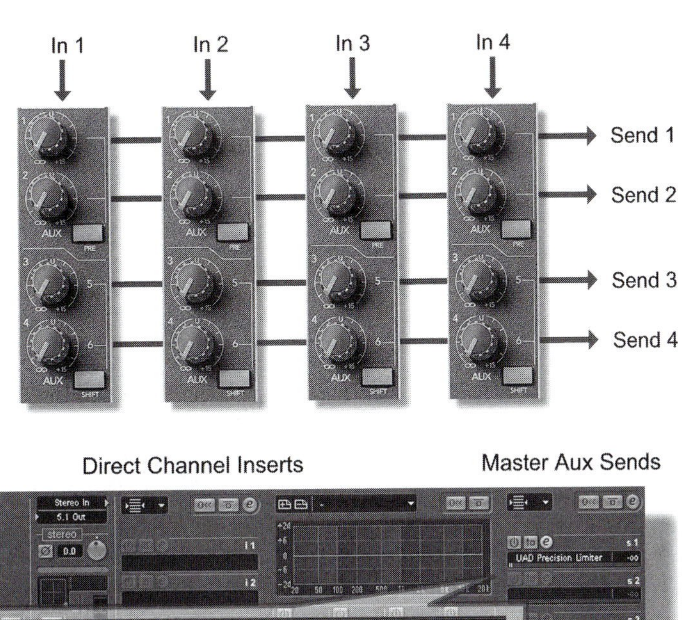

Figure 12.11. *Although a hardware mixer's signal path generally flows vertically from top to bottom, an aux sends path flows in a horizontal fashion, in that the various channel signals are mixed together to feed a mono or stereo send bus. The combined sum can then be effected (or sent to another destination) or returned back to the main output bus. (Courtesy of Loud Technologies, Inc.; www.mackie.com.)*

Figure 12.12. *An effects plug-in can be inserted into an effects send bus, allowing multiple channels to share the same effects process. (Courtesy of Steinberg Media Technologies GMBH, www.steinberg.net, and Universal Audio, www.uaudio.com.)*

effects processing task that needs to be handled. How you deal a send is up to you, your needs, and your creativity.

In regard to a workstation, using an aux send is a way of keeping the processing load on the CPU to a minimum (Figure 12.12). For example, let's say that we wanted to make wide use of a reverb plug-in that's generally known to be a CPU hog. Instead of separately plugging the reverb into a number of tracks as inserts, we can greatly save on processing power by plugging the reverb into an aux send bus. This allows us to selectively route and mix audio signals from any number of tracks and then send the summed signals to a single plug-in that can then be mixed back into the master out bus. In effect, we've cut down on our power requirements by using a single plug-in to do the job of numerous plug-ins.

③ Equalization

The most common form of signal processing is *equalization* (EQ). The audio equalizer (Figure 12.13) is a device or circuit that lets us control the relative amplitude of various frequencies within the audible bandwidth. Put another way, it exercises tonal control over the harmonic or

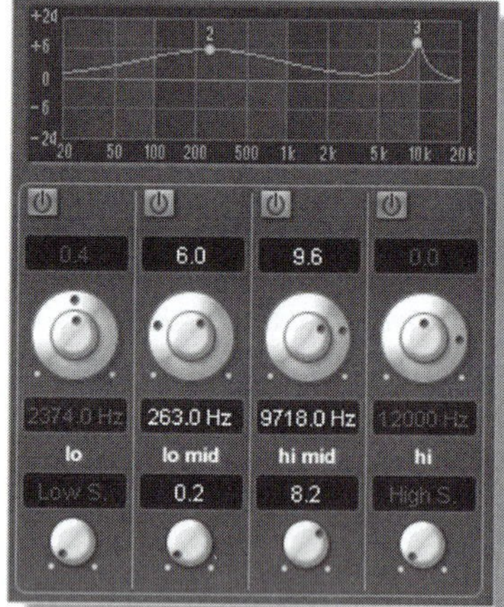

Figure 12.13. *Equalizer examples. (Courtesy of Loud Technologies, Inc., www.mackie.com, and Steinberg Media Technologies GMBH, www.steinberg.net.)*

timbral content of a recorded sound. EQ may need to be applied to a single recorded channel, to a group of channels, or to an entire program (often as a step in the mastering process) for any number of other reasons, including:

- To correct for specific problems in an instrument or in the recorded sound (possibly to restore a sound to its natural tone)
- To overcome deficiencies in the frequency response of a mic or in the sound of an instrument
- To allow contrasting sounds from several instruments or recorded tracks to better blend together in a mix
- To alter a sound purely for musical or creative reasons

When you get right down to it, EQ is all about compensating for deficiencies in a sound pickup or about reducing extraneous sounds that make their way into a pickup signal. To start our discussion on how to apply EQ, let's take a look at the "Good Rule," which states:

> Good musician + good instrument + good performance + good acoustics + good mike + good placement = Good sound

Let's say that at some point in the "good" chain something falls short—like, a mic was placed in a bad spot for a particular instrument during a session that's still under way. Using this example, we now have two options. We can change the mic position and overdub the track or rerecord the entire song—or, we can decide to compensate by applying EQ. These choices represent an important philosophy that's held by many producers and engineers (including myself): Whenever possible, EQ should not be used as a bandage to doctor a session after it's been completed. By this I mean that it's often a good idea to correct a problem on the spot rather than rely on the hope that you can fix it in the mix using EQ and other methods.

Although it's usually better to deal with problems as they occur, it isn't always possible. When a track needs fixing after it's completed, EQ is a good option when:

- There's no time or money left to redo the track.
- The existing take was simply magical … and too much feeling would be lost if the track were to be redone.
- You have no control over a track that's already been recorded during a previous session.

When EQ is applied to a track, bus, or signal, the whole idea is to take out the bad and leave the good. If you keep adding EQ to the signal, it'll degrade the gain structure and lead to a creeping up of volume. Thus, it's often a good idea to use EQ to take away a deficiency in the signal but not necessarily to boost the desirable part of the track (which would in effect serve to turn up the overall gain). Such a boost will often throw off a mix's overall balance and reduce its overall headroom. A couple of active examples would include:

- Reducing the high-end on a bass guitar instead of boosting its primary bass notes
- Using a peak filter to pull out the ring of a snare drum (a perfect example of a problem that should've been corrected during the session)

This use of EQ might not always be the best course of action; for example, to bring out an upper presence on a recorded vocal track, it might be best to use a peak curve to slightly boost the upper mid-range. Just like life, nothing's ever absolute.

(4) Monitor Section

During the recording phase, since the audio signals are commonly recorded to tape or DAW at their optimum levels (without regard to the relative musical balance on other tracks), a means for creating a separate monitor mix in the control room is necessary in order to hear a musically balanced version of the production. A separate monitor section (as well as an aux bus) can be used to provide control over each input's level, pan, effects, etc., as well as to route this mix to the control room's mono, stereo, or surround speakers (Figure 12.14). The approach and techniques of monitoring tracks during a recording will often vary from mixer to mixer (as well as among individuals). Again, no method is right or wrong compared to another. It simply depends on what type of equipment you're working with, and on your own personal working style.

During the overdub and general production phases, this monitor phase can be easily passed over in favor of mixing the signal levels directly at the main faders in a standard mixdown environment. The straightforward system lets us pass go and collect $200 by setting up a rough mix all through the production phase, allowing us to finesse the mix during production. By the time the final mix rolls around, many or most of the "mix as you go" and automation kinks will have been worked out of the mix.

(a)

(b)

Figure 12.14. *Monitor mix sections: (a) Mackie 8-bus console (courtesy of Loud Technologies, Inc.; www.mackie.com); (b) monitor section within the Cubase/Nuendo virtual mixer (courtesy of Steinberg Media Technologies GMBH, www.steinberg.net).*

⑤ Channel Fader

Each input strip contains an associated *channel fader* (which determines the strip's bus output level) (Figure 12.15) and *pan pot* (which is often designed into or near the fader and determines the signal's left/right placement in the stereo and/or surround field). Generally, this section includes a solo/mute feature, which performs the following functions:

- *Solo*—When pressed, the monitor outputs for all other channels will be muted, allowing the listener to monitor only the selected channel (or soloed channels) without affecting the multitrack or main stereo outputs during the recording or mixdown process.
- *Mute*—This function is basically the opposite of the solo button, in that when it is pressed the selected channel is cut or muted from the main and/or monitor outputs.

Depending upon the hardware mixer or audio interface design, the channel fader might be motorized, allowing automation moves to be recorded and played back in the physical motion of moving faders. In the case of some of the high-end console and audio interface/controller designs, a *flip fader* mode can be called up that literally reassigns the control of the monitor section's fader to that of the main channel fader. This "flip" allows the monitoring of levels during the recording process

Figure 12.15. Input channel fader. (Courtesy of Loud Technologies, Inc., www.mackie.com.)

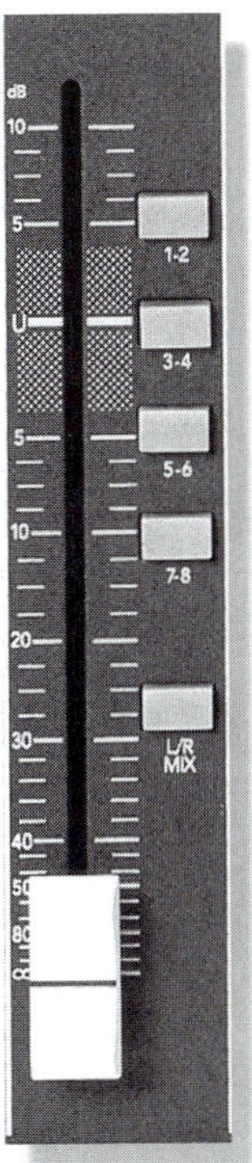

to be controlled from the larger, long-throw faders. In addition to swapping monitor/channel fader functions, certain audio interface/controller designs allow a number of functions such as panning, EQ, effects sends, etc., to be swapped with the main fader, allowing these controls to be finely tuned under motorized control.

(6) Output Section

In addition to the concept of the signal path as it follows through the chain, there's another important signal path concept that should be understood: *output bus*. From the above input strip discussion, we've seen that a channel's audio signal by and large follows a downward path from its top to the bottom; however, when we take the time to follow this path, it's easy to spot where audio is routed off the strip and onto a horizontal output path. Conceptually, we can think of this path (or bus) as a single electrical conduit that runs the horizontal length of a console or mixer (Figure 12.16). Signals can be inserted onto or routed off of this bus at multiple points. Much like a city transit bus, this signal path follows a specific route and allows audio signals to get on or off the line at any point along its path.

Aux sends, monitor sends, channel assignments, and main outputs are all examples of signals that are injected into buses for routing to one or more output destinations; for example, the aux send

Figure 12.16. *Example of a stereo output bus, whereby multiple channels are mixed and routed to a master output fader. (Courtesy of Loud Technologies, Inc.; www.mackie.com.)*

controls are horizontally duplicated across a console's surface. These gain controls are physically tied to an auxiliary send bus that routes the mixed levels to an output destination. The main stereo or surround buses feed to the channel faders and pan positioners and then onto the mixers main output buses, which are combined with the various effects return signals and routed to the monitor speakers and/or recording device.

Grouping

Many consoles and professional mixing systems allow any number of input channels to be organized into *groups*. Such groupings allow the overall relative levels of a series of channels to be interlinked into organized groups according to instrument or scene change type. This important feature makes it possible for multiple instruments to retain their relative level balance while offering control over their overall group level from a single fader or stereo fader pair.

The obvious advantage to grouping channels is that it makes it possible to avoid the dreaded and unpredictable need to manually change each channel volume individually. Why try to move 20 faders when you can adjust their overall levels from just one? For example, the numerous tracks of a string ensemble and a drum mix could each be varied in relative level by assigning them to their own stereo or surround sound groupings—ahhhh … much easier!

On an analog mixer, grouping is simply accomplished by assigning the various channels in the desired group to their own output bus (Figure 12.17). During mixdown, each instrument group

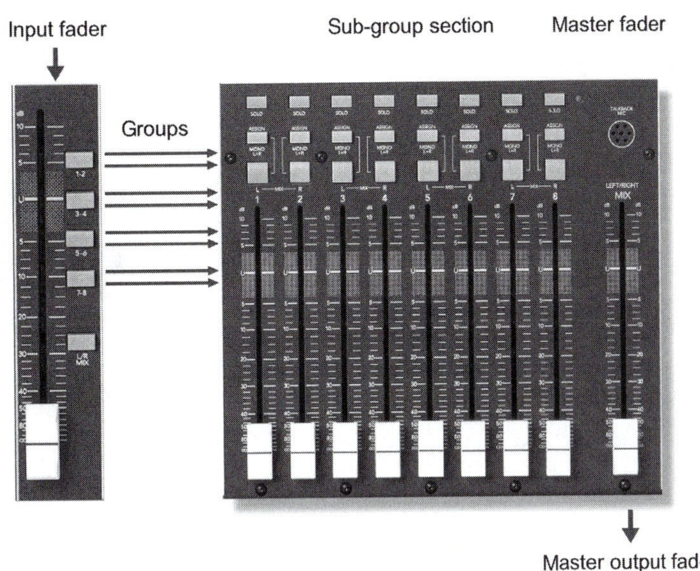

Figure 12.17. *Simplified anatomy of the output section on the Mackie 8-bus analog console. (Courtesy of Loud Technologies, Inc.; www.mackie.com.)*

bus can be summed into the main stereo or surround output through the use of pan pots or L/R assignment buttons.

In the digital or workstation domain, the bussing process is actually easier in that any number of channels can be grouped together by assigning their relative fader level control to a single digital control signal. This level can then be controlled from a single fader or on-screen automation controls. Alternatively, a digital or DAW mixer allows any number of faders to be simply linked by allowing their relative levels to be connected without needing to be physically assigned to a group.

Hardware and Virtual Effects in Action

The following sections offer some insight into the basics of effects processing and how they can be integrated into a recording or mixdown. It's a forgone conclusion that the power of these effects can be harnessed in hardware or software plug-in form. The important rule to remember is that there are no rules; however, there are a few general guidelines that can help you get the sound that you want. When using effects, the most important asset you can have is experience and your own sense of artistry. The best way to learn the art of processing, shaping, and augmenting sound is through experience; gaining experience takes time, a willingness to learn, and lots of patience.

EQ

Although most equalization is done by ear, it's helpful to have a sense of which frequencies affect an instrument in order to achieve a particular effect (Figure 12.18). On the whole, the audio spectrum can be divided into four frequency bands: low (20–200 Hz), low-mid (200–1000 Hz), high-mid (1000–5000 Hz), and high (5000–20,000 Hz). When the frequencies in the

Figure 12.18. *EQ plug-ins. (a) Pro Tools 4-band EQII (courtesy of Digidesign, A Division of Avid Technology, Inc.; www.digidesign.com); (b) Cubase SX/Nuendo (courtesy of Steinberg Media Technologies GmbH, A Division of Yamaha Corporation; www.steinberg.net).*

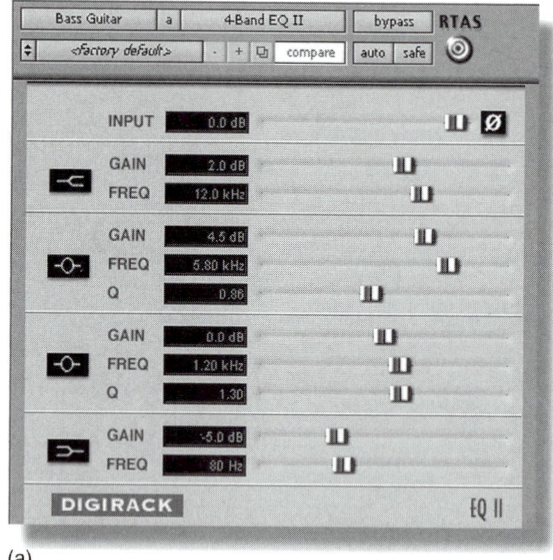

(a)

(b)

20- to 200-Hz (low) range are modified, the fundamental and the lower harmonic range of most bass information will be affected. These sounds are often felt as well as heard, so boosting in this range can add a greater sense of power or punch to music. Reducing the level in this range will weaken or thin out the lower frequency spectrum.

Table 12.1. Instrumental Frequency Ranges of Interest

Instrument	*Frequencies of Interest*
Kick drum	Bottom depth at 60–80 Hz, slap attack at 2.5 kHz
Snare drum	Fatness at 240 Hz, crispness at 5 kHz
Hi-hat/cymbals	Clank or gong sound at 200 Hz, shimmer at 7.5–12 kHz
Rack toms	Fullness at 240 Hz, attack at 5 kHz
Floor toms	Fullness at 80–120 Hz, attack at 5 kHz
Bass guitar	Bottom at 60–80 Hz, attack/pluck at 700–1000 Hz, string noise/pop at 2.5 kHz
Electric guitar	Fullness at 240 Hz, bite at 2.5 kHz
Acoustic guitar	Bottom at 80–120 Hz, body at 240 Hz, clarity at 2.5–5 kHz
Electric organ	Bottom at 80–120 Hz, body at 240 Hz, presence at 2.5 kHz
Acoustic piano	Bottom at 80–120 Hz, presence at 2.5–5 kHz, crisp attack at 10 kHz, "honky tonk" sound (sharp Q) at 2.5 kHz
Horns	Fullness at 120–240 Hz, shrill at 5–7.5 kHz
Strings	Fullness at 240 Hz, scratchiness at 7.5–10 kHz
Conga/bongo	Resonance at 200–240 Hz, presence/slap at 5 kHz
Vocals	Fullness at 120 Hz, boominess at 200–240 Hz, presence at 5 kHz, sibilance at 7.5–10 kHz

The fundamental notes of most instruments lie within the 200- to 1000-Hz (low-mid) range. Changes in this range often result in dramatic variations in the signal's overall energy and add to the overall impact of a program. Because of the ear's sensitivity in this range, a minor change can result in an effect that's very audible. The frequencies around 200 Hz can add a greater feeling of warmth to the bass without loss of definition. Frequencies in the 500- to 1000-Hz range could make an instrument sound hornlike, while too much boost in this range can cause listening fatigue.

Higher-pitched instruments are most often affected in the 1000- to 5000-Hz (high-mid) range. Boosting these frequencies often results in an added sense of clarity, definition, and brightness. Too much boost in the 1000- to 2000-Hz range can add a tinny effect on the overall sound, while the upper mid-frequency range (2000–4000 Hz) affects the intelligibility of speech. Boosting in this range can make music seem closer to the listener, but too much of a boost often tends to cause listening fatigue.

The 5000- to 20,000-Hz (high-frequency) region is composed almost entirely of instrument harmonics. Boosting frequencies in this range will often add sparkle and brilliance to, say, a string or woodwind instrument. Boosting too much might produce sibilance on vocals and make the upper range of certain percussion instruments sound harsh and brittle. Boosting at around 5000 Hz has the effect of making music sound louder. A 6-dB boost at 5000 Hz, for example, can sometimes make the overall program level sound as though it's been doubled in level; conversely, attenuation can make music seem more distant. Table 12.1 shows how instruments can interact at various frequencies and EQ settings.

One way of using an equalizer (especially a parametric one) to zero in on a particular frequency is to accentuate or attenuate the EQ level and then vary the center frequency until the desired range is found. The level should then be scaled back until the desired effect is obtained. If boosting in one instrument range causes you to want to do the same in other frequency ranges, it's likely that you're simply raising the program's overall level. You should be careful, as it's easy to get caught up in the "bigger! better! more!" syndrome of wanting an instrument to sound louder. If this "EQ-gain creep" continues to happen on a mix, it's likely that one of the frequency ranges of an instrument or ensemble will become too dominant and require attenuation.

With regard to laying down a recorded track using EQ, there are a number of different situations in which EQ might be used—and a number of differing opinions regarding its use:

- Some use EQ liberally to make up for placement and mic deficiencies, whereas others might use it sparingly, if at all. One example where EQ is used sparingly is when an engineer knows that someone else will be mixing a particular song or project. In this situation, the engineer who's doing the mix might have a very different idea of how an instrument should sound. If large amounts of EQ were recorded to tape during the session, the mix engineer might have to work very hard to counteract the original EQ settings.
- If everything was recorded flat, the producer and artists might have difficulty passing judgment on a performance or hearing the proper balance during the overdub phase. Such a situation might call for equalization in the monitor mix, while leaving the recorded tracks alone.
- In situations where several mics are to be combined onto a single tape track, the mics can be individually equalized only during the recording phase. In situations where a project is to be engineered, mixed, and possibly even mastered, the engineer might want to know in advance the type and amount of EQ that the producer or artist would want. Obviously, this comes under the "art" category of recording and comes with experience.
- Above all, it's wise that any sound shaping be determined and discussed with the producer or artist before the sounds are committed to tracks.

Whether you choose to use EQ sparingly, as a right-hand tool for correcting deficiencies, or not at all, there's no getting around the fact that an equalizer is a powerful tool. When used properly, it can greatly enhance or restore the musical and sonic balance of a signal. Experimentation and experience are the keys to proper EQ usage, and no book can replace the trial-and-error process of "just doing it!"

Dynamic Range Processors

The overall dynamic range of music is potentially between 120 to 140 dB (Figure 12.19), whereas the overall dynamic range of a compact disc is often 80 to 90 dB, and analog magnetic tape is on the order of 60 dB (excluding the use of noise-reduction systems, which can improve this figure by 15 to 30 dB). However, when working with 20- and 24-bit digital word lengths, a system, processor, or channel's overall dynamic range can actually approach or exceed the full range of hearing. Even with such a wide dynamic range, unless the recorded program is played back in a noise-free environment, either the quiet passages would get lost in the ambient noise of the listening area (35–45 dB SPL for the average home and much worse in a car) or the loud passages would simply be too loud to bear. Similarly, if a program of wide dynamic range were to be played through a medium with a limited

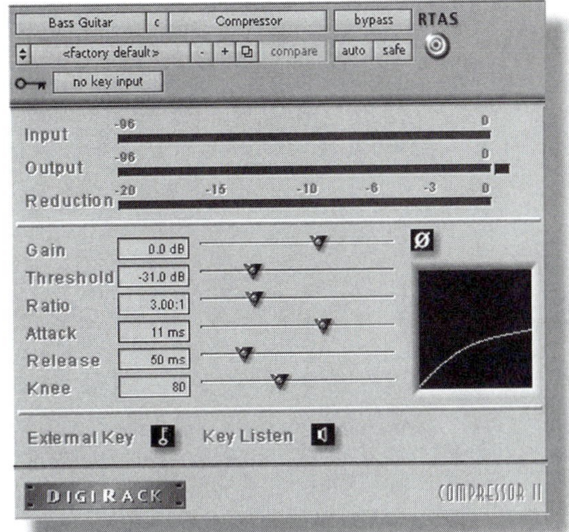

(a)

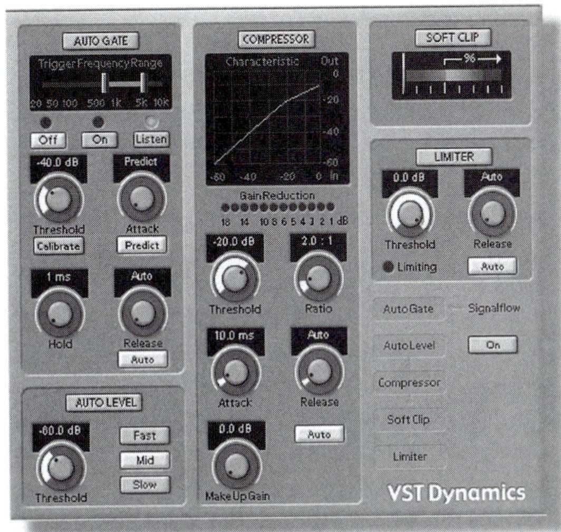

(b)

Figure 12.19. *Dynamic range plug-ins: (a) Pro Tools Compressor II (courtesy of Digidesign, A Division of Avid Technology, Inc.; www.digidesign.com); (b) VST Dynamics for Cubase SX/Nuendo (courtesy of Steinberg Media Technologies GMBH, www.steinberg.net).*

dynamic range (such as the 20- to 30-dB range of an AM radio or the 40- to 50-dB range of FM), a great deal of information would get lost in the background noise. To prevent these problems, the dynamics of a program can be restricted to a level that's appropriate for the reproduction medium (radio, home system, car, etc.). This gain reduction can be accomplished either by manually riding the fader's gain or through the use of a dynamic range processor that can alter the range between the signal's softest and loudest passages.

The concept of automatically changing the gain of an audio signal (through the use of compression, limiting, or expansion) is perhaps one of the most misunderstood aspects of audio recording.

This can be partially attributed to the fact that a well-done job won't be overly obvious to the listener. Changing the dynamics of a track or overall program will often affect the way in which it will be perceived (either unconsciously or consciously) by making it seem louder, by reducing its volume range to better suit a particular medium, or by making it possible for a particular sound to ride at a better level above other tracks within a mix.

Compression

A compressor (Figure 12.20) is, in effect, an automatic fader that's used to proportionately reduce the gain of a signal that rise above a user-definable level (known as the threshold) to a lesser volume range. This process is done so that:

- The dynamics can be managed by the electronics or amplifiers in the signal chain.
- The range is appropriate to the overall dynamics of a playback or broadcast medium.
- An instrument better matches the dynamics of other recorded tracks within a song or audio program.

Since the signals of a track, group, or program will be automatically turned down (hence, the terms *compressed* or *squashed*) during a loud passage, the overall level of the newly reduced signal can now be amplified. In other words, since the dynamics have been reduced, the overall level can now be boosted such that the range between the loud and soft levels is less pronounced. We've not only raised the louder signals back to a prominent level, but we have also turned up the softer signals that would otherwise be buried in the mix or ambient background noise.

The most common controls on a compressor (and most other dynamic range devices) include input gain, threshold, output gain, slope ratio, attack, release, and meter display:

- *Input gain*—This control is used to determine how much signal will be sent to the compressor's input stage.

Figure 12.20. *Fairchild 670 compressor plug-in for the UAD-1 effects processing card. (Courtesy of Universal Audio, www.uaudio.com.)*

- *Threshold*—This setting determines the level at which the compressor will begin to proportionately reduce the incoming signal. For example, if the threshold is set to −20 dB, all signals that fall below this level will be unaffected, while signals above this level will be proportionately attenuated, thereby reducing the overall dynamics. On some devices, varying the input gain will correspondingly control the threshold level. In this situation, raising the input level will lower the threshold point and thus reduce the overall dynamic range. Most quality compressors offer a hard and soft knee threshold option. A soft knee widens or broadens the threshold range, making the onset of compression less obtrusive, while the hard knee setting causes the effect to kick in quickly above the threshold point.
- *Output gain*—This control is used to determine how much signal will be sent to the device's output. It's used to boost the reduced dynamic signal into a range where it can best match the level of a medium or be better heard in a mix.
- *Ratio*—This control determines the slope of the input-to-output gain ratio. In simpler terms, it determines the amount of input signal (in decibels) that's needed to cause a 1-dB increase at the compressor's output (see Figure 12.26). For example, with a ratio of 4:1, for every 4-dB increase at the input there will be only a 1-dB increase at the output; an 8-dB input increase will raise the output by 2 dB, while a ratio of 2:1 will produce a 1-dB increase in output for every 2-dB increase at its input. Get the idea?
- *Attack*—This setting (which is calibrated in milliseconds; 1 msec = 1 thousandth of a second) determines how quickly or slowly the device will turn down signals that exceed the threshold. It is defined as the time it takes for the gain to decrease to a percentage (usually 63%) of its final gain value. In certain situations (as might occur with instruments that have a long sustain, such as the bass guitar), setting a compressor to instantly turn down a signal might be audible (possibly creating a sound that pumps the signal's dynamics). In this situation, it would be best to use a slower attack setting. On the other hand, such a setting might not give the compressor time to react to sharp, transient sounds (such as a hi-hat). In this case, a fast attack time would probably work better. As you might expect, you'll need to experiment to arrive at the fastest attack setting that won't audibly color the signal's sound.
- *Release*—Similar to the attack setting, release (which is calibrated in milliseconds) is used to determine how slowly or quickly the device will restore a signal to its original dynamic level once it has fallen below the threshold point (defined as the time required for the gain to return to 63% of its original value). Too fast a setting will cause the compressor to change dynamics too quickly (creating an audible pumping sound), while too slow a setting might affect the dynamics during the transition from a loud to a softer passage. Again, it's best to experiment with this setting to arrive at the slowest possible release that won't color the signal's sound.
- *Meter display*—This control changes the compressor's meter display to read the device's output or gain reduction levels. In some designs, there's no need for a display switch, as readouts are used to simultaneously display output and gain reduction levels.

As was previously stated, the use of compression (and most forms of dynamics processing) is often misunderstood, and compression can easily be abused. Generally, the idea behind these processing systems is to reduce the overall dynamic range of a track, music, or sound program or to raise its overall perceived level, without adversely affecting the sound of the track itself. It's a well-known fact that over-compression can actually squeeze the life out of a performance by limiting the dynamics and reducing the transient peaks that can give life to a performance. For this reason, it's important to be aware of the general nuances of the controls we've just discussed.

Limiting

If the compression ratio is made large enough, the compressor will actually become a limiter. A limiter (Figure 12.21) is used to keep signal peaks from exceeding a certain level in order to prevent the overloading of amplifier signals, recorded signals onto tape or disc, broadcast transmission signals, and so on. Most limiters have ratios of 10:1 (above the threshold, for every 10-dB increase at the input there will be a gain of 1 dB at the output) or 20:1, although some have ratios that can range up to 100:1. Since a large increase above the threshold at the input will result in a very small increase at its output, the likelihood of overloading any equipment that follows the limiter will be greatly reduced. Commonly, limiters have two basic functions:

- *To prevent signal levels from increasing beyond a specified level*—Certain types of audio equipment (often those used in broadcast transmission) are often designed to operate at or near their peak output levels. Significantly increasing these levels beyond 100% would severely distort the signal and possibly damage the equipment. In these cases, a limiter can be used to prevent signals from significantly increasing beyond a specified output level.
- *To prevent short-term peaks from reducing a program's average signal level*—Should even a single high-level peak exist at levels above the program's rms average, the average level can be significantly reduced. This is especially true whenever a digital audio file is normalized at any percentage value, as the peak level will become the normalized maximum value and not the average level. Should only a few peaks exist in the file, they can easily be zoomed in on and manually reduced in level. If multiple peaks exist, then a limiter should be considered.
- *To prevent high-level, high-frequency peaks from distorting analog tape*—When recording to certain media (such as cassette and videotape), high-energy, transient signals actually don't significantly add to the program's level, relative to the distortion that could result from their presence (if they saturated the tape) or from the noise that would be introduced into the program (if the signal was recorded at such a low level that the peaks wouldn't distort).

Unlike the compression process, extremely short attack and release times are often used to quickly limit fast transients and to prevent the signal from being audibly pumped. Limiting a

Figure 12.21. *Universal Audio 1176LN limiter plug-in for the UAD-1 effects processing card. (Courtesy of Universal Audio, www.uaudio.com.)*

signal during the recording and/or mastering phase should only be used to remove occasional high-level peaks, as excessive use would trigger the process on successive peaks and would be noticeable. If the program contains too many peaks, it's probably a good idea to reduce the level to a point where only occasional extreme peaks can be detected.

The Noise Gate

One other type of expansion device is the noise gate (Figure 12.22). This device allows a signal above a selected threshold to pass through to the output at unity gain and without dynamic processing; however, once the input signal falls below this threshold level, the gate acts as an infinite expander and effectively mutes the signal by fully attenuating it (see Figure 12.34). In this way, the desired signal is allowed to pass while background sounds, instrument buzzes, leakage, or other unwanted noises that occur between pauses in the music aren't. Here are a few reasons why a noise gate might be used:

- To reduce leakage between instruments—Often, parts of a drum kit fall into this category; for example, a gate can be used on a high-tom track in order to reduce excessive leakage from the snare.
- To eliminate noise from an instrument or vocal track during silent passages.

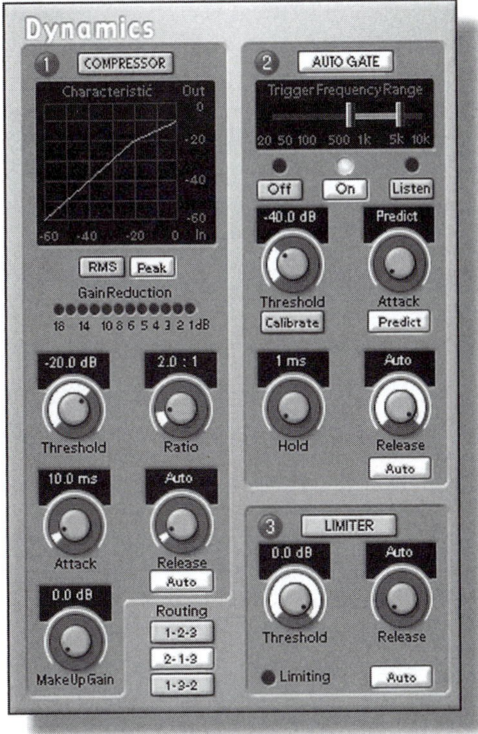

Figure 12.22. *Noise gates are commonly included within many dynamic plug-in processors. (Courtesy of Steinberg Media Technologies GMBH, www.steinberg.net.)*

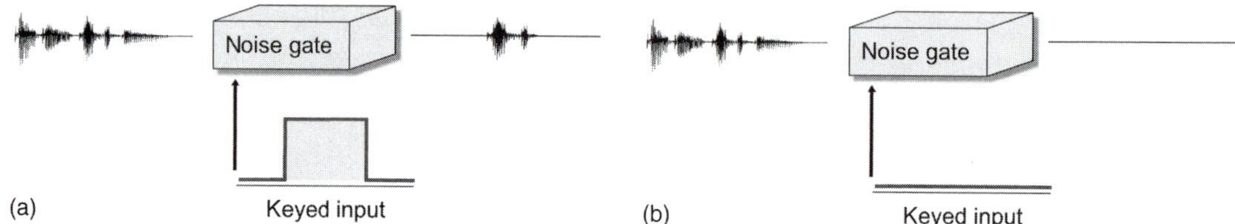

Figure 12.23. *Diagram of a key sided-chain input to a noise gate. (a) The signal is passed whenever a signal is present at the key input. (b) No signal is passed, when no signal is present at the key input.*

The general rules of attack and release apply to gating as well. Fortunately, these settings are a bit more obvious during the gating process than with any other dynamic tool. Improperly set attack and release times will often be immediately obvious when you're listening to the instrument or vocal track (either on its own or within a mix) as the sound will cut in and out at inappropriate times.

Commonly, a key input (Figure 12.23) is included as a side-chain path to a noise gate. A key input is an external control that allows an external analog signal source (such as a miked instrument, synthesizer, or oscillator) to trigger the gate's audio output path. For example, a mic or recorded track of a kick drum could be used to key a low-frequency oscillator. Whenever the kick sounds, the oscillator will be passed through the gate. By combining the two, you can have a deep kick sound that'll make the room shake, rattle, and roll.

Time-Based Effects

Another important effects category that can be used to alter or augment a signal revolves around delays and regeneration of sound over time. These time-based effects often add a perceived depth to a signal or change the way we perceive the dimensional space of a recorded sound. Although a wide range of time-based effects exist, they are all based on the use of delay (and/or regenerated delay) to achieve such results as:

- Time-delay or regenerated echoes
- Chorus
- Flanging
- Reverberation

Delay

One of the most common effects used in audio production today alters the parameter of time by introducing various forms of delay into the signal path. Creating a delay circuit is a relatively simple task to accomplish digitally. Although dedicated delay devices (often referred to as *digital delay lines*, or DDLs) are readily available on the market, most multifunction signal processors and time-related plug-ins are capable of creating this straightforward effect (Figure 12.24). In its basic form, digital delay is accomplished by storing sampled audio directly into RAM. After a defined length of time

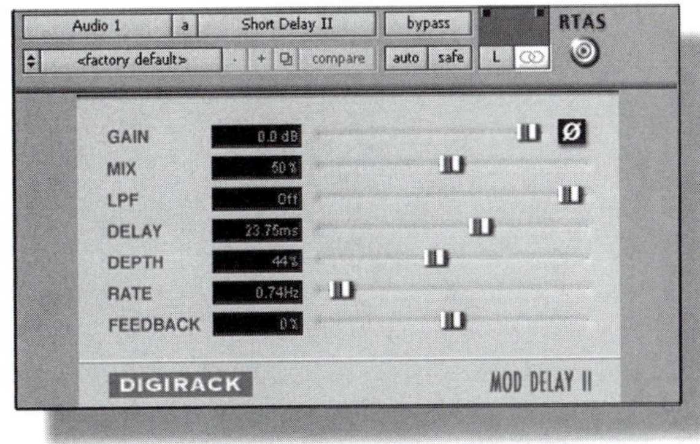

Figure 12.24. *Pro Tools Mod Delay II. (Courtesy of Digidesign, A Division of Avid Technology, Inc.; www.digidesign.com.)*

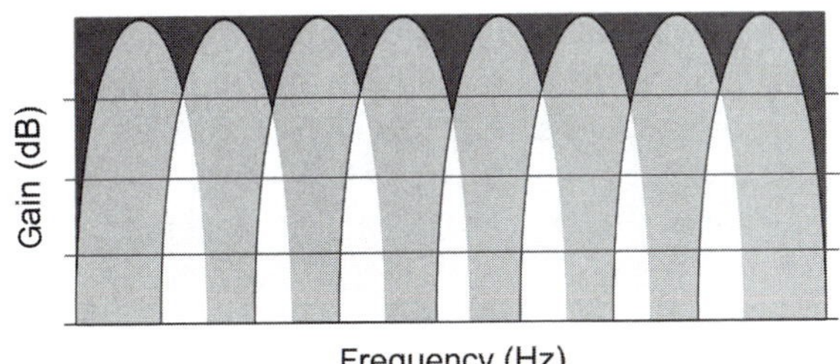

Figure 12.25. *Peaks and dips in a signal's frequency response (as shown in the blue areas) result from the combination of several short-term delays that shift over time to create the effects of phasing or flanging.*

(usually measured in milliseconds), the sampled audio can be read out from memory for further processing or direct output. Using this basic concept, a wide range of effects can be created simply by assembling circuits and program algorithms into blocks that can introduce delays or regenerated echo loops. Of course, these circuits will vary in complexity as new blocks are introduced.

Delay in Action!

Probably the best place to start looking at the delay process is at the sample level. By introducing delays downward into the microsecond (one millionth of a second) range, control over a signal's phase characteristics can be introduced to the point where selective equalization actually begins to occur. This is actually how EQ is carried out in the digital domain! Whenever delays that fall below the 15-msec range are slowly varied over time and then are mixed with the original undelayed signal, an effect known as *combing* is created. Combing is the result of changes that occur when equalized peaks and dips appear in the signal's frequency response (Figure 12.25). By either manually or automatically varying the time of one or more of these short-term delays, a constantly shifting series of effects known as phasing or flanging can be created. Depending on the application, this effect (which makes a unique "swishing" sound that's often heard on guitars or vocals) can range from being relatively subtle (phasing) to having moderate to wild shifts in time and pitch (flanging).

By combining two identical (and often slightly delayed) signals that are slightly detuned in pitch from one another, an effect known as *chorusing* can be created. Chorusing is an effects tool that's often used by guitarists, vocalists, and other musicians to add depth, richness, and harmonic structure to their sound. Increasing delay times into the 15- to 35-msec range will create signals that are spaced too closely together to be perceived by the listener as being discrete delays.

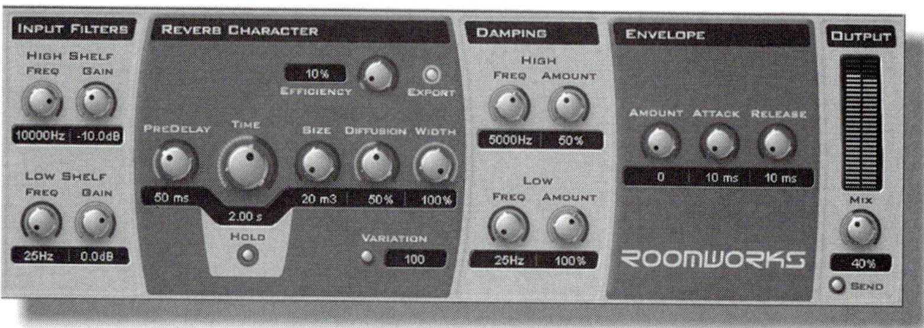

Figure 12.26. *Roomworks Ververb plug-in for Cubase SX/Nuendo. (Courtesy of Steinberg Media Technologies GMBH, www.steinberg.net.)*

Instead, these closely spaced delays create a *doubling* effect when mixed with an instrument or group of instruments. In this instance, the delays actually fool the brain into thinking that more instruments are playing than actually are. The effect, at least subjectively, increases the sound's density and richness. This effect can be used on background vocals, horns, string sections, and other grouped instruments to make the ensemble sound as though it has doubled (or even tripled) its actual size. This effect also can be used on foreground tracks, such as vocals or instrument solos, to create a larger, richer, and fuller sound.

Some delay devices introduce slight changes in delay times in order to create a more natural, humanized sound. Should time or budget be an issue, it's also possible to create this doubling effect by actually recording a second pass to a new set of tracks. Using this method, a ten-piece string section could be made to sound like a much larger ensemble. In addition, this process automatically gives vocals, strings, keyboards, and other legato instruments a more natural effect than the one you get by using an electronics effects device. This having been said, these devices can actually go a long way toward duplicating the effect. Some delay devices even introduce slight changes in delay times in order to create a more natural, humanized sound. As always, the method you choose will be determined by your style, your budget, and the needs of your particular project.

Reverb

In professional audio production, natural acoustic reverberation is an extremely important tool for the enhancement of music and sound production. A properly designed acoustical environment can add a sense of space and natural depth to a recorded sound that'll often affect the performance as well as its overall sonic character. In situations where there is little, no, or substandard natural ambience, a high-quality reverb device or plug-in (Figure 12.26) can be extremely helpful in filling the production out and giving it a sense of dimensional space and perceived warmth. In fact, reverb consists of closely spaced and random multiple echoes that are reflected from one boundary to another within a determined space (Figure 12.27). This effect helps give us perceptible cues as to the size, density, and nature of a space (even though it might have been artificially generated). These cues can be broken down into three subcomponents:

- Direct signal
- Early reflections
- Reverberation

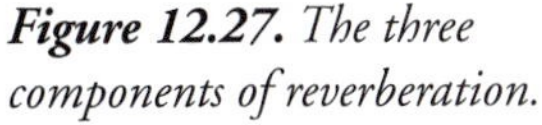

Figure 12.27. The three components of reverberation.

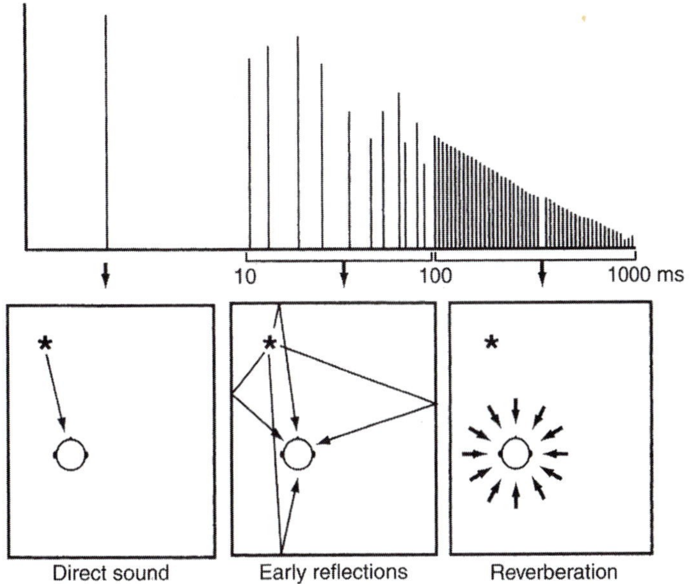

The direct signal is heard when the original sound wave travels directly from the source to the listener. *Early reflections* is the term given to those first few reflections that bounce back to the listener from large, primary boundaries in a given space. Generally, these reflections are the ones that give us subconscious cues as to the perception of size and space. The last set of reflections makes up the signal's reverberation characteristic. These sounds are comprised of zillions of random reflections that travel from boundary to boundary within the confines of a room. These reflections are so closely spaced in time that the brain can't discern them as individual reflections, so they're perceived as a single, densely decaying signal.

Reverb Types

By varying program and setting parameters, a digital reverb device can be used to simulate a wide range of acoustic environments, reverb devices, and special effects. A few popular categories include:

- *Hall*—Simulates the acoustics of a concert hall. This is often a diffuse, lush setting with a longer RT60 decay time (the time that's required for a sound to decay by 60 dB).
- *Chamber*—Simulates the acoustics of an echo chamber. Like a live chamber, these settings often simulate the brighter reflectivity of tile or cement surfaces.
- *Room*—As you might expect, these settings simulate the acoustics of a mid- to large-sized room. It's often best suited to intimate solo instruments or a chamber atmosphere.
- *Live (stage)*—Simulates a live performance stage. These settings can vary widely but often simulate long early-delay reflections.

- *Spring*—Simulates the low-fidelity "boingyness" of yesteryear's spring reverb devices.
- *Plate*—Simulates the often-bright diffuse character of yesteryear's metallic plate reverb devices. These settings are often used on vocals and percussion instruments.
- *Reverse*—These backward-sounding effects are created by reversing the decay trail's envelope so that the decay increases in level over time and is quickly cut off at the tail end, yielding a sudden break effect. This can also be realistically created in a DAW by reversing a track or segment, applying reverb, and then reversing it again to yield a true backward reverb trail.
- *Gate*—Cuts off the decay trail of a reverb signal. These settings are often used for emphasis on drums and percussion instruments.

Pitch Shifting

Ever had a perfectly good vocal take that was spoiled by just one or two flat notes? Or had a project come in the door with a guitar track that was out of tune? Or needed to change the key on a 30-second radio spot? It's times like these that pitch shifting can save your day! Pitch shifting can be used to vary the pitch of a signal or sound file (either upward or downward) in order to transpose the relative pitch of an audio program without affecting its duration. This process can take place in either real time or non-real time. Pitch shifting works by writing sampled audio data to a temporary memory, where it's resampled to either a higher or a lower sample rate (according to the desired final pitch). Once this is done, the processor either adds interpolated samples to (lowers the pitch) or subtracts them from (raises the pitch) the resampled data to return it back to the original output rate, while keeping the altered pitch intact. Figure 12.28 gives two basic examples of how this is often carried out.

A degree of caution should be used when changing the pitch of a program or audio segment. Whenever uneven or minute interval changes are made, the interpolation of samples doesn't always fall perfectly into place. This can lead to digital artifacts that add unacceptable amounts of harmonic distortion. If the track is in the background, there shouldn't be a problem; however, care should be taken with upfront instruments and vocals. It's important to keep in mind that large pitch changes might be more noticeable. As always, your ears are the best judge.

Time and Pitch Changes

By combining variable sample rates and pitch shifting techniques, it's possible to create three different variations:

- *Time change*—A program's length can be altered, without affecting its pitch, by raising or lowering its playback sample rate)
- *Pitch change*—A program's length can remain the same while pitch is shifted either up or down.
- *Both*—Both a program's pitch and length can be altered by means of simple resampling techniques.

These functions have become an important part of the signal processing and music production arsenal that are used by the audio-for-video, film, and broadcast industries. These tools help give producers control over the running time of film video and audio soundtracks while maintaining the original, natural pitch of voice, music, and effects. For example, using a DAW, we could add a 5-second trailer onto the end of an existing 30-second public service radio spot simply by time compressing the 30-second spot to 25 seconds (while keeping the pitch intact) and then adding the trailer.

In addition to the basic time/pitch techniques that are commonly used in music production (most often by electronic musicians), this technology has allowed for the huge explosion in

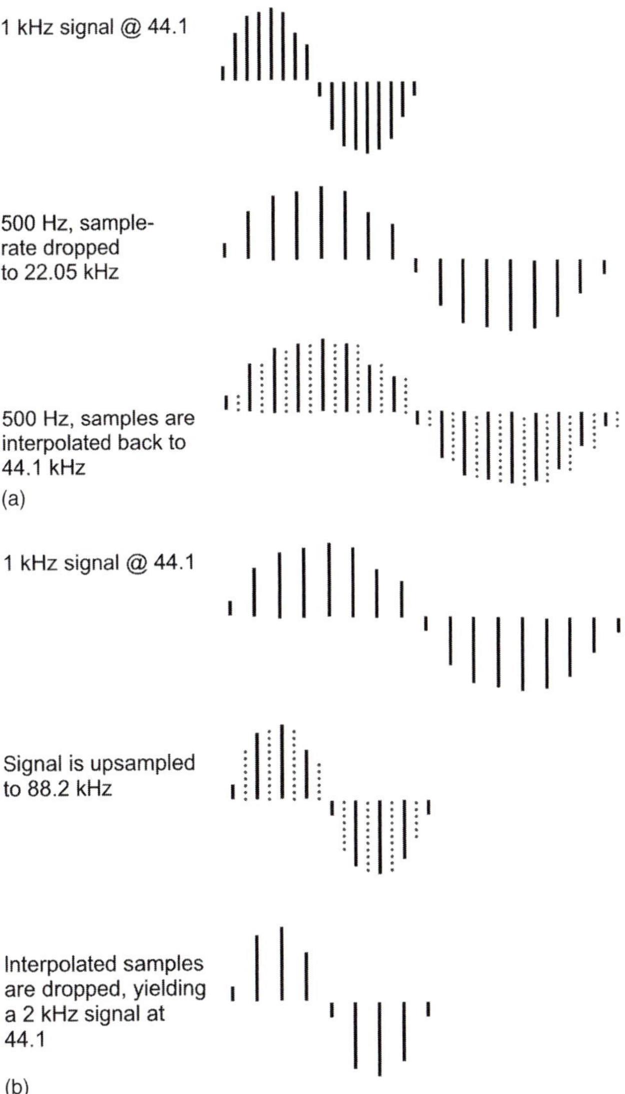

Figure 12.28. *Two pitch shift examples with an initial 1-kHz digital signal and a sample rate of 44.1 kHz. (a) The signal can be halved in pitch (to 500 Hz) by internally downsampling to a new rate of 22.05 k. In order to return the output rate to 44.1 (while retaining the 500-Hz pitch), new sample points must be added into each dropped position. (b) The signal can be doubled in pitch (to 2 kHz) by internally upsampling to a new rate of 88.2 k. In order to return the output rate to 44.1 (while retaining the 2-kHz pitch), every other sample point must be dropped.*

loop-based music composition and production. These popular programs and music plug-ins involve the use of recorded sound files that are encoded with headers that include information on their native tempo and length (in both samples and beats). When you set the loop program to a master tempo (or a specific tempo at that point in the score), a loop segment, once imported, can go about the process of recalculating its pitch and tempo to match the current session tempo and—*voila*! The file's in sync with the song! Should you wish to check it out for yourself, a number of loop-based demo programs can be found on the web.

Effects Automation

One of the great strengths of the digital age is how easily all of the mix and effects parameters can be automated and recalled within a mix. A computer-based DAW is particularly strong in this area, allowing levels, panning, effects, and control over any parameter within a project to be completely recorded, edited, and played back under complete automation during all phases of the production. The beauty of being able to write, read, and recall a mix is that it allows you to build the mix over time and even save multiple mix versions, allowing you to go back in time and explore other production avenues. In short, the job of mixing becomes much less of a chore, allowing you pursue less of the technology of mixing, and more of the art of mixing. What could be bad about that?

Although terminologies and functional offerings will differ from one system to the next, control over the basic automation functions will be carried out in one of two operating modes (Figure 12.29):

- Write mode
- Read mode

Write Mode

Once the mixdown process has gotten underway, the process of writing the automation data into the system's memory can begin (actually, that's not entirely true, as basic mix moves will often begin during the recording or overdub phase). When in the *write mode*, the system will begin the process of encoding mix moves for the selected channel or channels in real time. This mode is used to record all of the settings and moves that are made on a selected strip or strips (allowing track mixes to be built up individually) or on all of the input strips (in essence, storing all of the mix moves, live and in one pass). The first approach can help us to focus all of our attention on a difficult or particularly important part or passage. Once that channel has been finessed, another channel or group of channels can then be written into the system's memory ... and then another, until an initial mix is built up.

Often, modern automation systems will allow previously written automation data to be automated by simply grabbing the fader (either on-screen or on the console/controller) and moving it to the newly desired position. Once the updated move has been made, the automation will remain at that level or position until a previously written automation move is initiated, at which point the values will revert to the existing automation settings.

Whenever data is updated over previously written automation on an older analog mixer or a DAW controller that isn't equipped with moving faders, the concept of matching current fader level or controller positions to their previously written position becomes important. For example, let's say that we needed to redo several track moves that occur in the middle of a song. If the current controller positions don't match up with their previously written points, the mix levels could jump during the updated transition (and thus during playback).

Read Mode

Automated mixer or DAWs that have been placed into the *read mode* will play the mix information from the system's automation data, allowing the on-screen and moving faders to follow the written mix moves in real time. Once the final mix has been achieved, all that's needed is to press play, sit back, and listen to the mix. Matching the actual fader level to its current automated position is performed manually, and the engineer will need some form of indicator in order to do it.

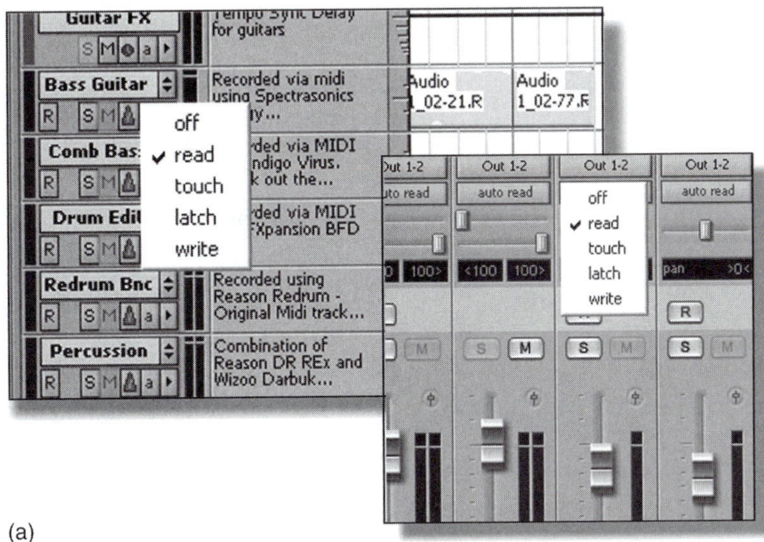

Figure 12.29. *Automation mode selections: (a) Pro Tools showing auto selectors within the edit and mixer screens (Courtesy of Digidesign, A Division of Avid Technology, Inc.; www.digidesign.com). (b) Cubase SX3/Nuendo3 showing auto selectors within the edit and mixer screens (courtesy of Steinberg Media Technologies GMBH, www.steinberg.net).*

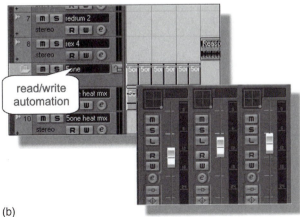

Figure 12.30. Automation rubberbands: (a) Pro Tools (courtesy of Digidesign, A Division of Avid Technology, Inc.; www.digidesign.com); (b) Cubase SX/Nuendo (courtesy of Steinberg Media Technologies GMBH, www.steinberg.net).

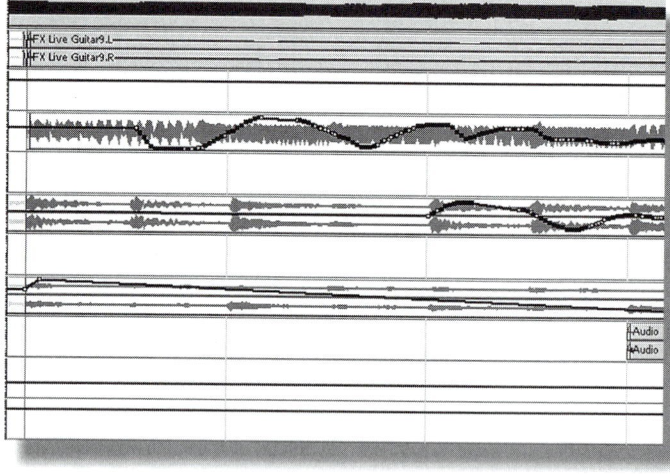

(a)

(b)

This match-up is often done with the aid of nulling indicator lights that read the difference and indicate whether the setting is higher, lower, or equal to the current mix position. After a level match has been achieved, the engineer can update the current mix data without fear of a sudden jump in level. Of course, subtle differences in and flavors of automation modes will exist from workstation to workstation (for example, Pro Tools has at least six different automation modes). As always, it's best to refer to your DAW manual and take a session for an automated spin in order to make full use of its important benefits.

Drawn (Rubberband) Automation

In addition to physically moving on-screen and controller faders under read/write automation control, one of the most accurate ways to control various automation parameters is through the drawing and editing of on-screen *rubberbands*. These useful tools offer a simple, graphic form of automation that lets us draw fades and complicated mix automation moves over time. This user interface is so-named because the graphic lines (that represent the relative fade, volume, pan, and other parameters) can be bent, stretched, and contorted like a rubberband (Figure 12.30).

Commonly, all that's needed to define a new mix point is to click on a point on the rubberband (at which point a box handle will appear) and drag it to the desired position. You can change a move simply by clicking on an existing handle (or range of handles) and moving it to a new position.

Hardware Effects Automation and Editing via MIDI

In addition to automating a production using a DAW, you can use MIDI program-change commands to automate and control external hardware devices during a mix. In the same way that a favorite sound patch can be stored into an instrument's memory for later recall, most hardware effects devices will let you store effects types as a patch that can be saved to its internal memory. These patch changes can be automated from a MIDI sequencer or, in the case of a live stage performance, from a MIDI controller by transmitting a program change command. Often, making such a scene change is as simple as transmitting the desired program change number (from a sequencer or controller) on a port and MIDI channel number that's being sent to the device.

In addition to changing program numbers and basic controllers (such as volume and pan) in real time, any number of effects parameters such as delay, reverb time, depth, or EQ can often be changed through the use of system-exclusive messages (see Chapter 2). Control over these messages is generally carried out through the use of a hardware controller or software application. By assigning physical controllers to an individual function, it's possible to tweak the effects (or electronic instrument) settings in real time.

The Art of Mixing

Actually, the topic of the art of mixing could easily fill a book (and I'm sure it has); however, I'd simply like to point out the fact that it is indeed an art form—and, as such, is a very personal process. I remember the first time that I sat down at a console (an older Neve 1604). I was truly petrified and at a loss as to how to approach the making of a mix. Am I over-equalizing? Does it sound right? Will I ever get used to this sea of knobs? Well, folks, as with all things, the answers come to you when you simply sit down and mix, mix, mix! It's always a good idea to watch others in the process of practicing their art and to take the time to listen to the work of others (both the known and the not so well known). With practice, it's a foregone conclusion that you'll begin to develop and master your own sense of the art and style of mixing—which, after all, is what it's all about. It's all up to you. Dive into the deep end—and have fun!

Mixing and Balancing Basics

Once all of the tracks of a project have been recorded, assembled, and edited, it's time to put the above technology to use to mix the tracks of your project into their final media forms. The goal

Figure 12.31. *Graphic representation of the stereo playback soundfield.*

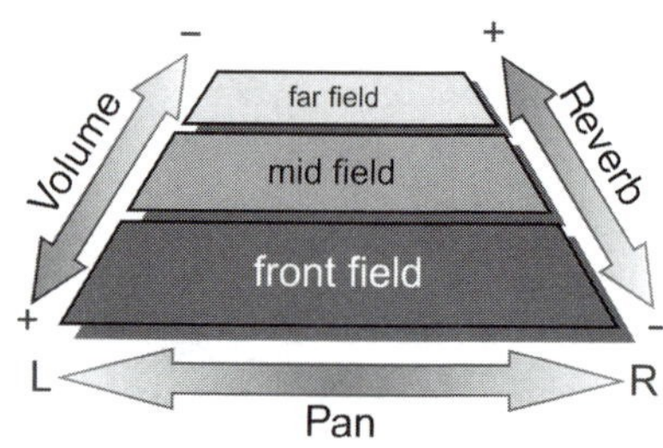

of this process is to combine audio, MIDI, and effects tracks into a pleasing form that makes use of such traditional tools as:

- Relative level
- Spatial positioning (the physical panned placement of a sound within a stereo or surround field)
- Equalization (affecting the relative frequency balance of a track)
- Dynamics processing (altering the dynamic range of a track, group, or output bus to optimize levels or to alter the dynamics of a track so it fits within a mix)
- Effects processing (adding reverb-, delay-, or pitch-related effects to a mix in order to augment or alter the piece in a way that is natural, unnatural, or just plain interesting

Figure 12.31, for example, shows how sounds can be built up and placed into a sonic stage through the use of natural, psycho-acoustic, and processed signal cues to create a pleasing, interesting, and balanced soundscape. Now, it's pretty evident that volume can be used to move sound forward and backward within the soundfield and that relative channel levels can be used to position a sound within that field. It's less obvious that changes in timbre (often but not always introduced through EQ), delay, and reverb can be used to move sounds within the stereo or surround soundscape. All of this sounds simple enough; however, the dedication that's required to hone your skills within this evolving art is what mixing careers are made of.

OK … let's take a moment to walk through a fictitious mix. Remember, there's no right or wrong way to mix as long as you watch your levels along the signal path. There's no doubt that, over time, you'll develop your own sense of style. The important thing is to keep your ears open, care about the process, and make it sound as good as you can.

- Let's begin building the mix by setting the output volumes on each of the instruments to a level that's acceptable to your main mixer or console. From a practical standpoint, you might want to set your tracks to unity gain or to some easily identifiable marking.
- The next step would be to begin playing the project and change the fader levels for any instrument or voice until they begin to blend in the mix. Once done, you can play the tracks to see how the overall mix levels hold up over the course of the song.

- Should the mix need to be changed at any point from its initial settings, you might turn the automation on for that track and begin building up the mix. You might want to save your mix at various stages of its development under a different name (MyMix 001, MyMix 002, etc.). This makes it easier to return to a point where you began to take a different path.
- You might want to group (or link) various instrument sections, so that overall levels can be automated. For example, during the bridge of a song you might want to reduce the volume on several tracks by simply grabbing a single fader … instead of moving each track individually.
- This calls to mind a very important aspect of most types of production (actually it's the basic tenant of life): Keep it simple! If there's a trick you can use to make your project go more smoothly, use it. For example, most musicians interact with their equipment in a systematic way. To keep life simple, you might want to explore the possibility of creating a basic mixing template file that has all of the instruments and general assignments already programmed into it.
- Once you've begun to build your mix, you might want to create a rough mix that can be burned to CD. Take it out to the car, take it to a friend's studio, put it in a your best/worst boom box, have a friend critique it. Take notes and then revisit the mix after a day. You'll be surprised at what you might find out.

Need More Inputs?

It's no secret that electronic music has had a major impact on the physical requirements of mixing hardware and software. This is largely due to the increased need for the large number of physical inputs, outputs, and effects that are commonly encountered in the modern project and MIDI production facility. For those who are using a large number of hardware instruments and who are doing their production and mixing work on an analog or digital hardware mixer, it's easy to understand how we might run out of physical inputs when you're faced with synths that have four outputs, drum machines that have six, and samplers that have up to eight. Although you might not plug all of these outputs into your mixer or console at once, you can still see how a system might easily outgrow your connection needs, leaving you with the unpleasant choice of either upgrading or dealing with your present system as best you can. This is why it's always wise to anticipate your future mixing requirements when buying a console or mixer.

One way to keep from running out of inputs on your main mixer or console is to use an outboard line mixer (take another look at Figures 12.1 and 12.2). These rack-mountable mixers (also known as *submixers*) are often equipped with extra line-level inputs, each having equalization, pan, and effects send capabilities that can be mixed down to either two or four channels. These channels can then be used to free up a multitude of inputs on your main mixing device.

For those who prefer to mix (or submix) within a DAW, the problems associated with the mixing of virtual instruments basically come down to a need for raw computing speed and power. In a

Figure 12.32. Octane eight-channel preamp with ADAT lightpipe. (Courtesy of M-Audio, A Division of Avid Technology, Inc.; www.m-audio.com.)

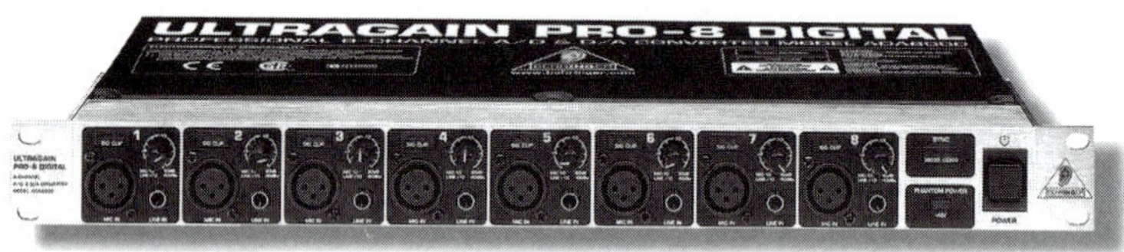

Figure 12.33. Behringer UltraGain Pro-8 eight-channel preamp and digital converter. (Courtesy of Behringer International GMBH, www.behringer.com.)

virtual environment, any number of software instruments can be plugged and played into a workstation in real time, or the tracks can be transferred to disk and then mixed into the project as soundfile tracks (making sure to save all of the original instrument settings and MIDI files). These tracks can continue to add up within a session, but as long as the CPU "busy bar" indicator shows that the system is able to keep up with the processing and disk demands, none of this should be a problem.

When using a workstation-based system, there are often a number of ways to tag on additional analog inputs. For example, a number of interface designs offer eight or more analog inputs; however, a number of these systems offer ADAT lightpipe I/O. In short, this means that a number of mic preamp designs (usually having eight mic pres) that include ADAT lightpipe outputs, can be plugged into a system, thereby adding eight more analog inputs to the interface (Figures 12.32 and 12.33).

Exporting a Mixdown to File

Speaking of mixdown, most DAWs systems are able to *export* (Digidesign uses the traditional recording term "bouncing") all or part of a session to a single file or set of soundfiles (Figure 12.34). This is because of the system's ability to mix the entire session to one (mono), two (stereo), or multiple (typically 5.1 surround sound tracks that can be *interleaved*, combined together into a single file in a L–R–L–R or multichannel fashion, or individual, non-interleaved) channel files. Certain programs allow the session to be exported in non-real-time fashion (a faster than real time process that can include all mix, plug-in effects, automation, and virtual instrument calculations) or in real time (a slower process that, on certain DAWs, can send and receive real-time analog signals through the audio interface so as to allow for the insertion of external effects devices, etc.). A session can be mixed down to a number of chosen final soundfile and bit-/sample-rate formats, while certain DAWs allow third-party plug-ins to be inserted into the final (master) output section, allowing for the export of a session to a specific output file format. For

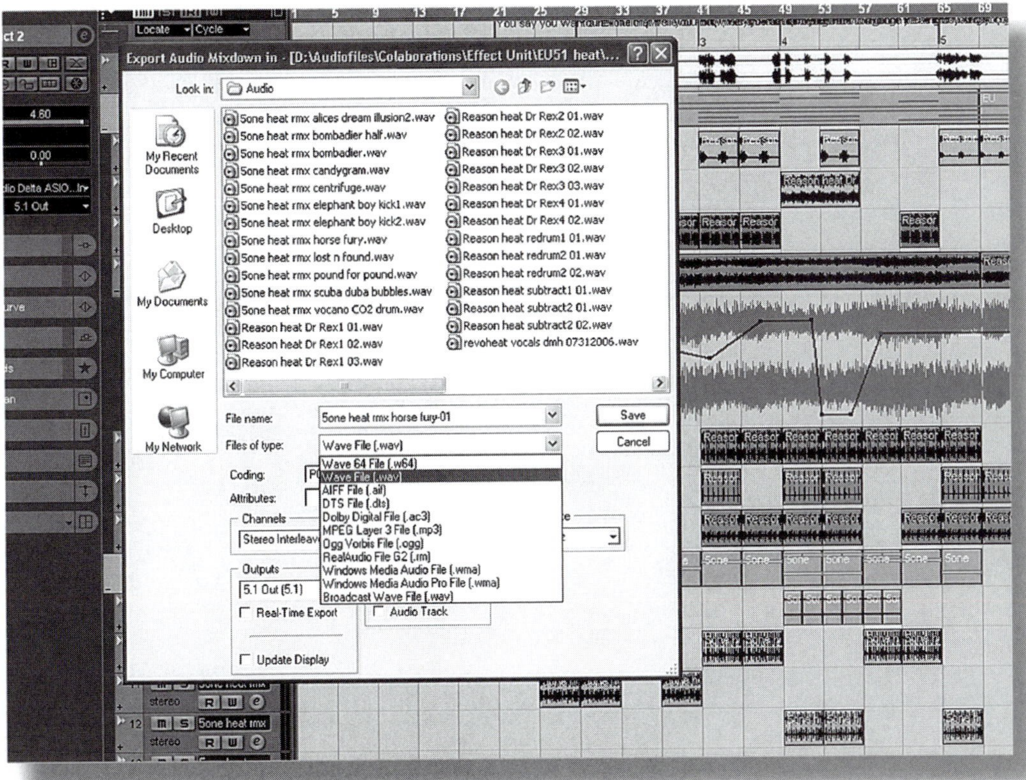

Figure 12.34. Many DAWs are capable of exporting a session soundfiles, effects, and automation to a final set of mixdown tracks. (Courtesy of Steinberg Media Technologies GMBH, www.steinberg.net.)

example, a discrete surround mix could be exported to a two-channel Dolby ProLogic surround-sound encoded file, or the same file could be rendered as a Dolby® Digital 5.1 or DTS file for insertion into a DVD video soundtrack.

In the final analysis, mixing in a project or professional studio environment isn't only fun—it's often a lesson in understanding the power, nuances, and subjectivity of music and sound. Once all the songs in a music project are finished, they can be edited into a final, sequenced order that might (or might not) be sent to a mastering engineer or manufactured into a commercially salable product (which, by the way, is when the REAL work begins).

Studio Tips and Tricks

As we near the end of this book, I'd like to take some time to offer some tips that can help make a session go more smoothly, both in the project and in the professional studio environment (Figures 13.1 and 13.2). Many of these deal with the physical layout of the project/mixdown space itself because without a decent monitoring and production environment some of the steps you've taken could easily and unnecessarily miss their intended mark. It's important to remember that the hallmark of both a good production and a good production facility is their investment in the details. Basically, the glory goes not so much to those who simply do the job, but to those who take the time to get the details right. OK, let's take some time to look at some of the details that can help your projects shine.

Studio Acoustics

While many professional studios are built from the ground up using standard acoustic and architectural guidelines, most budget-minded production and project studios are limited by their own unique building, space, and acoustic considerations. Even if a budget, project, or bedroom

Figure 13.1. *Workin' out the kinks at home. (Courtesy of Yamaha Corporation of America, www.yamaha.com.)*

Figure 13.2. *AIX Media Group Studio (Los Angeles) with Euphonix System 5 console and R-1 recorder. (Courtesy of AIX Media Group, www.aixmediagroup.com, Euphonix, Inc., www.euphonix.com.)*

control room can't have an acoustically perfect design, certain ground rules of acoustical physics should still be followed in order to create a proper working environment.

Of these ground rules, the one that's arguably more important than control over the room's overall sound and frequency response is the need for symmetrical reflections on all axes within the design of a control room or single-room project studio. In short, the *acoustic imaging* within a room (ability to discriminate placement and balance in a stereo or surround field) is best when the listener, speakers, walls, and other acoustical boundaries are symmetrically centered around the listener's position. In a rectangular room, the best low-end response can be obtained by orienting the console and loudspeakers into the room's long dimension (Figure 13.3).

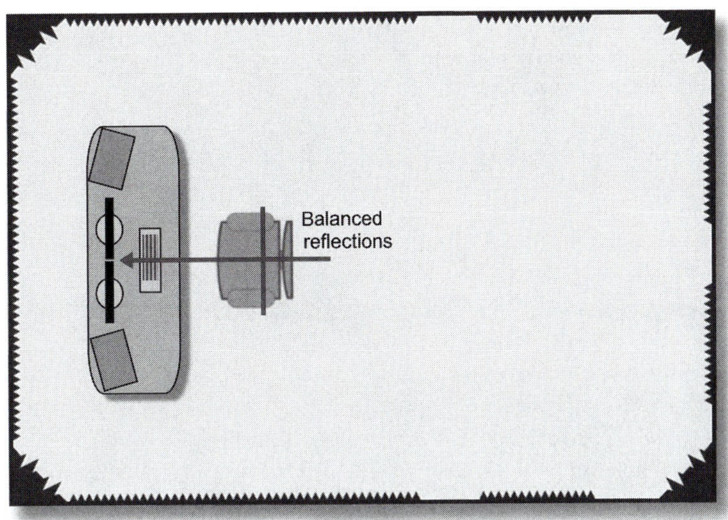

Figure 13.3. *Orienting a control room along the long dimension can extend its low-end response.*

Should any primary boundaries of a control room (especially wall or ceiling boundaries near the mixing position) be asymmetrical from side to side, sounds heard by one ear will receive one combination of direct and reflected sounds, while the other ear will hear a different acoustic balance (Figure 13.4a). This condition alters the sound's center image characteristics, so when a sound is actually panned between the two monitor speakers the sound will appear to be centered; however, when the sound is heard in another studio or standard listening environment the imaging may be off center. In order to avoid this problem, care should be take to ensure that both the side and ceiling boundaries (as well as the materials on these boundaries) are as symmetrical as possible with respect to each other (Figure 13.4b).

If general symmetry isn't followed, one speaker might sound louder or have a different frequency balance, and you might be tempted either to pan the instruments toward the far speaker or boost that entire side of the mix to balance the overall sound. As with asymmetrical control room boundaries, the resulting mix would sound properly centered when played in that room, but in another environment it might be off-center. As a quick check against this, the engineer should always make sure that the audible volume difference between speakers is accompanied by a corresponding visual difference on the main output display meters. Another guard against off-center levels is to monitor pink noise (or a test tone signal) from each speaker in the soundfield to check that they're equally loud (either by doing a quick audible check or by placing a mic in the center listening position). In the latter case, the output level from each speaker can be read and matched using a sound pressure level (SPL) meter or level meter on a spare console input.

It's generally wise to place nearfield and all other speaker enclosures at points that are symmetrically balanced around the listening position (in both the stereo or surround soundfields). Whenever possible, speaker enclosures should be placed 1 to 2 feet away from the nearest wall or corner. This helps avoid bass buildups that acoustically occur at boundary and corner locations. In addition to strategic speaker placement, homemade or commercially available isolation pads

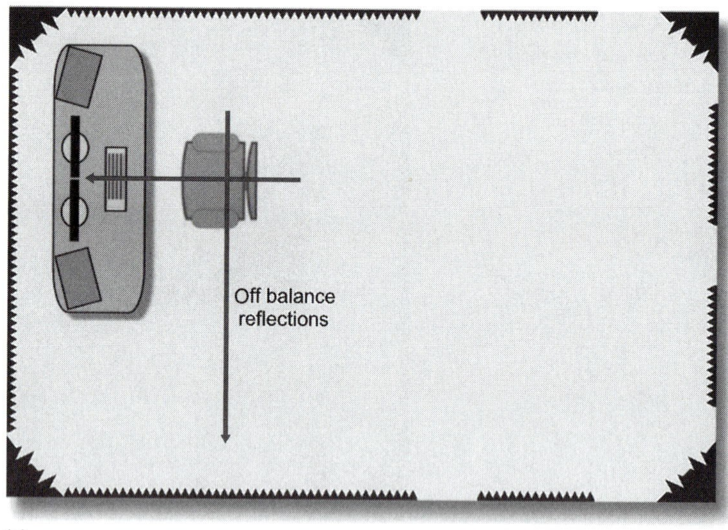

(a)

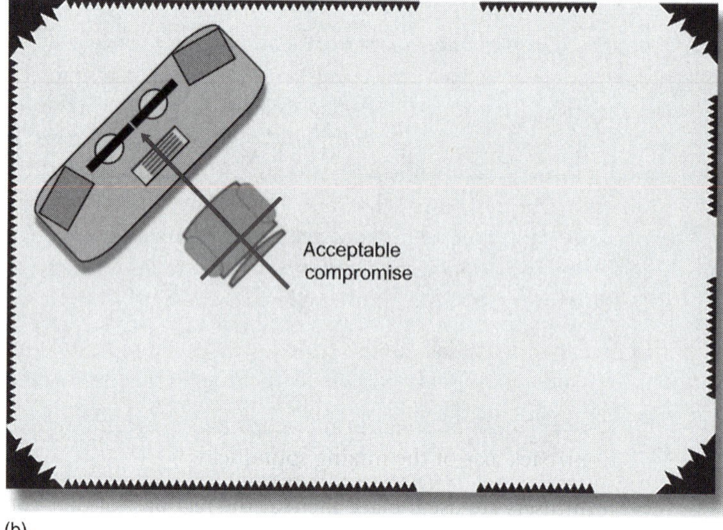

(b)

Figure 13.4. *Various symmetries in a monitoring environment. (a) Placing the monitoring environment off-center and in a corner will affect the audible center image; placing one speaker in a 90° corner can cause an off-center bass buildup and adversely affect the mix's imagery. Shifting the listener/monitoring position into the center will greatly improve the left/right imagery, as can be seen in Figure 13.3. (b) Centering the listener/ monitoring position at a 45° angle within a symmetrical corner is another example of how the left/right imagery can be improved over the first example.*

(Figure13.5) can be used to reduce resonances that often occur whenever enclosures are placed directly onto a table or flat surface.

Monitoring

When mixing, it's important that the engineer be seated as closely as possible to the symmetrical center of the soundfield (making allowances for the producer, musicians, and others who are also

Figure 13.5. MoPad speaker isolation pads. (Courtesy of Auralex Acoustics, Inc.; www.auralex.com.)

doing their best to be in the sweet spot) and that all the speaker volumes are adjusted equally, with no compensation EQ (unless required to balance out the speaker sound) or fancy special effects settings.

Here are a few additional pointers that can help you get the best sound from your control room monitors:

- Make sure that the room's reverb time is both low and smooth over the audible range (absorptive and diffusive materials can help).
- Keep large reflections to a minimum within the room (absorptive and diffusive materials can help reduce reflections to levels that are at least 20 dB down from the direct signal).
- Keep all room boundaries and reflections as symmetrical as possible along the L/R and front/back axis of the mixing soundfield.
- If diffusers are used, place them at the rear part of the room.
- Angle the monitors symmetrically toward the listening position in both the horizontal and vertical planes.

Monitor Volume

Before continuing, volume is another extremely important monitoring factor. During the record and mixdown stage, it's important to keep in mind that the Fletcher–Munson curves (Figure 13.6) will always have a direct effect on the frequency balance of a mix. These curves relate to the ear's sensitivity at various sound pressure levels. As the diagram illustrates, our ears are less sensitive to high and low frequencies at lower listening levels.

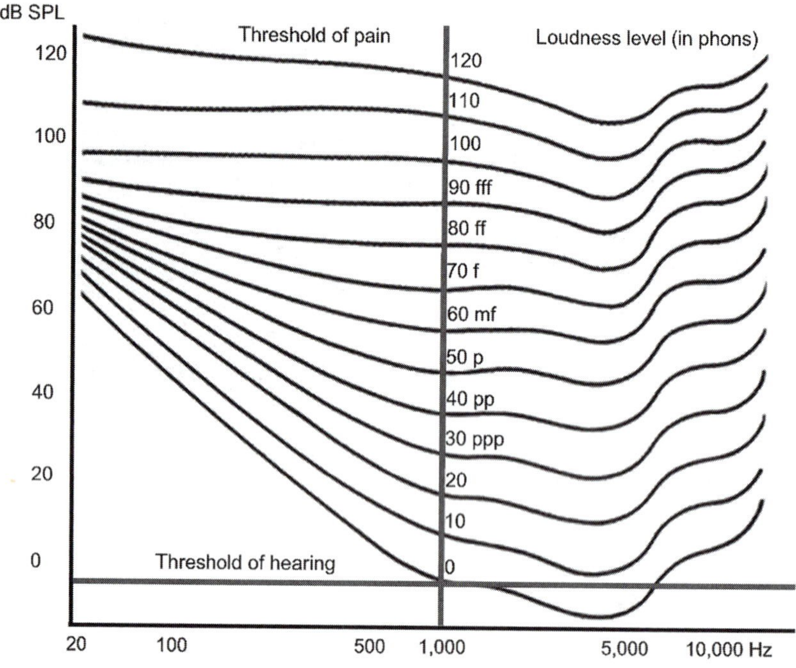

Figure 13.6. *The Fletcher–Munson curve shows an equal loudness contour for pure tones as perceived by humans that have an average hearing acuity. These perceived loudness levels are charted relative to sound-pressure levels at 1000 Hz.*

Our ears perceive recorded sound differently at various monitoring levels. When monitoring at loud levels our ears will easily perceive the extreme high and low frequencies in the mix (sounds good doesn't it?). However, when the mix is played back at lower levels (such as over the radio, TV, or computer), our ears will be much less sensitive to these frequencies, and the bass and extreme highs will probably be deficient (leaving the mix sounding distant and lifeless).

> Unlike during the 1970s, when excruciatingly high SPLs tended to be the rule in most studios, recent decades have seen the reduction of monitor levels to a more moderate 75- to 90-dB SPL. A good rule of thumb is that if you have to shout to communicate in a room you're monitoring too loudly.

These moderate levels offer a good compromise for mixing, as they more accurately represent listening levels that are likely to be encountered in the average home (*e.g.*, the Fletcher–Munson curves will be more closely matched). By mixing at these lower levels, industry professionals can also avoid the ear fatigue and potential ear damage that are caused by prolonged exposure to high SPLs. For more information on safe monitor levels and hearing conservation, contact the House Ear Institute at www.hei.org.

Figure 13.7. Event Studio Precision 8 and Precision 6 reference monitors. (Courtesy of Event Electronics, www.event1.com.)

Monitor Speaker Types

In order to obtain the best possible compromise in sound balance, several alternative monitor options are often available as a reference during a session or mix; therefore, it's wise to listen to your rough and final mixes over several types of speaker configurations. In fact, it's usually a good idea to take a rough mix on the road—listen to it at a friend's studio, play it back on your buddy's favorite ghetto blaster, chill to the beat in your car ... you get the picture. As for your own studio space, the most common speaker types include nearfield, small-speaker, and headphone monitoring.

Nearfield Monitoring

The term *nearfield* refers to the placement of small to medium-sized bookshelf speakers on each side of a desktop working environment or on (or slightly behind) the metering bridge of a production console. These speakers (Figures 13.7 through 13.9) are generally placed at closer working distances, allowing us to hear more of the direct sound and less of the room's overall acoustics. In recent times, nearfields have become an accepted standard for monitoring in almost all areas that relate to audio production for the following reasons:

- Quality nearfield monitors more accurately represent the sound that would be reproduced by the average home speaker system.
- The placement of these speakers at a position closer to the listening position reduces unwanted room reflections and resonances. In the case of an untuned room, this helps create a more accurate monitoring environment.
- These moderate-sized speaker systems cost significantly less than their larger studio reference counterparts (not to mention their reduced amplifier cost, as less wattage is needed).

Figure 13.8. *Mackie HR 624 high-resolution studio monitor. (Courtesy of Loud Technologies, Inc.; www.mackie.com.)*

Figure 13.9. *Studiophile BX8a 130-watt bi-amplified studio reference monitors. (Courtesy of M-Audio, A Division of Avid Technology, Inc.; www.m-audio.com.)*

Many of the more popular nearfield types that are in use today incorporate an active-powered amplifier into its design. These cost-effective systems have become widely accepted by the professionals and project communities due to their:

- Compact design
- High-quality sound (often these systems are bi- or tri-amplified)

- Expandability (additional speakers can be cost-effectively added for surround-sound monitoring)
- Lack of a need for an external power amplifier

For these reasons, these systems are often ideal for project- and DAW-based facilities, and are steadily increasing in popularity.

With so many variables to consider in a speaker and room combination, it quickly becomes clear that there's no such thing as the ideal monitor system. The choice is often more of a matter of personal taste and current marketing trends than one of subjective room and system measurements. Monitors that are widely favored over a long period of time tend to become regarded as the industry standard; however, this can easily change as preferences vary. Again, the best judge of what works best for you should be your own ears and personal working style.

Small Speakers

As radio, television, and web airplay are major forces in audio production and help in the distribution and sales of recordings, it's often a good idea to monitor your final mix through a small, inexpensive speaker set that mimics the nonlinearities, distortion, and poor bass response of these media. Before listening to a mix over such small speakers, take a break to allow your ears and your brain to recover from listening to larger speakers (possibly at higher sound levels).

Headphones

Headphones are also an important monitoring tool in that they remove us from the room's acoustic environment. Headphones also offer excellent spatial positioning, as they let the artist, engineer, or producer place a sound source at critical positions within the stereo field without reflections or other environmental interference from the room. Because they're portable, you can take your favorite headphones with you to quickly and easily check out a mix in an unfamiliar environment. It should be noted that, while headphones eliminate the acoustics of a room from the monitoring situation, they don't always give a true representation of how sounds will behave through loudspeakers (especially with regard to imaging). Monitoring through headphones will also often emphasize low-level sounds like reverb and other effects in a room more than loudspeakers. As a result, it's usually best to listen to a mix over both monitor types. Monitoring over headphones in the studio is by far the most common way to monitor a mix during a session. When recording, it's generally best to use sealed headphones (Figure 13.10) to prevent or minimize the monitor feed from leaking back into the newly recorded track.

Your Car

Last, but not least, your (or any other) car can be a big help in determining how a mix might sound in one of modern society's most popular listening environments. Go ahead, take your mix for a spin on a basic car system—or try it on a souped-up, window-shakin' bass bomb.

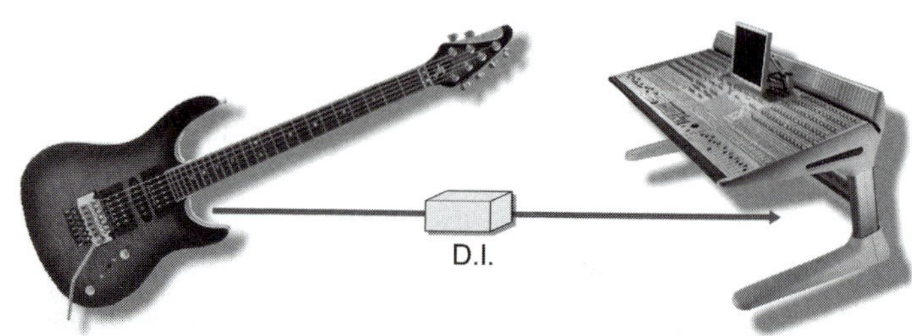

Figure 13.10. *A guitar or line-level instrument can be recorded direct to a DAW or recording medium, bypassing the need for a mic and acoustic environment.*

Is It Live or Is It MIDI?

For those who want to dive into a whole new world of experimentation and sonic surprises, here's a new adage for ya: If the instrument supports MIDI, record the performance to a sequence track. For example:

- You might record the MIDI out of a keyboard performance to its own MIDI track. If a note conflicts with the composition later in its evolution you can simply change the note and re-record it. If you want to change the sound, simply pick another sound.
- Record the sequenced track of an electronic drum controller kit to a track. If you do a surround mix at a later time, simply re-record each drum part to new tracks and pan the drums around the soundfield.
- If the MIDI guitar needs some tweaking to fill out the sound, double the MIDI track an octave down.
- Acoustic drum recordings can benefit from MIDI through the use of a trigger device that can accept audio from existing tracks and output MIDI messages to a sampler or instrument, in order to replace the bad tracks with samples that rock the house. By the way, don't forget to record the trigger outputs to a MIDI track on a DAW or sequencer, just in case you want to edit or change the sounds at a later time. You could even use replacement triggering software to replace the tracks on a DAW with killer sounds. It's all up to you—as you might imagine, surprises can come from experiments like these.

Reamping It in the Mix

Another way to alter the sound of a recorded track or to inject a new sense of acoustic space into an existing take is to *reamp* a track. As most of you know, the process of recording a guitar direct to a tape or digital audio workstation (DAW) track involves using a direct insertion (DI) box that takes the high-impedance output of a guitar/bass and converts it to a low-impedance mic or line signal. This signal is then recorded directly to a track, in a clean-sounding manner that completely bypasses the guitar/amp/room acoustics miking process (Figure 13.10).

In its most basic form, the "reamp" process (originally conceived in 1993 by recording engineer John Cuniberti; www.reamp.com.) works in roughly the reverse fashion, in that it lets us record

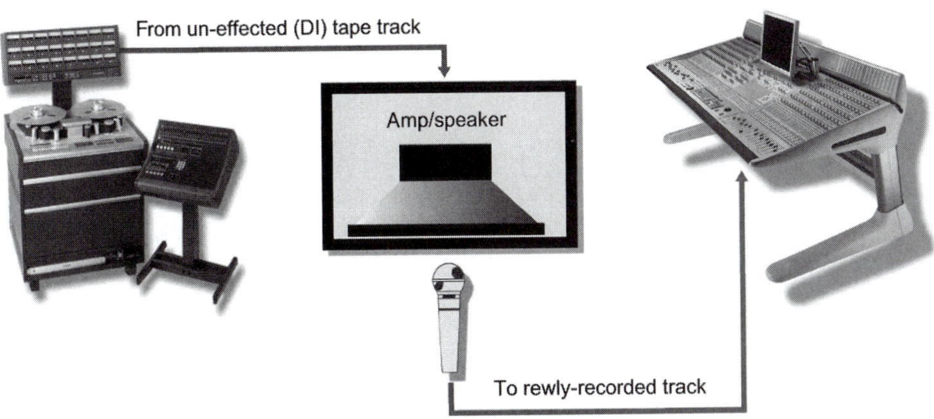

Figure 13.11. *Example of how a direct recording can be reamped in a studio, allowing for complete tonal, mic placement, and acoustical control … after the fact!*

a guitar's signal directly to a track during the recording session and then play this cleanly recorded track back through a miked guitar amp/speaker, allowing it to be re-recorded to new tracks at another time (Figure 13.11).

The re-recording of an instrument that has been recorded directly gives us total flexibility for changing the final, recorded amp and mic sound at a later time. For example, it's well known in the recording community that it's far easier to add an effect during mixdown than to attempt to remove an effect after it's been printed to track. Whenever reamping is used, it becomes possible to audition any number of amps, using any number of effects and/or mic settings, until the desired sound has been found. This process lets the musician concentrate solely on getting the best recorded performance, without having to spend extra time getting the perfect guitar, amp, mic, and room sound. Leakage problems in the studio are also reduced, as no mikes are used in the process.

Although the concept of recording an instrument directly and playing the track back through a miked amp at a later time is relatively new, the idea of using a room's sound to fill out the sound of a track or mix isn't. For example, mics and playback monitors have been used in the studio for decades to fill out the sound of a track or mix (in essence, the studio space becomes an extra acoustic reverb/room chamber). The reamp concept takes this idea a bit further by letting you go as wild as you like. For example, you could use the process to re-record a single, close-miked guitar amp and then go back and layer a larger stack at a distance. Electronic guitarist could even take the process a step further by recording their MIDI guitar both directly and into a sequenced MIDI track. In this way, the reamp and patch combinations would be virtually unlimited.

Probably my favorite example of using MIDI to acoustically reamp a track has to do with a grand piano intro that I had created for a major project. Because the track was an important part of the piece, I contacted a studio that has a Yamaha Disklavier MIDI grand piano, played the sequence into the piano, and recorded the acoustic tracks. Since the project was to be mixed in surround sound, we used both close and distant mics and recorded to four tracks. Unfortunately, the close-mic technique didn't sound as good as the sampled grand, so we ended up using the samples for the front speakers and placed the recorded ambient tracks into the rear. It sounded great!

Transferring MIDI to Audio Tracks

When mixing down a sequence in a professional recording studio, many folks prefer not to mix the sequence in the MIDI domain. Instead, they'll play the sequenced track through its intended instrument, which is then recorded as an audio track within the project. Here are a few helpful hints that can make the process go more smoothly:

- If any mix-related moves have been programmed into the sequence, strip out these volume, pan and other controller messages (this can be done manually; however, most DAW/sequencer packages include features for stripping controller data from a MIDI track).
- Set the main volume and velocities to a reasonably high level.
- Solo the MIDI and instrument's audio input track and take a listen, making sure to turn off any reverb or other effects that might be on that instrument track (you might need to pull out your instrument's manual to flip through these settings). If you really like the instrument effect, of course, go ahead and record it; however, you might consider recording the track both with and without effects.
- Record the instrument to a new stereo or surround track. By soloing the MIDI and input tracks, you can either record the instrument in real time or export (bounce) the soloed mix in real-time. On certain DAWs, plug-in instruments can be exported at greater than real-time speeds.
- With certain instruments having multiple outputs or instrument voices, you might consider making multiple passes to export each output or stereo output pair to their own tracks. Obviously, this will allow for the greatest degree of flexibility during a mixdown.

> *Important note:* It's *always* wise to save the original MIDI track or file within the session. This makes future changes in the composition infinitely easier. Failure to save your MIDI files will often limit your future options or result in major production headaches down the road.

Helpful Production Hints

The following section is meant as a tricks and tips guide to organizing and carrying out your sessions and mixdowns. Many (but not all) of these hints relate to working with a DAW; however, the general overriding concept of organization and data protection is a production and work ethic that can help you to take control of your sessions and your career.

 Project Preparation

Probably the most important step that one can take to help ensure that a recording project will be successful—and that it has a chance of being marketable to its intended audience—is careful preparation and planning. By far the biggest mistake that a musician or group can make is to go into the studio, spend a lot of money and time, press a few thousand CDs, make a template website, and then sit back and expect an adoring audience to spring out of thin air! It ain't gonna happen! Beyond getting a good dose of business reality and added experience, these artists will have the dubious distinction of joining the throngs who have thousands of CDs sittings in their closet or basement.

Most professionals in the industry realize that music in the modern world is a business. Once you get to the phase of getting your or your client's band out to the buying public you'll quickly realize just how true this is. Building and maintaining an audience with an appetite for your product can easily be a full-time business—one in which you'll encounter both well-intentioned people and some who would think nothing of taking advantage of you or your project client.

Without a doubt, the best way to avoid pitfalls and to help get your, your client's, or your band's project off the ground is to discuss and outline the many factors and decisions that will affect the creation and outcome of that all-important final product. Just for starters, a number of basic questions need to be asked long before anyone presses the "REC" button:

- How are we going to recoup the production costs?
- How is it to be distributed to the public? Self-distribution? Indie? Record company?
- Will other musicians be involved?
- Do we need a producer or will we self-produce?
- How much practice will we need? Where and when?
- Should we record it in the drummer's project studio or at a commercial studio?
- Where will it be recorded and mixed?
- Who's going to keep track of the time and budget?
- Are we going to need a music lawyer with contacts and contracts? Do we know someone who can handle the job?

Also, carefully discuss the artist or group's artistic and financial goals and put them down on paper. Discuss budget requirements and possible rewards as early as possible in the game. These are but a few of the issues that should be addressed before tackling a project. Of course, they'll change from project to project and will depend on the final project's scope and purpose; however, in the final analysis, asking the right questions (or finding someone who can help you ask the right questions) can help keep you focused and on budget.

Now that you've answered the questions, here's a list of tasks that are often wise to tackle well before going into production:

- Create a *mission statement* for you or your group and the project. This can help clue your audience in as to what you are trying to communicate through your art and music and can greatly benefit your marketing goals. For example, you might want to answer such questions as: Who are you? What are your musical goals? How should the finished project sound? What emotions should it evoke? What is the budget for this project?

- How will it be sold? What are the marketing strategies?
- Start working on the project's artwork, packaging, and website ASAP.
- Copyright your songs. Form SR is used for the registration of published or unpublished sound recordings or a recorded performance. If you wish to register only the underlying musical composition or dramatic work, Form PA or Short Form PA should be used. Copies of and information on these and other forms can be found at http://www.copyright.gov/forms/.

2 Session Documentation

There are few things more frustrating than going back to an archived session and finding that no information exists as to what instrument patch, mic type, or outboard effect was used on a DAW session (or any session, for that matter). The importance of documenting a session in a separate written document or in the notepad apps within a DAW can't be overemphasized. The basic documentation that relates to the who, what, where, and when of a recording, mixdown, mastering, and duplication session should include such information as:

- Artists, engineers, and support staff that were involved with the project
- Session calendar dates and locations
- Session tempo
- Participants in the project and important dates (for future credits)
- Mic choice and placement (for future overdub reference)
- Outboard equipment types and their settings
- Plug-in effects and their settings or general descriptions (you never know if they'll be available at a future time, so a description can help you to duplicate it with another app)

The more information that can be archived with a session (and its backups), the better the chance that you'll be able to duplicate the session in greater detail at some point in the future. Just remember that it's up to us to save and document the music of today for the fans and future playback/mix technologies of tomorrow. Basic session documentation guidelines can be found in the Guidelines & Recommendation section at www.grammy.com/Recording_Academy/Producers_and_Engineers/.

3 Name That Track

When recording an audio or MIDI track to a DAW, it's always wise to name the track before the Record button is ever pressed. This simple habit lets you do the naming, instead of having the workstation assign an arbitrary track title. For example, giving the track a name of "johnsbass" will cause the DAW to name the recorded track as johnsbass 01 and subsequent takes as johnsbass 02, johnsbass 03, etc. If you don't name it at the outset the DAW might give it a name like "track 12-001." Obviously, finding the alternative take "johnsjazzbass9" would be easier to find in a crowded audio directory than "track 12-009." You might want to keep these names as short as possible (or keep the first few alphanumerics as descriptive as possible), as the track display on most workstations and controller readouts will allow only eight or so characters per name and drop the remaining ones.

4 Organize Your Colors

Almost every workstation will let you assign a color to a track or grouping of tracks from a selectable color palate (Figure 13.12). This handy feature makes it much easier to visually identify and organize your tracks according to instrument groups or types. I recommend that you create and save a document in which you list your favorite instrument groups and assign a color code to each. You could even create a desktop shortcut for quick and easy access.

5 Visual Groupings and Subfolders

Certain DAWs allow any number of tracks and/or media track types to be organized into physical track folders (Figure 13.13). In this way, tracks can be dragged into a folder that relates to a specific category (*e.g.*, drums, background vocals, strings). The obvious advantage is the ability to quickly and easily spot a track grouping. Certain DAWs (Steinberg's Cubase and Nuendo, for example) allow the tracks within the folder to be either shown or hidden. Whenever a track is hidden, a graphic representation of the parts and events within a collapsed folder will often be shown. An example of a hidden folder might be include any number of MIDI tracks that make up a complicated sequence. Lets say that the MIDI files have been transferred to audio tracks within a project. Of course, we'll want to keep and archive the original MIDI tracks with the session. By placing these MIDI tracks into a "MIDI Tracks" folder, a whole slew of MIDI tracks can be organized into a single folder that can be visually collapsed down to a single, representative

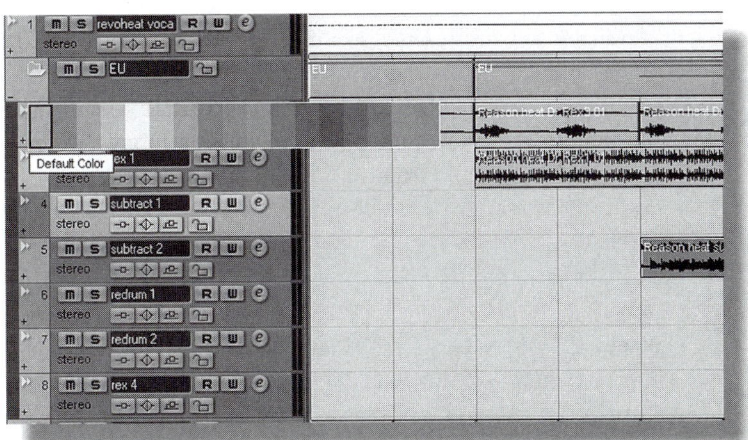

Figure 13.12. *Color can be a helpful organization tool within a DAW project. (Courtesy of Steinberg Media Technologies GmbH, A Division of Yamaha Corporation; www.steinberg.net.)*

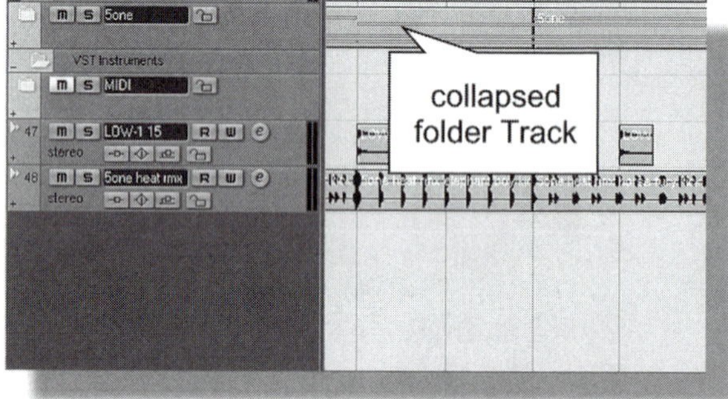

Figure 13.13. *This figure shows an expanded folder track (showing the included tracks) and the collapsed folder tracks (with the tracks hidden within the folder). (Courtesy of Steinberg Media Technologies GmbH, A Division of Yamaha Corporation; www.steinberg.net.)*

folder track. If you want to view the MIDI files, simply expand the folder to display all of the individual tracks—it's that easy!

(6) Recording and Mixing Templates

Most DAWs will allow a full setup arrangement (including I/O assignments, audio and MIDI track layouts, effects, etc.) to be saved as a template. The preprogramming of a recording or mix-down session makes it easy for complicated session setups to be instantly called up, with a minimum of muss or fuss. If you find a basic startup layout that you like, go ahead and save it as a template. Next time you start a session, you can call it up without going through the setup moves.

(7) Schedule a Pre-Mix Cleanup Session

It's important not to go overboard with cleaning up all of the ticks, pops, bad takes, and throat clearings that can accumulate during a session, but this process can help reduce time and frustration come mix time. If the session needs it, this simple and effective housecleaning can help you to better concentrate on the important stuff.

(8) Backup and Archive Strategies

The phrase "nothing lasts forever" is especially true in the digital domain of lost 1's and 0's, damaged media, dead hard drives, and lost data … you know, the "Oh @#$%! factor." It's a basic fact that you never quite know what lies around the techno bend—and, of course, it's extremely important that you protect yourself as much as is humanly possible against the inevitable. Of course, the answer to this digital dilemma is to back up your data in the most reliable (or redundant) way possible. Hardware and program software can (usually) be replaced; on the other hand, when valuable session soundfiles are lost, they're lost!

Backing up a session can be done in several ways, depending on the level of certainty that the files can be played back in the future with a minimum of hassle. Here are a few tips that can help you avoid the loss of data:

- As you might expect, the most straightforward backup system is to copy the session data, in its entirety, to the most appropriate media.
- In the longer run (5+ years), the most ironclad way to back up the track data of a session is to print each track as its own .wav or .aif file. Each track should always be recorded or exported as a contiguous file that flows from the beginning of the session (00:00:00:00 or appropriate begin point) to the end of that particular track. In this way, the individual track files could be loaded into any type of DAW, at the beginning point, for processing and mixdown.
- In such a track-by-track safety restoration situation, you might want to save two copies of a track that has a particular effect—one that contains the original and effected sound and another that simply contains the dry, unaltered sound.
- Those who want additional protection against the degradation of unproven digital media may also want to back each track (or group of tracks) to the individual tracks of an analog recorder.
- For those sessions that contain MIDI tracks, you should always keep these tracks within the session (*i.e.*, don't delete them). These tracks might come in very handy during a remix or future mixdown.
- Speaking of MIDI … it's always wise to export all of the MIDI tracks/data within a session as standard MIDI file. You should save all of the tracks as a type 1 file (where all of the multichannel track/data information is left intact) and, whenever possible, save it as both a type 0 and a type 1 (you never know what file format obstacles might haunt you in the future).
- Whenever possible, make multiple backups and store them in separate locations. Having a backup copy in your home as well as in the studio can save your proverbial butt in case of a fire or any other unforeseen situation.

> *A general backup rule of thumb:* Data is never truly backed up unless it's saved in three places!

As you might expect, all of these tracks can add up to a ton of data. For example, a 3-minute song that's been recorded onto 24 tracks at 16 bit/44.1 kHz would add up to 720 Mb (10 Mb min/track × 3 minutes × 24 tracks = 720 Mb). When you add up the session's tracks from multiple songs, original DAW session data, and alternative track takes, you could easily end up backing up huge amounts of data onto such media as:

- Hard disk—Large amounts of session/track data can be backed to a large removable drive (or duplicated onto multiple drives).
- DVD+R or DVD-R—This removable media lets you back up to 5 GB (gigabytes) of data onto a single disc. (My good buddy, Craig Anderton, recommends that you individually back each song onto its own disc. In case something happens to the media, just that song would be lost, not all of them.)
- CD-R—This removable media can hold up to 800 MB of data and is reliable, assuming that proper care is taken in labeling the disc with a water-based CD marker and that proper storage and handling precautions are taken.

Just think of how you would feel if all of your most precious, original session material was lost ... forever! All of the frustration and deep sense of loss can be avoided (or at least cushioned) by thinking ahead—and literally saving the day—by backing up your data.

9 Grounding Guidelines

Proper grounding is essential to maintaining equipment safety; however, within an audio facility, small AC voltage potentials between various devices in an audio system can leak into a system's grounding circuit. Although these potentials are small, they are sometimes large enough to induce noise in the form of hums, buzzes, or radiofrequency (RF) reception that can be injected (and amplified) directly into the audio signal path. These unwanted signals generally occur whenever improper grounding allows a piece of equipment access to two or more different paths to ground.

Power- and Ground-Related Issues

Because grounding problems arise as a result of electrical interactions among any number of equipment combinations, the use of proper grounding techniques and troubleshooting within an audio production facility are, by their very nature, situational. Therefore, the following procedures are simply a set of introductory guidelines for dealing with this age-old problem. There are a great number of technical papers, books, methods, and philosophies on grounding, and I recommend that you carefully research the subject further before tackling any ground-related problems. When in doubt, contact an experienced professional, and, as always, care should be taken not to sacrifice safety.

Keep all studio electronics on the same AC electrical circuit. Most stray hums and buzzes occur whenever parts of the sound system are plugged into outlets from different AC circuits. Plugging into a circuit that's connected to such noise-generating devices as air conditioners, refrigerators, light dimmers, neon lights, etc., will definitely invite stray noise problems. Because most project studio devices don't require a lot of current (with the possible exception of power amplifiers), it's usually safe to run all of the devices from a single, properly grounded line from the electrical circuit panel.

Keep audio wiring away from AC wiring. Whenever AC and audio cables are laid side-by-side, portions of the 60-Hz signal might be induced into a high-gain, unbalanced circuit as hum.

If this occurs, check to see if separating or shielding the lines helps reduce the noise. When all else fails:

- If you hear hum coming from a particular input channel, check that device for ground-related problems. If the noise still exists when the console or mixer is turned down, check the amp circuit or any device that follows the mixer. If the problem continues to elude you, then …
- Disconnect all of the devices (both power and audio) from the console, mixer, or audio interface, then systematically plug them back in one at a time (it's often helpful to monitor through a pair of headphones).
- Lifting the ground lug on a power line can sometimes solve a grounding problem, but it often isn't safe (care to put a 100-V potential across that vocalist's mic?). Power filter devices can help solve the problem while keeping the ground line intact.
- Check the cables for bad connections or improper polarity. It's also wise to keep the cables as short as possible (especially in an unbalanced circuit).
- Another common path for ground loops is through a chassis into a 19-inch rack and then into another chassis. Test this by removing the chassis from the rack. You can isolate the offending chassis from the rack with electrical tape and then insulate the rack screws by using nylon washers.
- Investigate the use of a balanced power source, if traditional grounding methods don't work.

Trouble-shooting a ground-related problem can be tricky, and finding the problem's source might be a needle in a haystack situation. When putting on your trouble-shooting hat, it's best to remain calm, be methodical, and consult with others who might know more than you do (or might simply have a fresh perspective).

Power Conditioning

Whether your facility is a project studio or a full-sized professional facility, it's often wise to regulate or isolate the voltage supply that's feeding one of your studio's most precious investments (besides you and your staff)–namely, the equipment! This discussion of power conditioning can basically be broken down into three topics:

- Voltage regulation
- Eliminating power interruptions
- Keeping the lines quiet

In an ideal world, the power that's being fed to your studio outlets will be very close to the standard reference voltage of the country you are working in (*e.g.*, 120 V, 220 V, 240 V). The real fact of the matter is that these line voltages regularly fluctuate from this standard level, resulting in voltage sags (a condition that can seriously underpower your equipment), surges (rises in voltage

Figure 13.14. *Furman P-8 Pro Series II power conditioner. (Courtesy of Furman Sound, Inc.; www.furmansound.com.)*

that can harm or reduce the working life of your equipment), transient spikes (sharp, high-level energy surges from lightning and other sources that can do serious damage), and brown-outs (long-term sags in the voltage lines). Through the use of a voltage regulator (Figure 13.14), high-level, short-term spikes and surge conditions can be clamped, thereby reducing or eliminating the chance that the mains voltage will rise above a standard, predetermined level.

Certain devices that are equipped with voltage regulation circuitry are able to deal with power sags, long-term surges, and brown-outs by electronically switching between the multiple voltage level taps of a transformer so as to match the output voltage to the ideal mains level (or as close to it as is possible). One of the best approaches for regulating voltage fluctuations both above and below its nominal power levels is to use an adequately powered uninterruptible power supply (UPS). In short, a quality UPS works by using a regulated power supply to constantly charge a rechargeable battery or bank of batteries. This battery supply is again regulated and used to feed sensitive studio equipment (such as a computer, a bank of effects devices, etc.) with a clean and constant voltage supply.

Just as the "uninterruptible" part of the name implies, should the power be momentarily interrupted or give out altogether, a good UPS can draw upon its battery supply to see you through a momentary power loss or to give you enough time to safely shut your system down without losing data during a total power failure. The most important thing here is to buy a UPS that has enough power reserves to adequately and continuously power the equipment during normal operation and give you enough time to save your session and file data and then safely shut the system down during an outage. In short, if you're looking to buy a UPS, make sure that it has a continuous power rating that's high enough for your supply needs.

Household Tips

Producers, musicians, audio professionals, and engineers spend a great deal of time in the control room and studio. It only makes sense that this environment should be laid out in a manner that's esthetically, functionally, and acoustically pleasing from a *feng shui* point of view. Creating a good working environment that's conducive to making good music is the goal of every professional and project studio owner. Beyond the basics of creating a well-designed facility from an acoustic and electronic standpoint, a number of basic concepts should be kept in mind when building or designing a recording facility, no matter how grand or humble. Here are a few helpful hints:

- Given the fact that an engineer spends a huge amount of time sitting on his or her bum, it's always wise to invest in both your and your clients' posture and creature comforts by

Figure 13.15. *The venerable Herman Miller Aeron® chair. (Courtesy of Herman Miller, Inc.; www.hermanmiller.com.)*

having comfortable, high-quality chairs around for both the production team and the musicians (Figure 13.15).

- Velcro™ or tie-straps can be used to organize studio wiring bundles into groups that can be laid our in a way that reduces clutter, improves organization (color-coded straps can be used), and makes the studio look more professional.
- Most of us are guilty of cluttering up our workspace with unused gear, papers … you know, junk! I know it's hard, but a clean, uncluttered working environment tells your clients a lot about you, your facility, and your work habits.
- Unused cables, adapters, and miscellaneous stuff can be sorted into plastic storage boxes with snap-on lids and stacked for easy storage.
- Important tools and items that are used every day (such as screwdrivers, masking tape, or markers) can be stored in a rack-mounted drawer that can be easily accessed without cluttering up your space.
- Portable label printers can be used to identify cable runs within the studio, identify patch points, I/O strip instrumentation … you name it.

(11) Musician's Tools

By now it's probably painfully obvious to most musicians that producing the music is only the first step toward building a career in the business. It takes hard work, perseverance, blood, sweat, tears, and laughter. For every person that makes it, a large number don't. There are a lot of people waiting in line to get into what is perceived by many to be a glamorous biz. So, how do you get to the front of the line? Well, folks, here are some keys to help you through the golden gate.

- A ton of self-motivation
- Good networking skills
- A good, healthy attitude
- Realize that showing up is *huge*!

The business of art (the techno-arts of recording and music production being no exception) is one that's generally reserved for self-starters. Even if you get a degree from XYZ College or recording school, there's absolutely no guarantee that a label will be knocking on the door with a contract in hand (if they do, get a lawyer, quick!). It takes a large dose of perseverance, talent, and personality to make it. In fact, one of the best ways to get into the biz is to get down on your knees and knight yourself on the shoulder with a sword (figuratively or literally—it doesn't matter) and say: "I am now a __________!" Whatever it is you want to be, become it … Shazammm! Make up a business card, start a business, begin contacting artists to work with, or make the first step toward becoming the artist you want to be.

There are many ways to get to the top of your own personal mountain. You could get a diploma from a school of education or the school of hard knocks (it usually ends up being from both), but the goals and the paths are up to you. Like a mentor of mine says: "Failure isn't a bad thing … not trying is!"

Another part of the success equation lies in your ability to network with other people. Like the venerable expression says: "It's not [only] what you know … it's who you know." Maybe you have an uncle or a friend in the business or a friend who has an uncle—you just never know where help might come from next. This idea of getting to know someone who knows someone else is what makes the business world go around. Don't be afraid to put your best face forward and start meeting people. If you want play gigs around your region (or beyond), get to know a promoter or venue manager and hang out without being in the way. You never know—the music maven down the street might know someone who can help get you in the proverbial door. The longer you stick with it, the more people you'll meet, thus making a bigger and stronger network than you thought could be possible.

Like my own music maven always says, "Showing up is *huge*!" It's the wise person who realizes that being in the right place at the right time means being at the wrong place hundreds and hundreds of times. You just never know when lightning is going to strike—just try to be standing in the right field when it does.

Here are some more practical and immediate tips for musicians:

- Build a personal and/or band website—Making a site on www.MySpace.com or creating your own personal site helps to keep the world informed of your gigs, projects, bio, and general goings-on.
- Build a relationship with a music lawyer—Many music lawyers are open to building relations that can be kicked into gear at a future time. Take the time to find a solicitor

that's right for you. Does he or she understand your personal music style? If you don't have the bucks, is this person willing to work with you and your budget, as your career grows?

- The same questions might be asked of a potential manager, although this symbiotic relationship should be built with care, honesty, and safeguards (which is just one of the many reasons you want to know a music lawyer).
- Copyright your music—Always protect your music by registering it with the Library of Congress. It's easy and inexpensive and can give you peace of mind about knowing that the artistic property that you're sending out into the world is protected. Go to www.copyright.gov for more information. You'll want to look under Publications and then look for info and the actual forms relating to Form SR (www.copyright.gov/forms/formsr.pdf) and Form PA (www.copyright.gov/forms/formpa.pdf).

(12) Record Your Own Concerts and Practices

Although this topic doesn't seem to belong in this section, as a musician I've found that recording my concerts and putting them up on the web helps in many ways.

- Freely distributing concerts on your site helps to promote your music and provides a degree of good will that can go a long way with your fans.
- I've also found that they can be really helpful as a business card promo tool, in that a link can be sent to potential venues, allowing booking agents to hear and appreciate your music first-hand.
- These recordings can also be helpful as learning tools for the band and yourself. In general, these recordings don't lie and can help to point out shortcomings in your performance or the house mix.

This process doesn't have to be involved. For example, connecting a DAT or MiniDisc recorder to your audio interface or house mixer and hitting the record button before going on stage will often do the trick.

(13) A Word on Professionalism

Another subject that should be touched upon is the importance of a healthy professional demeanor. Without a doubt, the life and job of a producer or musician aren't always easy ones. The work often involves long hours and extended concentration with people who, more often than not, are new acquaintances. In short, it can be a high-pressure job. On the flip side, it's one that's often full of new experiences, with demands changing on an almost daily basis and opportunities, and it lets you be involved with exciting people who feel passionately about their art and chosen profession.

It's been my observation (and that of many I've known) that the best qualities that can be exhibited by anyone in the biz are:

- An innate willingness to experiment
- Openness to new ideas (flexibility)

- A sense of humor
- An even temperament (this often translates as patience)
- A willingness to communicate with others
- An ability to convey and understand the basic nuances of people from all walks of life and of many different temperaments

The best advice I can possibly give is to be open, be patient and, above all, be yourself. Also, be extra patient with yourself. If you don't know something, ask. If you make a mistake (trust me, you will—we all do), admit it and don't be hard on yourself. It's all part of the process of learning and gaining experience.

14 Protect your Investment

When you've spent several years amassing your studio through hard-earned sweat-equity and bucks, it's only natural that you'll want to take the necessary precautions to protect your investment. Obviously, the best way to protect your data is through a rigorous and straightforward backup scheme (the general rule is that something isn't backed up unless it's saved in three places—preferably with one of the backups being stored off-site). However, you'll also want to take extra steps to protect your hardware and software investments as well, by making sure that they're properly insured.

The best way to start the process of properly insuring your studio is to contact your trusted insurance agent or broker. If you don't have one, now's the time to get one. You might get some referrals from friends or people in your area and give them a call, set up some appointments, and get several quotes.

If you haven't already done so, sit down and begin listing your equipment, their serial numbers, and replacement values. Next you might consider taking pictures or a home movie of your listed studio assets. These steps will help your agent come up with an adequate replacement plan and will come in handy when filing a claim, should an unfortunate event occur. Being prepared isn't just for the Boy or Girl Scouts.

15 Update Your Software

Periodic software updates might help to solve some of those annoying problems that you've been dealing with in the studio. Many times the software that's been pressed onto a commercially available CD or DVD will be out of date by the time it reaches you. For this reason, it's a good idea to check the web regularly to see if there's a newer version that can be loaded at the outset.

16 Read, Read, Read...

As electronic musicians, it's important that we stay on top of the new gear, new technological advances, and new and old techniques. How can we best do that? For starters, you can get the best results from your gear by reading your manuals (a particularly difficult thing for me to do, but something that, as you well know by now, I advocate doing). Various recording and electronic music mags (available both for free and through paid subscription) can be placed in the bathroom as reading material (something that's always been a joy for me). Books can come in really useful. And, last, but not least, go to industry-related conferences—one of the smartest things you can do to broaden your network beyond your back yard! Keeping on top of the tools, toys, and techniques can definitely be fun! Get out there and get yourself some!

In Conclusion

Obviously, the above is just the beginning of an ever-changing list. The process of producing, recording, and mixing in any type of studio environment is an ongoing, lifelong pursuit. Just when you think you've gotten it down, the technology or the nature of the project changes under your feet—hopefully, you'll be the better for it and will learn from the process. Far more than just the technology, the process of coming up with your own production style and your own way of applying the tools, toys, and techniques to a production is what makes an artist—whether you're in front of the proverbial glass or behind it. Over time, your own list of studio tips and tricks will grow. Take the time to write them down and pass them on to others, and be open to the advice of your friends and colleagues. Use the trade mags, conventions, and web to lead you to new ideas. This way, you're opening yourself up to new insights to using the tools of your profession and to finding new ways of doing stuff. Learning is an ongoing process—have fun along the way!

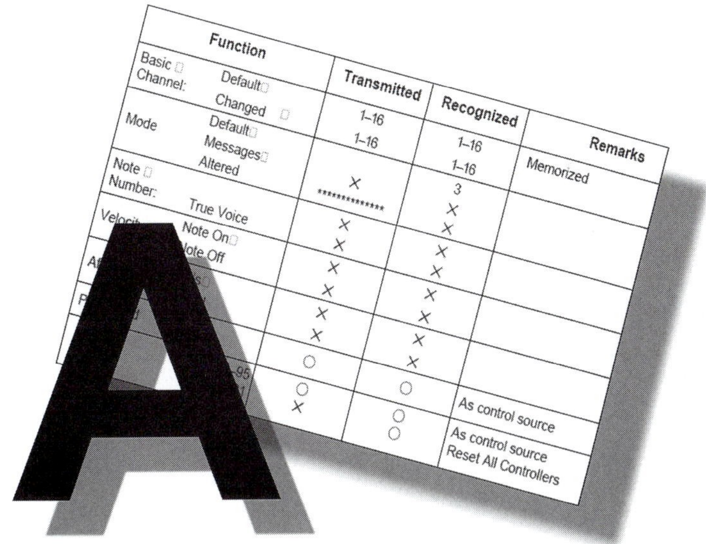

The MIDI Implementation Chart

It isn't always necessary for a MIDI device to transmit or receive every type of MIDI message defined by the MIDI specification, because certain messages might not relate to the device's function. For example, a synth module is only required to respond to Note On or Note Off messages. Because the device does not have a built-in keyboard, there's no need to transmit these messages. Other devices might limit certain MIDI messages due to factors such as design limitations or cost effectiveness. For example, no amount of keyboard banging on a velocity-sensitive keyboard controller will vary the output volume on a synth that does not respond to velocity messages.

To ensure that two or more MIDI devices will be able to communicate MIDI events effectively, the MMA (MIDI Manufacturers Association, www.midi.org) and the JMSC (Japan MIDI Standard Committee) have devised a MIDI Implementation Chart (Figure A.1) that lets the reader look at all the MIDI capabilities of a specific MIDI device at a single glance using a

Function …		Transmitted	Recognized	Remarks
Basic Channel	Default Changed	1–16 1–16	1–16 1–16	memorized
Mode	Default Messages Alterd	Mode 3, 4 OMNI OFF, MONO POLY * * * * * * * * *	Mode 3 ×	memorized
Note Number	True Voice	0–127 * * * * * * * * *	0–127 12–106	
Velocity	Note ON Note OFF	○ v = 1–127 × 9n v = 0	○ v = 1–127 ×	
After Touch	Key's Ch's	× ×	× ×	
Pitch Bender		○	○ 0–24 semitons	
Control Change	1	○	○	Modulation
	2–5	×	×	
	6	* *	* *	Data Entry MSB
	7	○	○	Volume
	8–15	×	×	
	16	×	○	General Purpose Control-1
	17–37	×	×	
	38	* *	×	Data Entry LSB
	39–63	×	×	
	64	×	○	Hold 1
	65–80	×	×	
	81	×	○	General Purpose Control-1
	82–99	×	×	
	100–101	* * (0)	* * (0)	RPC LSB, MSB
	102–120	×	×	
	121	○	○	Reset All Controllers
Prog Change	True #	○ 0–127 * * * * * * * * *	○ 0–127 0–127	
System Exclusive		○	○	
System Common	Song Pos Song Sel Tune	× × ×	× × ×	
System Real Time	Clock Commands	× ×	× ×	
Aux Message	Local ON/OFF All Notes OFF Active Sense Reset	× × ○ ×	× ○ ○ ×	
Notes		* Control Change messages from 0 to 95 which are recognized through Control channel are transmitted through all the channels which are used in Branches. However, General Purpose Control–1 and General Purpose Control – 6 are converted into the same functions as the FC-100 EV-5 assign and the FC-100 Switch assign in the System Setup, and are transmitted. * * RPC = Registered Parameter Control Number RPC # 0: Bender Range The value of parameter is to be determined by entering data.		

Mode 1: OMNI ON, POLY   Mode 2: OMNI ON, MONO   ○: Yes
Mode 3: OMNI OFF, POLY   Mode 4: OMNI OFF, MONO   ×: No

Figure A.1. *Example of a MIDI implementation chart.*

standardized printed format. From the user's standpoint, it is always wise to compare implementation charts with other devices within the existing MIDI system when considering a new piece of equipment. This will ensure that the device will recognize existing messages and add to your system's current capabilities.

Guidelines for Using the Chart

The MMA specifies that the MIDI implementation chart (Figure A.1) be printed the same size using a standardized spreadsheet format consisting of 4 columns by 12 rows. The first column lists the MIDI function in question. The second lists information relating to whether (or how) the device transmits this function's data. The third lists whether (or how) the device recognizes (receives) this data, and the final column is used for additional remarks by the manufacturer.

Despite efforts at standardization, slight inconsistencies within the chart's specifications allow for variations in the symbols, abbreviations, spelling, etc., that can be used by different manufacturers. The following guidelines provide a basic understanding of these differences.

- In general, the symbol 0 is used to indicate that a MIDI function is implemented, while an X is used to show that the function is not implemented. However, some charts may use an X to equal a "yes" and an 0 to equal a "no." This is usually indicated within a key at the lower right-hand corner of the chart.
- OX or an asterisk (*) is used to indicate a selectable function. Further information on the range or type of selectability will be placed within the remarks column.
- MIDI modes are listed as follows: mode 1 (Omni on, Poly), mode 2 (Omni on, Mono), mode 3 (Omni off, Poly), and mode 4 (Omni off, Mono). The modes will often be listed at the bottom of the chart. Occasionally abbreviations of these modes (*e.g.*, Omni on/off, Omni on, or Poly) may be used by a manufacturer.

Detailed Explanation of the Chart

The following paragraphs provide a detailed explanation of the various functions and their related categories that are found within the chart.

Header

The header provides the user with the model number, brief description, date, and version number of the device.

Basic Channel

Basic channel indicates which MIDI channels are used by the device to transmit and receive data. The subheadings for this function are *Default* and *Changed*:

- Default—This indicates which MIDI channel is in use when the device is first turned on.
- Changed—This indicates which of the MIDI channels can be addressed after the device is first turned on.

Mode

Mode indicates which of the MIDI modes may be used by the device. The subheadings for this function are *Default, Messages*, and *Altered*:

- Default—This indicates which of the four MIDI modes is active when the device is first turned on.
- Messages—This describes which of the four MIDI modes can be transmitted or recognized by the device.
- Altered—This refers to mode messages that cannot be recognized by the device. It may be followed by a description of the mode that the device automatically enters into upon receiving a request message for an unavailable mode.

Note Number

The transmitted note number indicates the range of MIDI note numbers that are transmitted by a device. The maximum possible range spans from 0 to 127, while 21 to 108 corresponds to the 88 keys of an extended keyboard controller. Should the note number be greater than the actual number of keys on a keyboard device, a key transposition feature is indicated. The recognized note number indicates the range of MIDI note numbers that can be recognized by a device. MIDI notes that are out of this range are ignored by this device. A second note number range, known as true voice, indicates the number of notes the device can actually play. Recognized notes that are out of the actual voice range are transposed up or down in octaves until they fall within this range.

Velocity

This category indicates whether the device is capable of transmitting or receiving attack and release velocity messages. The subheadings for this function are *Note On* and *Note Off*.

- Note on—This indicates if the device is capable of transmitting and responding to variable-velocity (attack) messages. Not all dynamically controllable devices respond to the full velocity range (1–127). Some devices, such as drum machines, respond to a finite number of velocity steps.

- Note off—This indicates whether the device is capable of transmitting and responding to variable release velocity messages. Many devices use a message (note-on velocity = 0) to indicate a note-off condition. This is often indicated in the chart by 9NH v = O or $9n 00, which is the hexadecimal equivalent for this message.

Aftertouch

Aftertouch indicates how pressure data is transmitted or received. The subheadings for this function are *Key's* and *Ch's*.

- Key's—This indicates if the device will transmit or receive independent polyphonic-pressure messages for each key.
- Ch's—This indicates whether the device is capable of transmitting or receiving channel-pressure changes (a common after-touch mode, providing one pressure value for an entire MIDI channel).

Pitch Bender

Pitch Bender indicates if the device is capable of transmitting or receiving pitch-bend information. If so, the remarks column will often give information as to the pitch-bend range and resolution.

Control Change

Control Change indicates whether the device is capable of transmitting or receiving continuous-controller messages. The chart will often list which of these messages are supported in addition to providing a detailed breakdown of their parameters within the remarks column.

Program Change

This category indicates if the device is capable of transmitting or receiving program-change messages. True # indicates the message numbers that are actually supported by the device's program-change buttons.

System Exclusive

This indicates if the device is capable of transmitting and receiving system exclusive data. The remarks column will often give general information as to which type of SysEx data is supported; however, more detailed data will generally be provided within the device's manual.

System Common

This indicates whether the device is capable of transmitting or receiving the different types of system common messages, such as SPP, MIDI time code, song select, and tune-request messages.

System Real Time

This category indicates whether the device can transmit or receive system real-time messages. The subheadings for this function are *Clock* and *Commands*:

- Clock—This refers to the device's ability to receive or transmit MIDI clock messages. A device that can transmit MIDI clock messages may be used to provide master timing information within a MIDI system, while a device capable of receiving clock data may only be slaved to other MIDI devices.
- Commands—This indicates whether the device is capable of transmitting or responding to Start, Stop, and Continue messages.

Auxiliary Messages

This indicates if a device is capable of transmitting or receiving local Control-On/Control-Off, All Notes-Off, Active-Sensing, and System-Reset messages.

Notes

This area is used by the manufacturer to comment on any function or implementation particular to the specific MIDI device.

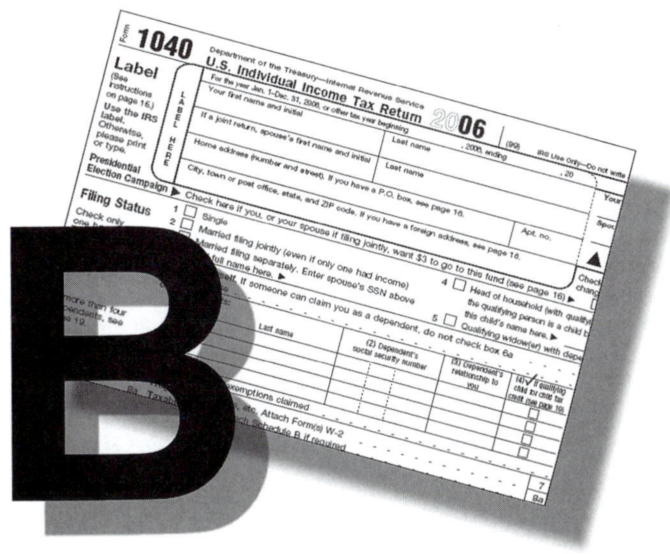

Tax Tips for Musicians

Jeffrey P. Fisher

Everybody complains about taxes, but few do anything about it. Well, there are several actions you can take to improve your tax situation right now. If you're making even the tiniest amount of music-related scratch, there's no reason to pay more taxes than you have to. To reap the most tax benefits, start running your music career as a legal small business. The IRS loves small businesses. According to the Small Business Administration, there are 25 million small businesses in the United States today, and a large percentage of them are sole proprietorships—one-person shops. As a sole proprietor, you report your music business income as part of your personal income using Schedule C (and a few other forms).

A Business Of Your Own

It makes real financial sense to run your project studio as legal business. Follow these basic steps:

- Set up your business by choosing its legal structure (sole-proprietorship, partnership, corporation, etc.). Consult with a tax adviser for details about the financial aspects of each form. Contact a legal adviser for answers to liability issues.

- Get legal. Make sure you meet any regulations that pertain to running a business in your town. For example, you may have to get a business license from your local clerk's office.
- File a doing-business-as (dba) with your local government if you call your business something other than your legal name. You may need a separate tax ID for your business and some states require a sales tax ID number.
- Open a business checking account. Deposit your project studio income into it and pay your business expenses using checks drawn on it. Also, use a credit card only for business purchases and pay it off on time from your business checking account.
- Use bookkeeping software to track all your business income and expenses. This makes tax preparation and monitoring your financial situation easier. Understand the various tax consequences of your business, too. You'll probably need to make quarterly tax payments in addition to yearly tax preparation.
- Protect yourself through health and property insurance. Also, consider life, disability, and liability insurance if it makes sense for your situation.

It all comes down to income and expenses—the money you make and the money you spend. The more you make, the more you pay in taxes. Simple, right? Even the most convoluted of IRS instructions make that point painfully clear. That means the inverse is also true. Since the IRS only taxes your business profits, cut back on the profit and pay less taxes.

"But, Dude, I Gotta Eat."

Whoa, there. I'm not saying that you should earn less. Instead, look for all the possible ways to convert some of your everyday expenses into legitimate business deductions. Even some personal expenses may be deductible against the business. The more expenses you have, the more you reduce your taxable income. And, since you were going to spend the money anyway, you might as well realize some tax benefits, too.

Write-Offs

Basically, all the expenses you incur to run your little music business are deductible. To be fully deductible, however, these business expenses must be "ordinary and necessary" according to the IRS. That's just fuzzy enough to be dangerous. Ordinary means the expenses must be typical for the business. Buying a new guitar could apply; buying a dishwasher wouldn't. Necessary just means the expense is vital to the success of your business. Office equipment, postage, phone charges, graphic design charges, recording studio fees, duplication, dues, magazine subscriptions, and other such related items are definitely necessary for the success of the typical music business. Here is a basic list of business deductions for musicians:

- Advertising and promotion costs
- Car and truck expenses

- Commissions and fees you pay to other people/businesses
- Depreciation and Section 179 deduction
- Insurance
- Interest (on business loans)
- Legal and professional fees (hire a lawyer and deduct the charge)
- Office expenses
- Rent/lease payments
- Repairs and maintenance
- Supplies
- Taxes and licenses such as a business license
- Travel costs
- Meals and entertainment
- Utilities
- Wages, salaries paid

For every $100 you earn, you pay approximately $45.30 in taxes (if you're in the 27% tax bracket, pay the 15.3% self-employment tax and send an additional 3% to your state). Of course, that also implies that for every legitimate $100 business expense you incur, you also save the $45.30 you would otherwise pay in taxes. Hey, that's like getting everything you buy at a discount.

Why does the IRS let you deduct all these expenses? They want you to succeed. So, they let you invest money (spend it) in your business as incentive for you to earn more. And the more money you make, the more you'll pay in taxes. You see they have an ulterior motive; they ain't jus' bein' neighborly.

But there's a caveat (isn't there always?). You need to be gainfully engaged in making a buck. You must turn a profit in your business three out of every five years or your business will be classified as a hobby, and you forfeit the expense deductions. Bottom line: very bad news and very high tax bill. (And one dollar in profit those three years ain't gonna make you popular down at the Treasury department neither.)

Since the burden of proof falls solely to you, it's vital that you record all your music business income and expenses diligently. A shoebox full of receipts does not a bookkeeping system make. Get help setting up your books or look for a software solution to help document your business financial transactions.

Another important gotcha: If you're just launching your music business, startup expenses can't be deducted all at once. You must amortize them over 5 years by taking 20% portions of the total expenses and deducting them over five consecutive years.

Gear Lust = Tax Savings

Did you know that the gear you buy for making your music magic could be a sweet tax deduction? Under Section 179 of the tax code you can deduct or expense up to $100,000 of tangible property and write it all off when you prepare your taxes. For tangible property think expensive, long-lasting items, such as a new computer. This amount can be above and beyond many other normal business expenses you might incur. If you've had a particularly strong earnings year, you might want to offset some of that gain by deducting all the cost of large purchases in one year (up to the limit). Alternatively, you can choose to depreciate what you buy and deduct a portion of those costs over the next several years.

Home Sweet Home

If you do the majority of your music work in your home office, you can deduct a portion of the same expenses that now do little or nothing to lessen your tax burden. You can write off rent or mortgage interest, property taxes, utilities (gas, electricity, water/sewer), insurance, repairs, and depreciation. First, dedicate a portion of your home entirely for your music business. Keep it free of personal items and make it your primary business location. Be aware that if do most of your work elsewhere (gigging, for instance) and only use this home office occasionally, your deduction may be limited or entirely verboten.

Here's how to figure your deductions. Total up the square footage of this exclusive and principal place and compare it to the total square footage of your crib. Say your math works out to 10% (100 square feet of a 1000 square foot home—you need a bigger place!). You can then deduct 10% of the aforementioned expenses using Form 8829, Expenses for Business Use of Your Home. The total deduction then flows through to your Schedule C reducing your income, and therefore your taxes, considerably.

There is a recapture clause (which doesn't apply to renters). If you sell your home and make a profit, those profit dollars become taxable business income at the same percentage rate as your deduction. Score a $50,000 gain (good for you!), and, following the above example, $5,000 of it belongs to the business (subject to self-employment tax and regular income tax, of course). It's important to note that the personal income you make from a house sale is generally not taxed, though! Stop taking the home-office deduction for 2 tax years prior to the home sale and this recapture clause doesn't apply.

Self-Employment Tax

Yes, we self-employed have a special tax just for us. Actually every worker pays the same tax—funding for Social Security and Medicare—it's just a little different when you're on your own. You must contribute both the employee and employer contributions which total up to a whopping 15.3%. Yep, just over 15 pennies on every buck you earn goes right into the Social Security kitty. This is, of course, before you start paying any regular income taxes. Ouch! And you have to pay the self-employment taxes (along with income taxes) quarterly. You need to predict what you are going to earn this year, and the taxes that would be due on that dollar amount. Then, you send in 25% of that money on April 15, June 15, September 15, and January 15 of the next year. These estimated tax payments are important, because if you don't pay enough, there's a penalty due the next April 15.

Health Insurance

You can deduct the premiums paid for health insurance, too. This doesn't come off the Schedule C but is actually a front page deduction on your personal 1040. Self-employed individuals can deduct 100% of the premiums they pay (unless Congress changes its collective mind). Other typical medical costs are deductible on Schedule A (if you qualify).

Eat, Drink, and Be Merry

When you entertain your clients, the money you spend is another write-off; however, these meals and entertainment are subject to a 50% limitation. Spend a $100 on a pizza party, and take $50 off on Schedule C. Give clients gifts, up to $25 per client, and you can take that as a full deduction, though. When you travel as part of your music business, those expenses are deductible including airfare, lodging, and meals. You must support your travel and lodging deductions with receipts. However, instead of keeping track of your meals, you can take the government's standard per diem allowance of $30 for meals and incidentals. Other cities may have higher amounts so check the official website (www.policyworks.gov/perdiem). Meals on the road are, of course, still subject to the 50% limit. The IRS figures you gotta eat anyway, so they limit the expense. Bummer.

Vehicular Reductions

Yes, that old beater is worth money! Keep track of actual vehicle expenses (gas, repairs, etc.) or take the standard mileage rate (which changes every year; check with the IRS). In either case, you must document the miles you drive for business, date, and purpose of trips, along with expenses incurred. A dedicated notebook/diary earns a gold star from the IRS. Even if you use your ride for business and personal use, the business portion of your expenses is still deductible. Determine your business percentage by dividing your business miles by the total miles driven (2500 business/10,000 total = 25%). If you just use the standard mileage rate, multiply your business miles driven by that rate (2500 × $/mile = deduction) to arrive at your deductible expense amount. You can also deduct the full cost of tolls and parking fees incurred while on business. And the loan interest on the car is deductible (subject to the business use percentage, if it applies).

Feed The Nest Egg and Save, Too

You also save money by contributing to a qualified retirement plan. The IRS makes it easy to sock away some cash for a rainy day and rewards you with a nice, fat deduction each year. This is another 1040 deduction, not Schedule C. IRAs are the first method that pop up. However, they're limited to $2250 (which changes regularly). With a Simplified Employee Pension (SEP) you can deduct as much as 15% of your business income topping out at $30,000 total per year. The more you put away, the more you save. And, since you're really helping yourself down the road, it's a smart way to manage your taxes and your retirement. For some of us, a Roth IRA may be more prudent. Roths give you no up-front deduction, but the earnings are tax free. Talk to a financial planner.

End-of-the-Year Tax Tips

At the end of each year, you have another opportunity to reduce your tax burden: accelerate expenses and decelerate income. First, spend some cash on business expenses. Don't just blow the wad. Make sensible purchases this year that will reduce your taxable income. Ideal last-minute purchases include postage, equipment, general office supplies, and promotions. You can also pay your mortgage and health insurance premium before the year-end to realize some other tax savings on the personal side. Second, put off collecting money this December to January by billing your clients a little later. Though you'll have to pay taxes on the money eventually, you defer that payment for a whole year.

Get Help From the IRS

Surf on over to the always exciting IRS website (www.irs.gov) and download the free guides that explain the specific tax benefits for small business owners:

- #334, Tax Guide for Small Business
- #463, Travel, Entertainment, Gift, and Car Expenses
- #533, Self-Employment Tax
- #535, Business Expenses
- #583, Starting a Business and Keeping Records
- #587, Business Use of Your Home.

Final Word

While we all have to pay taxes, we are only required to pay our fair share. Make sure you are not throwing money out the window. Take advantage of these and all the other tax breaks available to you. And put more music money in your pocket, where it belongs!

Author Bio

Jeffrey P. Fisher provides audio, video, music, writing, training, and media production services. He's published ten books, including *Cash Tracks: Compose, Produce, and Sell Your Original Soundtrack Music and Jingles* (Thomson, 2005), *Ruthless Self-Promotion in the Music industry, Second Edition* (Thomson, 2005), *The Voice Actor's Guide to Home Recording* (with Harlan Hogan, www.artistpro.com, 2005), and *Profiting From Your Music and Sound Project Studio* (Allworth Press, 2001). He teaches at the College of DuPage in Glen Ellyn, IL. Also, Jeffrey co-hosts the Acid, Sound Forge, and Vegas forums on Digital Media Net (www.dmnforums.com.). For more information visit his website at www.jeffreypfisher.com or contact him at jpf@jeffreypfisher.com.

Musician's Guide to Saving for Retirement

Jeffrey P. Fisher

Don't sit idly by and wait to control your retirement destiny. There are proactive alternatives to just collecting government checks in your golden years. A personal Individual Retirement Account (IRA) lets you contribute up to $4,000 for the years 2006–2007 and $5000 a year starting in 2008. If you are 50 or above, it's $5000 for 2006 and 2007 and $6000 in 2008. You direct how and where to invest your money.

The traditional IRA lets you deduct your contribution amount from your income each year, thus reducing the taxes you may owe. Money deposited into the account also grows tax-deferred, and compound growth can be substantial. Instead, you pay taxes when you withdraw the funds at retirement, including taxes due on the account's investment growth. This tax-advantaged account is based on the principle that your tax bracket may be lower in the future.

Conversely, the Roth IRA doesn't let you take the contribution amount off your taxes today; rather, the Roth gives you tax-free withdrawals. Roth IRAs are very attractive for younger investors. Your investment account could grow considerably, and your withdrawals, including the account's investment growth, would be tax free. People who already have funds in a traditional IRA need to investigate what conversion costs would be (*e.g.*, having to pay taxes today on previously tax-sheltered contributions).

A personal IRA should be part of everyone's retirement planning. In addition, self-employed musicians and those with incorporated companies should investigate three additional retirement options: SEP, SIMPLE, and Keogh plans.

SIMPLE SEP Steps

Simplified Employee Pension (SEP) and Savings Incentive Match Plan for Employees (SIMPLE) are the most common retirement plans for businesses. A SEP is a good choice for sole proprietors; SIMPLE plans are more suited to incorporated companies. Both plans are easy to set up and administer and allow contributions of up to 25% of income with a 2005 total cap of $42,000 based on a $210,000 income maximum (I know, that math doesn't add up; see box). Like traditional personal IRAs, contributions are tax deductible with tax-deferred growth. The assets in either plan must be managed by a financial institution (bank, investment firm, etc.), but plan members have control over their account's specific investments. A company must offer its SEP/SIMPLE plan to all employees and let them deduct the same percentage as the owner. The SIMPLE plan also lets the company match employee contributions. Since both plans are based on income, the deductible amount can vary from year to year. When you have an especially strong earnings year, contribute more. If times are tough, scale back.

For SEP, SIMPLE, and Keogh plans, the maximum allowable contribution is 25% of income; however, the math is a somewhat convoluted. Essentially, you can contribute the percentage of your income only after you reduce your income by that same percentage amount. Huh? First, determine your business profits (net after paying self-employment taxes). Second, divide that number by 1 plus the percentage you want to contribute (*e.g.*, 1.25). Finally, multiple the answer to the second step by the percentage you want to take. The final figure is your contribution amount. For example: 100,000/1.25 = 80,000 × .25 = $20,000.

Keogh Dough

Another retirement option for sole proprietors and partnerships (but not incorporated businesses) is the Keogh plan. Again, this plan is tax advantaged (deduct now, defer growth, and pay taxes later) but may be set up in several ways. A defined-benefit plan pays a fixed benefit amount to retirees. The profit-sharing model is similar to the SIMPLE plan with employer-matched contributions. The money purchase Keogh plan requires establishing a contribution percentage and sticking to it every year with strong penalties for noncompliance. Keogh plans require more work to set-up and run and therefore require the help of a professional pension manager.

Whether you choose a SEP, SIMPLE, or Keogh retirement plan, you can save a substantial amount of your earnings (at far higher levels than personal IRAs and light years beyond what's proposed for Social Security personal investment accounts), direct the growth options, and then retire on your own tidy little nest egg.

Don't ignore this advice, as saving for the future is an important part of running both your music business and your life. For more information on this subject, pick-up IRS Publication 560, Retirement Plans for Small Business from www.irs.gov or 1-800-TAX-FORM.

Author Bio

Jeffrey P. Fisher provides audio, video, music, writing, training, and media production services. He's published ten books, including *Cash Tracks: Compose, Produce, and Sell Your Original Soundtrack Music and Jingles* (Thomson, 2005), *Ruthless Self-Promotion in the Music industry, Second Edition* (Thomson, 2005), *The Voice Actor's Guide to Home Recording* (with Harlan Hogan, www.artistpro.com, 2005), and *Profiting From Your Music and Sound Project Studio* (Allworth Press, 2001). He teaches at the College of DuPage in Glen Ellyn, IL. Also, Jeffrey co-hosts the Acid, Sound Forge, and Vegas forums on Digital Media Net (www.dmnforums.com.). For more information visit his website at www.jeffreypfisher.com or contact him at jpf@ jeffreypfisher.com.

Continued Education

As the equipment used by electronic musicians places an ever-increasing emphasis on technology, education must play a greater role in the understanding of basic industry skills. Education can take many forms, ranging from a formal education to simply keeping abreast of industry directions from the many industry magazines.

Beyond getting a hands-on education by playing with all of the tools and toys, one of the best ways to get a better handle on the pulse of this industry is to read. Read all the magazines, books, and articles that you can get your hands on. In addition to reading, many schools offer courses in electronic music production. If this approach works for you, I'd recommend that you carefully check the institution and department out before potentially laying down your hard-earned cash to further your education in electronic music technology.

An expanded and continually updated listing of current music industry-related magazines, trade shows, and press-related information can be obtained by contacting either the mediamanager site (www.mediamanager.com) or the Music Yellow Pages (www.musicyellowpages.com).

The Web

For those MIDI professionals and enthusiasts out there who don't have access to the World Wide Web, I really encourage you to get your hands on a modem (28.8 kbps or faster), find an Internet provider that fits your needs and budget, and start surfing the "Net."

One of the best ways to begin surfing the web is to log on to a popular search engine, such as Google (www.google.com) or Yahoo (www.yahoo.com). Or at the search prompt, simply type "midi" and watch the world of information—such as SysEx patches, standard MIDI files, product reviews and tons more—open up before your very eyes.

If you want to access a company or organization directly (in order to get product information, specs, pricing information, or even to download product manuals electronically), you can often enter the name (using lowercase letters) in the following form: www.companyname.com or www.organizationname.org. For example, in order to log onto Yamaha's home site, you simply type in www.yamaha.com.

Users can also use the web to communicate using public and private forums (allowing the users to voice their comments and ideas to other users and manufacturers) in addition to contacting the company directly using electronic mail services … Happy surfing!

Index

D

E

N

O

P

Q

R

T